Hydroxyapatite Coatings for Biomedical Applications

Advances in Materials Science and Engineering

Series Editor

Sam Zhang

Advances in Materials Science and Engineering Series

Hydroxyapatite Coatings for Biomedical Applications

Edited by
Sam Zhang

CRC Press
Taylor & Francis Group
Boca Raton London New York

CRC Press is an imprint of the
Taylor & Francis Group, an **informa** business

CRC Press
Taylor & Francis Group
6000 Broken Sound Parkway NW, Suite 300
Boca Raton, FL 33487-2742

© 2013 by Taylor & Francis Group, LLC
CRC Press is an imprint of Taylor & Francis Group, an Informa business

No claim to original U.S. Government works

Printed on acid-free paper
Version Date: 20130220

International Standard Book Number-13: 978-1-1380-7616-7 (Paperback)
International Standard Book Number-13: 978-1-4398-8693-9 (Hardback)

Library of Congress Cataloging-in-Publication Data

Hydroxyapatite coatings for biomedical applications / editor, Sam Zhang.
    pages cm. -- (Advances in materials science and engineering)
    Includes bibliographical references and index.
    ISBN 978-1-4398-8693-9 (hardback)
    1. Biomedical materials. 2. Hydroxyapatite coating--Therapeutic use. I. Zhang, Sam.

R857.M3H944 2013
610.28'4--dc23   2013004203

Visit the Taylor & Francis Web site at
http://www.taylorandfrancis.com

and the CRC Press Web site at
http://www.crcpress.com

Contents

Contents

Series Preface

Advances in Materials Science and Engineering

Materials form the foundation of technologies that govern our everyday life, from housing and household appliances to handheld phones, drug delivery systems, airplanes, and satellites. Development of new and increasingly tailored materials is key to further advancing important applications with the potential to dramatically enhance and enrich our experiences.

The Advances in Materials Science and Engineering series by CRC Press/ Taylor & Francis Group is designed to help meet new and exciting challenges in materials science and engineering disciplines. The books and monographs in the series are based on cutting-edge research and development and are thus up to date with new discoveries, new understanding, and new insights in all aspects of materials development, including processing and characterization and applications in metallurgy, bulk or surface engineering, interfaces, thin films, coatings, and composites, to name a few. The series aims at delivering an authoritative information source to readers in academia, research institutes, and industry. The publisher and its series editor are fully aware of the importance of materials science and engineering as the foundation for many other disciplines of knowledge. As such, the team is committed to making this series the most comprehensive and accurate literary source to serve the whole materials world and the associated fields.

As series editor, I would like to thank all the authors and editors of the books in this series for their noble contributions to the advancement of materials science and engineering and to the advancement of humankind.

Sam Zhang

Preface

This book, *Hydroxyapatite Coatings for Biomedical Applications*, is dedicated to hydroxyapatite (HA) coatings (films) for biomedical applications. HA coatings are very important in the biological and biomedical coatings fields, especially in the current era of nanotechnology and bioapplications, given the advances that society has made in recent years. Stainless steel implants or titanium alloy implants are not biotoxic, but they are not bio-active either—they are only bioinsert. For a fast healing effect, a bioactive surface is needed to promote bone growth. HA coatings or its variations can be applied to otherwise bioinactive implants to make their surface bioactive, thus achieving fast healing and recovery. HA has a bone-like structure and is not only biocompatible but is also bioactive. Thus, it can be rapidly integrated into the human body, with the body not being aware of the foreign, albeit friendly, invasion. HA bonds to bone and forms indistinguishable unions. Bulk HA materials (powder) have been used in dental treatments as refills. However, they do not have the strength needed to be used alone for most biomedical applications; thus, their coating finds great applications in biomedical implants as well as in drug delivery. HA coatings in biological and biomedical applications have been researched extensively and are still gaining momentum. The aim of this book is to capture developments in the processing and property characterization and applications of HA to provide timely information for researchers to refer to and for newcomers to learn from, thus contributing to the advancement of the technology.

This book is special in that it not only compiles data (tables and figures) but also covers material that will be useful for both novices and veterans. I have paid special attention in guiding the authors of each chapter to prepare their chapters in such a way that novices will find the book to be a useful stepping stone to their field and veterans will find it to be a rich source of information for their research. Researchers in the fields of nanoscience and nanotechnology, materials scientists, engineers, postgraduate students, especially those dealing with biomedical and medical research or studies, should also find this book useful.

This book consists of eight chapters written by experts in the field of HA from the United States, Japan, Singapore, and China, covering deposition processes (magnetron sputtering, electrochemical, and sol–gel), applications in implants and drug delivery, etc. Chapter 1 deals with magnetron sputtering; Chapter 2 focuses on electrochemical deposition; Chapters 3 and 4 discuss modifying properties of HA by sol–gel deposition to incorporate the other elements found in natural bones, such as Zn, Mg, and F; Chapters 5 and 6 describe pure HA in drug delivery applications; Chapter 7 deals with the application of shape memory alloy; and Chapter 8 deals with HA composite coatings (HA with CNT, HA with TiO_2, etc.) on the Ti alloy.

I would like to thank all the contributing authors for their painstaking work that finally resulted in this informative book. Special thanks go to the reviewers who patiently went through the long chapters (one or two reviewers for each chapter) to provide their professional critique, which enabled the authors to improve the chapters to their current form. Their professional contributions were the guarantee for the quality of the book. Last, but not least, the editor would also like to thank the staff at CRC Press, especially project managers Allison Shatkin and Jennifer Ahringer at Taylor & Francis Group, for their invaluable assistance throughout the endeavor that made the smooth publication of the book a reality.

Sam Zhang
Professor
Nanyang Technological University
Singapore, Singapore

Editor

Professor Sam Zhang Shanyong, better known as Sam Zhang, received his PhD in ceramics from the University of Wisconsin-Madison, Madison, Wisconsin, in 1991, and is a tenured full professor at the School of Mechanical and Aerospace Engineering, Nanyang Technological University. Professor Zhang serves as editor in chief for *Nanoscience and Nanotechnology Letters* and principal editor for *Journal of Materials Research* (United States).

After completing his PhD, Professor Zhang entered the field of cermets and engineering coatings/thin films. He has been in the processing and characterization of nanocomposite thin films and coatings for 20 years and has authored/coauthored more than 250 peer-reviewed international journal papers, with an average of more than 12 citations per paper, 8 books, 20 book chapters, and guest edited more than 20 journal volumes. His book *Materials Characterization Techniques* (CRC Press, Boca Raton, Florida, 2008) has been adopted as a textbook by eight American universities and one European university: Purdue University (Department of Materials Engineering), New York University (Department of Biomaterials and Biomimetics), Louisiana State University–Baton Rouge (Department of Mechanical Engineering), California Polytechnic State University (Department of Materials Engineering), University of Missouri (Department of Chemical and Biological Engineering), Rutgers University–Camden (Department of Physics), Johns Hopkins University (Department of Materials Science and Engineering), North Seattle Community College (Math, Science, and Social Sciences), and the University of Southern Denmark's Centre for Nanotechnology in Europe, was translated into Chinese and published by the China Science Press in 2010. His other books include *Nanocomposite Thin Films and Coatings: Processing, Properties and Performance* (Imperial College Press, London, United Kingdom, 2007); a three-volume set: vol. 1—*Nanostructured Thin Films and Coatings: Mechanical Properties*, vol. 2—*Nanostructured Thin Films and Coatings: Functional Properties*, and vol. 3—*Organic Nanostructured Thin Film Devices and Coatings for Clean Energy* (CRC Press, Taylor & Francis Group, Boca Raton, Florida, 2010); a two-volume set: vol. 1—*Biological and Biomedical Coatings Handbook: Processing and Characterization* and vol. 2—*Biological and Biomedical Coatings Handbook: Applications* (CRC Press, Taylor & Francis Group, Boca Raton, Florida, 2011); and *Aerospace Materials Handbook* (CRC Press, Taylor & Francis Group, Boca Raton, Florida, 2012). In addition, *Surface Technologies* (Springer, 2014) is in the pipeline.

Professor Zhang was conferred the title of honorary professor by the Institute of Solid State Physics, Chinese Academy of Sciences. He also holds a guest professorship at Shenyang University, Zhejiang University, Harbin

Institute of Technology, and the Central Iron and Steel Research Institute. He was featured in the first ever *Who's Who in Engineering Singapore* (2007) and also in the 26th and 27th edition of *Who's Who in the World*. He became a fellow of the Institute of Materials, Minerals and Mining, United Kingdom, in 2007. He has been invited to present invited or plenary keynote lectures more than 60 times at international conferences in Taiwan, Japan, the United States, France, Spain, Mainland China, Portugal, New Zealand, and Germany. He has also been invited by universities or industries to conduct short courses and workshops more than 20 times. He founded the Thin Films conference series in 2002 and has been the chair of this very successful conference series ever since. Professor Zhang is also the president of the Thin Films Society established in Singapore in 2009.

Professor Zhang's current research centers on the following four aspects: (1) hard yet tough nanocomposite coatings for tribological applications by physical vapor deposition; measurement of fracture toughness of ceramic films and coatings; (2) biological coatings and drug delivery application; (3) electronic thin films in optoelectronic applications, data storage, etc.; and (4) energy films and coatings for dye-sensitized solar cells. Details of Professor Zhang's research can be accessed at http://www.ntu.edu.sg/home/msyzhang.

Contributors

Yanli Cai
School of Materials Science and
  Engineering
Tianjin University
Nankai, Tianjin, People's Republic
  of China

and

School of Mechanical and
  Aerospace Engineering
Nanyang Technological University
and
Department of Orthopaedic Surgery
National University of Singapore
Singapore, Singapore

Minfang Chen
School of Materials Science and
  Engineering
Tianjin University
and
School of Materials Science and
  Engineering
Tianjin University of Technology
Nankai, Tianjin, People's Republic
  of China

Kui Cheng
State Key Laboratory of Silicon
  Materials
Department of Materials Science
  and Engineering
Zhejiang University
Hangzhou, Zhejiang, People's
  Republic of China

Cleo Choong
School of Materials Science and
  Engineering
Nanyang Technological University
Singapore, Singapore

Zhenduo Cui
School of Materials Science and
  Engineering
Tianjin University
Nankai, Tianjin, People's Republic
  of China

Travelle W. Franklin-Ford
Department of Biomedical
  Engineering
University of Wisconsin-Madison
Madison, Wisconsin

Kai Hu
School of Materials Science and
  Engineering
Tianjin University
Nankai, Tianjin, People's Republic
  of China

Zhao Jin
School of Materials Science and
  Engineering
Tianjin University
Nankai, Tianjin, People's Republic
  of China

Masakazu Kawashita
Graduate School of Biomedical
  Engineering
Tohoku University
Sendai, Japan

Jae Sung Lee
Department of Biomedical
  Engineering
and
Department of Orthopedics and
  Rehabilitation
University of Wisconsin-Madison
Madison, Wisconsin

Changyi Li
School of Materials Science and
  Engineering
Tianjin University
Nankai, Tianjin, People's Republic
  of China

and

School of Dentistry
Tianjin Medical University
Heping, Tianjin, People's Republic
  of China

Zhang Lijun
School of Chemistry and Chemical
  Engineering
Tian University of Technology
Nankai, Tianjin, People's Republic
  of China

Toshiki Miyazaki
Graduate School of Life Science and
  Systems Engineering
Kyushu Institute of Technology
Kitakyushu, Japan

William L. Murphy
Department of Biomedical
  Engineering
and
Department of Orthopedics and
  Rehabilitation
University of Wisconsin-Madison
Madison, Wisconsin

and

AO Foundation Collaborative
  Research Center
Davos, Switzerland

Soon-Eng Ong
Temasek Laboratories
Nanyang Technological University
Singapore, Singapore

Wei Qiang
School of Materials Science and
  Engineering
Tianjin University
Nankai, Tianjin, People's Republic
  of China

Lei Shang
School of Mechanical and
  Aerospace Engineering
Nanyang Technological University
Singapore, Singapore

Liu Shimin
Department of Precious Stones and
  Materials
Technology School of Business
Tianjin University of Commerce
Beichen, Tianjin, People's Republic
  of China

Darilis Suarez-Gonzalez
Department of Surgery
University of Wisconsin-Madison
Madison, Wisconsin

Eng San Thian
Department of Mechanical
  Engineering
National University of Singapore
Singapore, Singapore

Wilson Wang
Department of Orthopaedic
  Surgery
National University of Singapore
Singapore, Singapore

Wenjian Weng
State Key Laboratory of Silicon
  Materials
Department of Materials Science
  and Engineering
Zhejiang University
Hangzhou, Zhejiang, People's
  Republic of China

Xianjin Yang
School of Materials Science and
  Engineering
Tianjin University
Nankai, Tianjin, People's Republic
  of China

Liang Yanqin
School of Materials Science and
  Engineering
Tianjin University
Nankai, Tianjin, People's Republic
  of China

Xianting Zeng
Surface Technology Group
Singapore Institute of
  Manufacturing Technology
Singapore, Singapore

Sam Zhang
School of Mechanical and
  Aerospace Engineering
Nanyang Technological University
Singapore, Singapore

Zhe Zhang
School of Mechanical and
  Aerospace Engineering
Nanyang Technological University
Singapore, Singapore

Contributors

Xie Jin Yang
School of Materials Science and
Engineering
Tongji University
Nankai, Tianjin, People's Republic
of China

Liang Yanjin
School of Materials Science and
Engineering
Pohl University
Nankai, Tianjin, People's Republic
of China

Xianting Zeng
Surface Technology Group
Singapore Institute of
Manufacturing Technology
Singapore, Singapore

Sun Zhang
School of Mechanical and
Aerospace Engineering
Nanyang Technological University
Singapore, Singapore

Zhe Zhang
School of Mechanical and
Aerospace Engineering
Nanyang Technological University
Singapore, Singapore

1

Magnetron Sputtering Deposition of Chemically Modified Hydroxyapatite

Eng San Thian and Cleo Choong

CONTENTS

1.1 Introduction

Techniques for the reconstruction of bone defects are important in order to maintain the quality of life of patients since bones support our body and enable us to perform various motions. Although our bodies have the ability to self-reconstruct damaged tissues through the normal supply of nutrients, the loss of bone and its function may lead to loss of function of related tissues when the amount of damaged bone is too large for self-reconstruction. In such a case, the damaged bones are reconstructed using alternative materials. The most popular transplantation is autograft, which is transferred from other parts of the bones of the same patient. As the bone tissue is extracted from the patient, autograft has the problem of a limited amount of tissue being available. Another candidate is allograft, which is transferred from other people. Allograft has problems related to overcoming the foreign body reaction and infections when it is implanted into the recipient. There have been major advancements in the development of biomedical materials for bone replacements, where prosthetic components are strongly attached to the bone without using bone cement, during the past three decades.[1,2] Emphasis has now shifted toward the use of bioactive ceramic materials, particularly calcium phosphates (CaPs), due to their similarity with the mineral phase found in bone.[3] CaP is bioactive and, as follows, has the ability to encourage direct bone apposition, which distinguishes it from many other metallic or polymeric biomaterials.[4] However, their inherent mechanical properties, i.e., brittleness, poor tensile strength, and poor impact resistance, restrict their application and therefore cannot be used in load-bearing parts.

CaPs, in particular hydroxyapatite (HA), are often used in the form of coatings on metallic implants to confer the biological and mechanical advantages of both types of materials. The HA-coated metallic implants show high tensile strength and ductility of the metal and bioactivity of HA. The thickness of such an apatite coating must be a compromise among a number of limiting conditions. The thinner the coating, the better its mechanical properties, but in the first few months, some 10–15 μm of an apatite surface may dissolve during the process of acquiring bone union. On the other hand, a thick coating of 100–150 μm may suffer from fatigue failure under tensile loading. The required compromise leads to an ideal thickness of approximately 50 μm.[5] Currently, a standard protocol for producing HA coatings on implant surfaces does not exist. The methods of HA coating of metallic implants are summarized in Table 1.1. Various HA coatings have shown differences in coating properties through contradicting results in animal and clinical studies.[6,7] This is because there are many properties that determine the performance of HA coating both in vitro and in vivo such as chemical composition, crystallinity and purity, surface morphology, porosity, and thickness.[2,8] These properties differ from one technique to another, and some of them can be controlled by varying the operating parameters of each technique. Plasma-spraying technology has been used for HA coatings, in which it consists of partially dehydrated HA, with amorphous CaP and other soluble CaP phases such as tricalcium phosphate (TCP). This occurs during the high-temperature deposition plasma-spraying process. Plasma-spraying technology is able to deposit HA coating with a crystallinity of 30%–70%. The normal crystallinity for plasma-sprayed HA coatings during the deposition process is approximately 65%. The ratio of HA to TCP is crucial for osteoinduction, whereby undifferentiated cells are collected and developed into differentiated bone cells.[9,10]

Surfaces of HA-coated implants that are high in amorphous and TCP phases are susceptible to a faster dissolution. Cellular viability and proliferation were affected by the dissolution of HA composites.[16–18] It is important to note that bone–implant interfacial strength is greatly influenced by HA coating crystallinity.[8] Therefore, there is a need to study the interaction

TABLE 1.1

Comparison of Different Methods to Deposit Hydroxyapatite Coating

Technique	Thickness (µm)	Advantages	Disadvantages
Thermal spraying	30–200	High deposition rate Low cost	Line-of-sight technique High temperature Rapid cooling produces amorphous coating
Sputter coating	0.5–3	Uniform thickness Dense coating	Line-of-sight technique Time consuming Produces amorphous coatings
Electron-beam deposition	~1	Uniform thickness Dense coating	Line-of-sight technique Time consuming Produces amorphous coatings
Dip coating	50–500	Inexpensive Can coat complex substrates	High sintering temperatures Thermal expansion mismatch
Electrophoretic deposition	0.1–2.0	Uniform thickness High deposition rates Can coat complex substrates	High sintering temperatures Difficult to produce crack-free coating
Hot isostatic pressing	0.2–2.0	Dense coating	Expensive High temperature Thermal expansion mismatch Cannot coat complex substrates Elastic property differences
Biomimetic coating	<30	Bone-like apatite formation Can coat complex substrates	Time consuming Requires replenishment and constant conditions
Pulsed laser deposition	~0.05–10	Coating with crystalline and amorphous Dense and porous coating Control over phases and Ca/P ratio	Line-of-sight technique Expensive High substrate temperature
Dynamic mixing	0.05–1.3	High adhesive strength	Amorphous coatings
Sol–gel	<1	Low processing temperatures Can coat complex shapes Low cost as coatings are very thin	Some processes require controlled atmosphere processing Expensive raw materials

(continued)

TABLE 1.1 (continued)

Comparison of Different Methods to Deposit Hydroxyapatite Coating

Technique	Thickness (μm)	Advantages	Disadvantages
Ion beam-assisted deposition	~0.03–4	Uniform thickness Dense pore-free coating High reproducibility and controllability over microstructure and composition	Line-of-sight technique Expensive Amorphous coatings
Powder plasma spray (PPS)	~30–300	High deposition rates Micro-rough surface and porosity	Line-of-sight technique High temperature Rapid cooling produces crack Poor control of chemical and physical coating parameters Nonuniform thickness Poor control of biodegradation
Liquid plasma spray (LPS) and suspension plasma spray (SPS)	~5–50	High deposition rates Pure (only HA phase) Crystalline Nearly fully dense and highly porous coating can be obtained Excellent control on the coating microstructure	Line-of-sight technique Expensive High temperature Nonuniform thickness
RF magnetron sputtering	~0.04–3.5	Uniform thickness Dense pore-free coating Can coat heat-sensitive substrates High-purity films Control over coating structure (amorphous or crystalline) and Ca/P ratio	Line-of-sight technique Expensive Time consuming Low deposition rate
Electrochemical deposition	<95	Low processing temperature Highly crystalline deposit Can coat complex shapes Control over thickness composition and microstructure	Long deposition time causes uneven thickness and formation of crack and holes

Source: Albrektsson, T. et al., *Eur. Spine J.* 2(10 Suppl.), S96, 2001; Kurashina, K. et al., *Biomaterials*, 23, 407, 2002; Surmenev, R., *Surf. Coat. Technol.*, 206, 2035, 2012; Abdel-Aal, E. et al., *Saudi Med. J.*, 202, 5895, 2008; Kamitakahara, M. et al., *Biomed. Mater. (Bristol, England).*, 2, R17, 2007; Eliaz, N. and Sridhar, T. M., *Cryst. Growth Des.*, 8, 3965, 2008; Yang, Y. et al., *Biomaterials*, 26, 327, 2005.

between coating crystallinity and optimum dissolution properties in order to produce a better bone–implant interfacial strength.

However, HA, in comparison with bioactive glasses and glass ceramics, has a relatively low bone-bonding rate.[19] This effect will inevitably result in the requirement for a long postoperative rehabilitation program. One approach toward improving the osteointegration rate of HA is to chemically dope HA with small amounts of elements that are commonly found in human bone,[20] since the HA structure is very tolerant for ionic substitution.

The idea of developing silicon-substituted hydroxyapatite (SiHA) is largely based on the role of silicon (Si) ions in bone and the excellent bioactivity of silica-based glasses and glass ceramics. Carlisle[21–23] reported that chicks on dietary silicon showed enhanced bone growth and development. Significant upregulation of bone cell proliferation and gene expression was observed for bioactive glasses, due to the release of low levels of Si^{4+} ions into the physiological environment.[24–26] All of these studies suggested that Si incorporation into the HA structure would certainly lead to an enhancement of bioactive performance of HA, since researches have demonstrated a significant increase in the amount of bone apposition and the quality of bone repair, compared to HA ceramics.[27–29]

Based on the success of using dense SiHA, alternative techniques for the deposition of SiHA as thin films onto metallic surfaces were subsequently explored.

1.2 Sputter Deposition

The most common technique to deposit CaP on the surface of implants is by plasma spraying. Even though this technique has a high deposition rate, it still has numerous disadvantages such as nonstoichiometric, porous, and nonuniform coating, poor adhesion between the coating and the substrate, microcracks on coating surface, and relatively low crystallinity.[30,31] Moreover, the tendency of cracks to appear on the coating and titanium interface after implantation of the device leads to the dissociation of the coating from the titanium interface. As the dissociation increases, there is an increasing possibility of phagocytosis and inflammation to occur.[32] Hence, CaP coatings on titanium substrates must be able to adhere strongly, must not possess any defects, and must be thick enough, as this plays a very important role in osteogenesis and bone tissue replacement.

An alternative to plasma spraying would be sputtering. The advantage of using sputter coating to deposit CaP onto a surface is that it produces strong adhesion between the coating and the substrate, compact microstructure, and controlled elemental composition.[33] The strong bond between the coating and the substrate is due to mechanical interlocking and chemical bonding. Sputtering is usually carried out in a vacuum environment at room temperature. This process involves the removal and deposition of a coating material (target) onto an adjacent surface (substrate). Plasma or ion guns produce high-energy gaseous ions under high-voltage acceleration that will collide with the atoms or molecules on the surface of the target. As the collision occurs, the target's atoms or molecules are ejected from the surface and propelled toward the substrate. This would then result in the formation of thin film.

Sputtering can be considered as a versatile coating technique since it is able to coat a variety of substrates with thin films of electrically conductive and nonconductive materials. The use of direct current (DC), however, allows only the sputtering of conductive material. As opposed to conventional DC sputtering, radiofrequency (RF) sputtering, which uses alternating current (AC), can be utilized to sputter both conductive and nonconductive materials. Other modifications could also be made to fit certain types of material and applications. Reactive, ion beam, and magnetron sputtering techniques, for instance, are some well-known modified methods. From the modus operandi point of view, sputtering technique uses mechanical approach rather than chemical or thermal approach in coating a substrate.

Though magnetron sputtering possesses several advantages, it has its own limitations. As it is a line-of-sight process, it is not possible to coat complex or intricate shapes. Moreover, as-sputtered CaP coatings are generally amorphous unless heating is supplied to the substrates during deposition. Sputtering is also considered as a relatively time-consuming process.

1.2.1 Technical Knowledge behind Sputtering Deposition

The target material acts as a cathode and it is connected to the negative terminal of a DC or RF power supply and subjected to large negative potential. On the other hand, the substrate, also known as the anode, is placed facing the cathode and usually grounded, biased positively or negatively, heated, cooled, or any combination of these. Air is then pumped out of the chamber, which is followed by the injection of argon gas. Argon gas is normally used for sputtering since it produces high sputter yield for most metals, chemically inert, nontoxic, and relatively cheap. The gas pressure is usually set and maintained between 1 and 100 mTorr.

The sputtering process relies on the parallel plate glow discharge, which plays a role in the formation of the plasma. The plasma can be described as a low-pressure gas where a high-energy field is used to drive the creation of ions and free electrons by ionization. In ionization, collisions with energetic particles cause a loss or gain of electrons to atoms or molecules, thus causing them to be electrically charged. As such, mobile positive ions, negative ions, electrons, and radicals can be found in the space between the electrodes.[34] Since these ions are in high electrical field, the negatively charged ions are

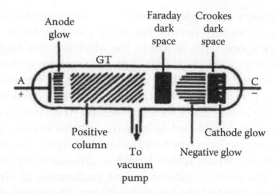

FIGURE 1.1

A discharge tube where both A and C are connected to a DC source. (From Gambhir, S. K. et al., *Foundations of Physics, Vol. II*, New Age International, Darya Ganj, Delhi, India, 1993.)

forced toward the grounded or positively charged substrate at high acceleration. As a result, this bombardment of ions contributes to the growing coating film. A small amount of current flows through due to the small number of initial charge carriers in the system.

As mentioned, glow discharge normally occurs when there is a current running through the gas at low pressure.[35] Hence, it is crucial for a discharge tube to be connected to a DC voltage at both ends of A and C in order to maintain the discharge (Figure 1.1). The Townsend discharge normally occurs during this process. Townsend discharge can be interpreted as a gas ionization process where a small initial amount of free electrons, accelerated by strong electric field, will increase the electrical conduction through a gas by avalanche multiplication (a form of current multiplication). As such, the discharge itself is self-sustainable with the right amount of electrons being generated to produce enough ions to regenerate the same number of initial electrons by positive ion bombardment of the cathode. As a result, a luminous glow is produced as the voltage drops followed by a sharp rise in the current. The increase in power signifies higher voltage and current density levels, indicating that it has reached the abnormal glow region, which is suitable for sputtering. At this stage, the ion bombardment over the entire substrate's surface would increase until the current density is almost uniform.

Different characteristic regions of a typical discharge tube connected to a DC source are presented schematically in Figure 1.1. The cathode glow is described as a highly luminous layer that can be found adjacent to the cathode. In the cathode glow region, the discharged ions and positive cathode ions are neutralized causing the secondary electrons to shift away and collide with nearby neutral gas atoms. The rise of cathode glow is caused by the decay in the excitation energy.[36] The region next to the cathode glow is known as the Crookes dark space, where the positive ion density is very high and a high drop in voltage occurs. Negative glow region is a region after the Crookes dark space. This is where the electrons are highly excited by the DC potential across the electrodes.[36] At the same time, the plasma is created due to the ionization of neutral gas molecules by the excited electrons. The electrons tend to lose their energy after several collisions and end up drifting toward the anode. Since they are lacking in excitation and ionization energy, this region is considered to be dark or also known as the Faraday dark space.[35] The region between the Faraday dark space and the anode glow is called the positive column, whose length is influenced by the length of the discharge tube. The positive column can be continuous or striated depending on the pressure. Beyond the positive column, the anode glow, where the electrons accelerated by the anode cause the gas to illuminate once again, can be found. Therefore, in order to obtain a better deposition rate, the substrate needs to be placed in the negative glow region, before the Faraday dark space.

1.2.2 Various Sputtering Deposition Techniques

1.2.2.1 DC Sputtering

The mechanism of DC sputtering has been explained in the previous section. DC sputtering is also known as diode or cathodic sputtering. The limitation of conventional DC sputtering is that it cannot be used for dielectric deposition or sputtering of insulating materials. This is because the positive charges acquired during sputtering leads to the neutralization of the target and this would prevent the necessary continuous discharge.

1.2.2.2 RF Sputtering

RF sputtering is typically used to sputter nonconductive materials. In RF sputtering, AC is used instead of DC. A sinusoidal wave generator is normally employed, and this would cause the target surface to accumulate positive charges during one cycle (Figure 1.2). Using a low frequency, which is normally less than 50 kHz, the ions can adequately mobile such that each electrode will take turns in being the anode and cathode. However, when a higher frequency is used (more than 50 kHz), two things will occur. First, the transfer of power from the cathode may occur through the insulating target by capacitive coupling. Second, the need to maintain secondary electrons in the cathode region will not be necessary, as ionizing collision will occur due to the energetic electrons that are oscillating in the glow region. RF frequency may range from 5 to 30 MHz, but the widely used frequency for RF sputtering is 13.56 MHz.[37,38]

RF sputtering works because target converts the insulating target to a negative potential, so that it may be sputtered by the bombarding positive ions. The electrons of negative target bias are known to be more mobile than

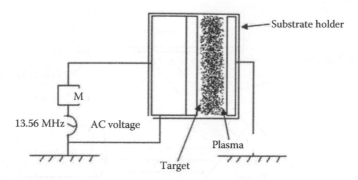

FIGURE 1.2
A schematic on RF sputtering where an AC with an RF of 13.56 MHz is used to coat the substrate.

ions and can easily adapt to the periodical change in the electric field. In this sputtering technique, the target is coupled to the RF generator. The ion flux is uniform throughout the plasma, thus causing the target of a small area to receive small ion current. The plasma potential being relative to the surface and independent of the surface area determines the electron current. This in turn will cause a negative potential due to the target receiving less ion current than electron current.

1.2.2.3 Reactive Sputtering

Reactive sputtering uses reactive gases (e.g., oxygen) instead of the inert gases like argon. This enables the deposition of insulating compounds such as oxides, nitrides, carbides, and sulfides by using DC sputtering with a metallic target.[39,40] By depositing oxides from metallic targets using reactive sputtering, a metallic target with high purity can be easily produced. Parameters such as gas pressure and sputter rate can cause the stoichiometric control to vary.

1.2.2.4 Ion Beam Sputtering

Ion beam sputtering is capable of producing thin, homogeneous, and adherent coating onto a surface. This process is carried out in a vacuum chamber where the target material will be coated onto the substrates in a cold plasma atmosphere. Unlike plasma spraying, ion beam sputtering is carried out under a high vacuum or in a cold plasma atmosphere. The target material is sputtered with inert gas ions and evaporated with electron beams (Figure 1.3). Argon is typically used in ion beam sputtering. The substrate is usually subjected to negative voltage ranging from a few hundred to a few thousand volts.[41] Plasma is produced due to glow discharge, and the ions from the target material will undergo ionization. Finally, due to the electric

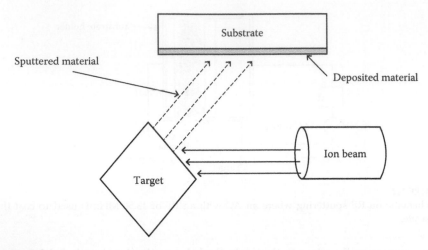

FIGURE 1.3
Ion beam (argon) is used to deposit the target material onto the substrate.

field in the system, the ions of the target material will be accelerated and deposited onto the surface of the substrate at high energies.

1.2.2.5 Magnetron Sputtering

Magnetron sputtering is a good deposition technique since it allows a magnetic field to be created above and parallel to the target surface and perpendicular to the electric field. Overall, this sputtering technique is good for thin film deposition, except that it does not utilize the target material efficiently.

By applying a magnetic field, the electrons are contained in orbits close to the cathode. The electrons are subjected to a force known as the Lorentz force, $F = -q[E + (v \times B)]$, where q is the electric charge of the particle, E is the electric field, v is the instantaneous velocity of the particle, and B is the magnetic field.[42] The free electrons are reoriented to the target surface by the magnetic field before losing their orientation to the anode. The discharging effectiveness is amplified with the increase in the collision of electrons with argon carrier molecules.

The planar magnetron is commonly the magnetic field of choice during sputtering. In this setup, magnets are positioned behind the target as shown in Figure 1.4. The magnetic field emanates at a right angle to the target and bends the magnetron component parallel to the target surface. Consequently, the electrons move in cycloid trajectory because of a force (Figure 1.4). Due to the arrangement of the magnets in the planar magnetron, the electrons are

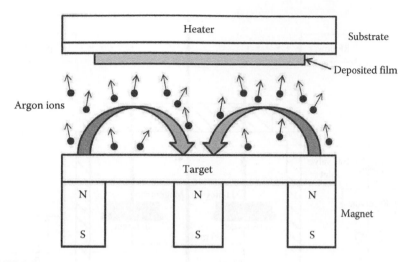

FIGURE 1.4
Deposited film onto the substrate with the use of magnetic fields in magnetron sputtering.

confined to a region called the "race track." The "race track" erodes greatly
due to the great ionization of the gas right above it.

The use of magnetron sputtering process at room temperature to deposit
SiHA thin films by using a combination of phase-pure HA and Si targets has
been reported to allow the formation of uniform well-adhered films with the
possibility of incorporating controlled amounts of Si. Thian et al.[43,44] reported
a systematic approach to deposit SiHA thin films. Figure 1.5 shows a typical setup for producing these films using a custom-built sputter deposition
system. A constant flow of argon gas at a working pressure of 0.6 Pa was supplied to the chamber at each deposition run. The composition of SiHA film
was controlled by the relative discharge power density supplied to each target. An RF (13.56 MHz) power density of 3.1 W cm^{-2} was supplied to the HA
target, and three different DC power densities of 0.2, 0.5, and 0.8 W cm^{-2} were
supplied to the Si target to achieve films with three different Si contents.

Following the successful deposition of SiHA thin films by magnetron
sputtering, there have been a number of recent reports on the fabrication of
SiHA films. Solla et al.[45,46] reported that SiHA films were obtained by pulsed
laser deposition technique since analyses revealed that Si was incorporated
into the HA structure. In a separate study, the authors found that osteoblasts
proliferate and differentiate well on these films.[47] Hijon et al.[48] produced
SiHA films with Si contents up to 2.1 wt.% using a sol–gel technique, and
these films supported cell proliferation in vitro.[49] SiHA thin films have also
been produced by electrophoretic deposition[50] and biomimetic[51,52] deposition
techniques, and the bioactive property of these films was enhanced by the
presence of Si.

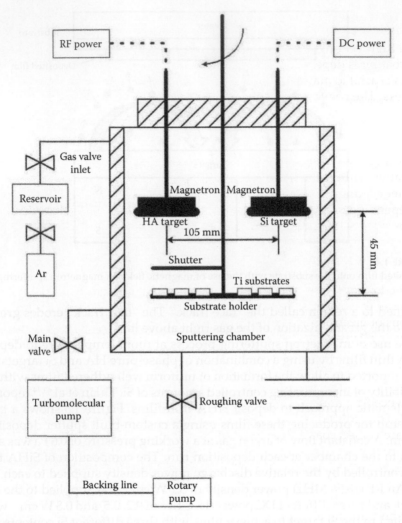

FIGURE 1.5
Magnetron sputtering deposits a layer of film onto the substrate with the use of magnetic fields. (With kind permission from Springer Science+Business Media: *J. Mater. Sci.*, Silicon-substituted hydroxyapatite (SiHA): A novel calcium phosphate coating for biomedical applications, 41, 2006, 709–717, Thian, E. et al.)

1.3 Physicochemical Properties

The mechanical properties of HA depend on two factors, namely, density (or porosity) and grain size of the ceramics. A dense HA with small grain size will constitute the high mechanical strength of HA.[53] Sintering temperature and time are crucial in improving density and governing the grain size. The usage of sputter coating technique is shown to produce coatings that have high coating–substrate bonding strength.[33] The surface roughness of the coating is dependent on the annealing temperature and time.[15,54,55]

It is crucial to maintain the integrity of HA coatings during the sputtering process. The conversion of amorphous to HA coatings according to the different degrees of crystallinity is carried out by controlling the heat-treatment temperatures and heating environment. Coating crystallinity is affected by the absence or presence of water vapor. The former will produce an increase in coating crystallinity when temperature is increased, while the presence of the latter will produce high crystallinity at a fixed temperature during the sputter coating.[54]

Depending on the type of sputtering system and deposition parameters, it is possible for the Ca/P ratio of the coating to differ from the target material. The Ca/P ratio sputter-coated onto the target surface can range from 1.6 to 2.6.[15,56] The range of Ca/P ratio between the HA target and coatings is influenced by the selective deposition of Ca ions and the loss of P ions. The cause for the loss of P ions might have been due to it being pumped away before reaching the substrate[57] or due to the weak P ion bonds being sputtered away by ions or electrons.[58]

Coating–substrate interface is another important factor that needs to be considered for an HA-coated bone implant. It has been reported that the average adhesion strength for sputter coating is 38 MPa.[59] The mechanical integrity and the adhesion strength of the coating–substrate interface are influenced by annealing and the types of annealing and surface condition used. The reduced strength of 17.9 and 9.0 MPa is caused by quenched and furnace cooled coatings respectively on passivated Ti surfaces.[59] The adhesion strength of 45.8 MPa was achieved only when the nonpassivated Ti surfaces were annealed first, followed by quenching. The difference in adhesion strength was due to the different substrate treatments used after annealing. The thickness of the oxide film and lateral oxide layer is caused by passivation and during annealing respectively.[60] The existing oxide layer is weakened due to the presence of the aforementioned lateral oxide layer. Therefore, the mechanical integrity of the coating–substrate interface is heavily influenced by the surface condition of the substrate.

The use of monolithic HA coating using RF magnetron sputtering by Ding et al. showed high bond strength of 59.9 MPa.[61] However, surface cracks and granular-shaped structures were formed with monolithic HA coating. Multilayered coating was developed to improve the bonding strength between the coating and the target material.[37]

Thian et al.[44] have fabricated SiHA films of thickness 0.7 μm. The Si composition in the film (0.8, 2.2, and 4.9 wt.%) was increased with the DC power density on the Si target. Following annealing at 600°C, the amorphous films transformed into phase-pure, polycrystalline films, with major peaks matching that of HA [Joint Committee on Powder Diffraction Standards

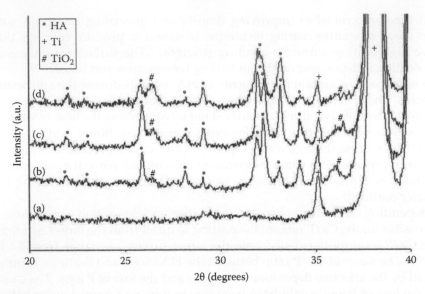

FIGURE 1.6
XRD patterns of SiHA films (a) as-deposited; (b) annealed, 0.8 wt.% Si; (c) annealed, 2.2 wt.% Si; and (d) annealed, 4.9 wt.% Si. (With kind permission from Springer Science + Business Media: *J. Mater. Sci.*, Silicon-substituted hydroxyapatite (SiHA): A novel calcium phosphate coating for biomedical applications, 41, 2006, 709–717, Thian, E. et al.)

(JCPDS) #9-432] was indicated by the x-ray diffraction (XRD) scans (Figure 1.6). There was no evidence of secondary phases such as TCP, tetracalcium phosphate, or calcium oxide present in the films. Also, the lattice plane reflections were found to become broader and less intense with increasing Si level.

Annealing led to several changes in the infrared (IR) spectra, with four typical absorption bands for HA being observed: one at 3571 cm⁻¹, which was assigned to the stretching mode of hydroxyl (O–H); the second at 1089 cm⁻¹, which was assigned to v_3 phosphate (P–O) stretching mode, became sharper and more intense; the third at 1038 cm⁻¹, which was also assigned to v_3 P–O stretching mode; and the last at 962 cm⁻¹, which was assigned to v_1 P–O stretching mode (Figure 1.7). Though all the IR spectra exhibited absorption bands associated with HA, the band intensities corresponding to O–H and P–O (most noticeably at 962 cm⁻¹) decreased with increasing Si level.

All these results signified the transformation of an amorphous film into a well-crystallized, phase-pure HA film after annealing. However, there has been an ongoing debate concerning the performance of amorphous and crystalline films. As crystalline HA has a considerably lower dissolution rate than amorphous HA,[62–64] it may thus be beneficial for long-term implant performance since it can lead to a decreased incidence of osteolysis.[63,65] On the other hand, amorphous HA was observed to enhance early bone apposition in vivo.[66] Phase purity is also a critical issue for thin films (thickness less

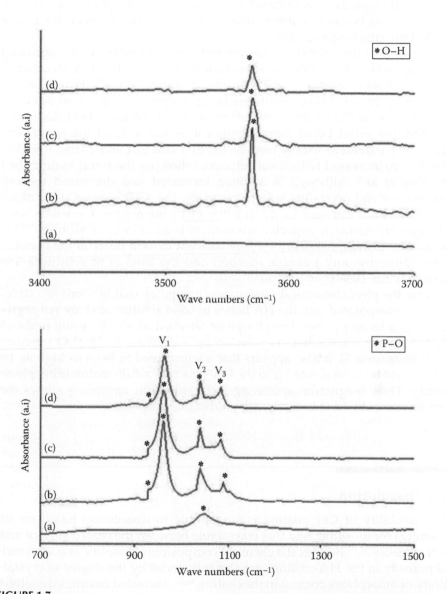

FIGURE 1.7
IR spectra of SiHA films (a) as-deposited; (b) annealed, 0.8 wt.% Si; (c) annealed, 2.2 wt.% Si; and (d) annealed, 4.9 wt.% Si. (With kind permission from Springer Science + Business Media: *J. Mater. Sci.*, Silicon-substituted hydroxyapatite (SiHA): A novel calcium phosphate coating for biomedical applications, 41, 2006, 709–717, Thian, E. et al.)

than 1 μm) since secondary CaP phases will dissolve faster than HA.[67] This effect will cause the film to dissolve completely into the physiological environment even before it is able to trigger the promotion of direct attachment at the bone/implant interface.

A change in the crystallite size and unit cell parameters of the annealed coatings was noted with the substitution of Si into the HA structure.[43] With Si addition from 0.8 to 4.9 wt.%, the crystallite size of SiHA decreased from 50 to 19 nm. These results are in agreement with those works done by Gibson et al.[68] and Arcos et al.,[69] for which Si addition inhibited grain growth, the effect being more significant as the Si level increased. The unit cell parameter's axes (*a* from 0.9436 to 0.9466 nm and *c* from 0.6915 to 0.6956 nm) increased with Si substitution, following the trend as described by Kim et al.[70] Although Si addition increased and decreased the cell parameters and crystallite size, respectively, the XRD scans did not show any significant changes in the relative peak intensities, for which such changes are normally expected when there is an atomic substitution. This is because Si and phosphorus (P) are adjacent to each other in the periodic table (differ by only 1 atomic number) and the level of Si substitution is relatively low (less than 5 wt.%).

From the physicochemical viewpoints, it appears that Si^{4+} ions are structurally incorporated into the HA lattice in solid solution and are not segregated as a second phase. The phosphate tetrahedral was basically replaced by the silicate tetrahedral, as reflected by a decrease in IR P–O intensity with increasing Si. It also appears that a substituted Si level as high as 4.9 wt.% could be incorporated into the HA structure while maintaining phase purity. Thus, magnetron sputtering technique with annealing allows the production of SiHA thin films with predictable properties.

1.4 Dissolution

The solubility of CaP coating is susceptible to dissolution based on its chemical composition, and this plays a role between the coated surface and cell adhesion.[62,63] Besides the chemical composition, quantity and structure of porosity in the HA coatings are also influenced by the degree of crystallinity or amorphous content in the coating.[64,65] Annealed coatings dissolved slower than amorphous coatings, and the crystallinity of the coatings will increase due to the rise in temperatures. Degradation of HA occurs when it dissolves readily in decreasing pH solutions[71] and also by the diffusion phenomenon caused by crystal defects such as impurities, vacancies, or dislocations.[72,73]

Solubility of HA coatings can be reduced using several ways such as increasing the crystal growth to a specific size,[74] using fluorine or hydroxyl ions as substitutes,[73,74] or adding sodium and carbonate ions to HA. Reports have shown that 100% amorphous plasma-sprayed coatings had lower dissolution rate compared to coatings with 40% amorphous and 60% crystallinity.[75] The decrease in porosity, surface area, and particle size had an effect on the decreasing solubility of HA coatings.[76,77] Moreover, the over- or undersaturation of the solution containing hydroxide, calcium, phosphate, and fluoride ions with the presence of proteins and amino acids in the solution influences the solubility of HA coatings.[78,79] The presence of amino acids and proteins have been reported to inhibit the formation of CaP on HA- or CaP-coated surfaces. It was also shown that osteoclasts also affected the dissolution rates of HA in hydrochloric acid.[71]

1.5 Biological Property: Acellular Testing

The development of a rapid means of ranking the bioactivity of materials has been the use of simulated body fluid (SBF), developed by Kokubo and coworkers.[80] As such, the rate at which a material is able to induce surface CaP precipitation from SBF-K9 solution is a measure of its level of bioactivity. Immersion testing using SBF-K9 solution demonstrated that annealed SiHA films possessed better bioactivity under physiological conditions, compared to as-deposited SiHA films or even as-deposited HA films and uncoated titanium (Ti) substrates.[43]

Dissolution of the as-deposited SiHA film, as evidenced by the appearance of submicrometer-sized pits on the surface, was observed after 1 day of immersion. This phenomenon seemed too exuberant by day 4 (Figure 1.8a). However, newly precipitated circular patches, comprising freshly nucleated crystallites, seemed to form on the surface by day 7 (Figure 1.8b). A homogeneous porous layer was observed after 14 days of immersion (Figure 1.8c). Surface cracks on the annealed SiHA film appeared to be exacerbated after immersion for 1 day. This effect possibly could be caused by the preferential dissolution along the crack regions. By day 4, freshly nucleated crystallites were observed, covering as much as 85% of the film surface (Figure 1.9a). A homogeneous, porous layer was then observed after day 7 (Figure 1.9b), which transformed into a dense layer by day 14 (Figure 1.9c). In contrast, porous layers consisting of short interconnecting rods (similar to Figure 1.9c) and loosely packed globules (Figure 1.10) were formed on the surface of as-deposited HA film and on the uncoated Ti substrate after 14 days of immersion, respectively.

Detailed characterization of the newly formed layer confirmed that it was rich in calcium and phosphorus, along with the incorporation of sodium and magnesium. Broad and diffuse XRD peaks ascribed to HA were observed, indicating that the CaP layer was apatitic in nature.

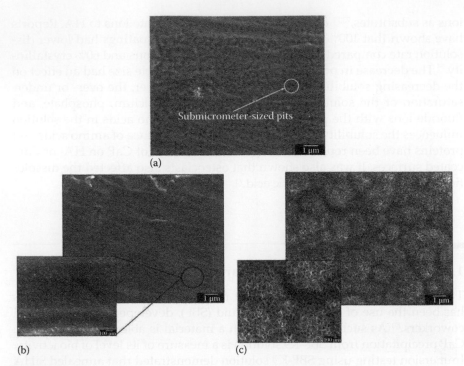

FIGURE 1.8
SEM images of as-deposited 0.8 wt.% SiHA film after immersion in SBF solution for (a) 4 days, (b) 7 days, and (c) 14 days. (From Thian, E. S. et al., *J. Biomed. Mater. Res. B Appl. Biomater.*, 76, 326, 2006.)

With prolonged immersion, the peak intensity corresponding to lattice plane reflection (002) increased, implying that this CaP layer is preferentially aligned along the *c*-axis. The appearance of carbonate (C–O) bands at 867 and 1416 cm^{-1} in the IR spectra confirmed the presence of carbonate-containing apatite layer. It was also noted that the hydroxyl (O–H) band intensity decreased while the intensities of the C–O bands became more pronounced, suggesting substitution of CO_3^{2-} ions for OH$^-$ ions in the HA structure.[83]

In the case of as-deposited SiHA film, the effect of precipitation process is attributed to the lack of coating crystallinity. Dissolution occurs in the initial stage where Ca^{2+} and P^{5+} ions are released into the SBF. These ions will increase the degree of supersaturation of SBF solution, thereby facilitating the initial formation of CaP nuclei on the film surface, by consuming the Ca^{2+} and P^{5+} ions in the SBF solution. CaP crystallite growth then proceeded along with the continued formation of additional nuclei.

As for the annealed SiHA film, the apatite nucleation process can be related to the structural-scale factor, which explains why annealed SiHA is

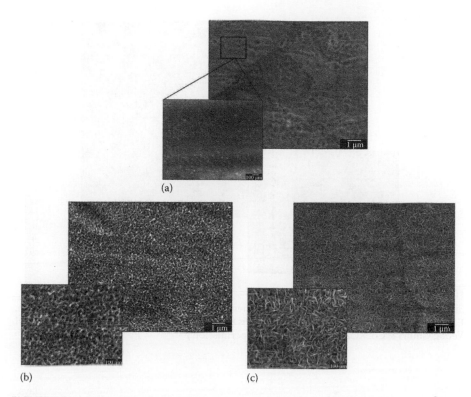

FIGURE 1.9
SEM images of annealed 0.8 wt.% SiHA film after immersion in SBF solution for (a) 4 days, (b) 7 days, and (c) 14 days. (From Thian, E. S. et al., *J. Biomed. Mater. Res. B Appl. Biomater.*, 76, 326, 2006.)

more bioactive than as-deposited SiHA. It is proposed that the nanocrystalline structure creates an increased grain boundary area, thereby providing an abundance of nucleation sites for CaP crystallites to form since crystallite formation usually commences at areas of low surface energy. Once formed, these crystallites grow spontaneously and transform into an apatite layer on the film.

When comparing SiHA and HA films, a dense carbonate-containing apatite layer was formed on SiHA by day 14, but this was not the case for HA, suggesting that SiHA was more bioactive than HA. As dissolution–precipitation is a surface-mediated process, the presence of surface silicate anions (due to the released Si^{4+} ions from SiHA) could have altered the surface properties of the material, thus attributing to the high bioactivity of SiHA. Botelho et al.[84] demonstrated that, with the presence of silicate anions, the surface charge of SiHA decreased, thereby inducing favorable sites for nucleation and crystallization of the apatite.

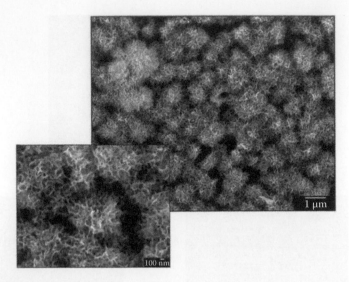

FIGURE 1.10

SEM image of uncoated Ti substrate after immersion in SBF solution for 14 days. (From Thian, E. S.: *Ceramic Nanocoatings and Their Applications in the Life Sciences*. 2011. Copyright Wiley-VCH Verlag GmbH & Co. KGaA. Reproduced with permission.)

1.6 Biological Property: Cellular Testing

The biological response of human osteoblast-like (HOB) cells to SiHA thin films, taking into account the effect of variation in Si substitution on the film stability and its bioactivity, was investigated using the alamarBlue™ assay. The results indicated that an increase in the growth of HOB cells with culture time was generally observed on all samples (Figure 1.11a).[85] HOBs growing on annealed 0.8 wt.% SiHA (S2) sample showed the highest growth at all time points. In addition, a significant difference was seen at days 3, 7, and 14 when comparing as-deposited 0.8 wt.% SiHA (S1) and S2 samples, or uncoated titanium (Ti) substrate and S2 samples. Thian et al.[86] also found that the metabolic activity of HOB cells varied with different levels of Si substitution (Figure 1.11b). Generally, all samples were able to support cell growth, with annealed 2.2 wt.% SiHA (S3) samples exhibiting the highest growth throughout the culture period. A statistically significant increase in cell growth was observed at day 14 on annealed 4.9 wt.% SiHA (S4) as compared to S2 samples. Furthermore, a significant increase was noted for S3 when compared to S2 at days 4 and 7.

HOBs attaching on S1 and S2 samples displayed numerous well-developed actin filaments (being stained green) and distinct vinculin focal adhesion

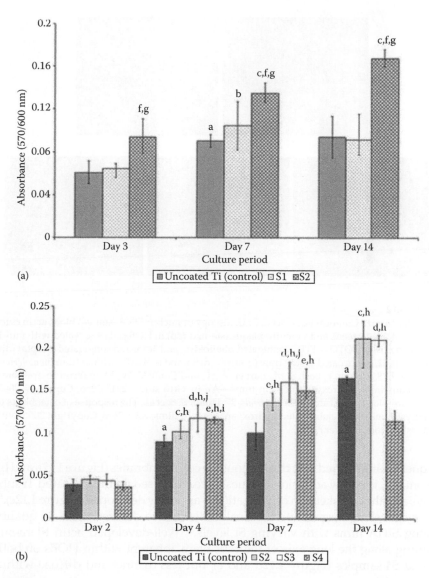

FIGURE 1.11

Growth of HOB cells with culturing time. [a]$p < 0.05$: higher on uncoated Ti substrate (control) between groups; [b]$p < 0.05$: higher on S1 between groups; [c]$p < 0.05$: higher on S2 between groups; [d]$p < 0.05$: higher on S3 between groups; [e]$p < 0.05$: higher on S4 between groups; [f]$p < 0.05$: higher on S2 than control within groups; [g]$p < 0.05$: higher on S2 than S1 within groups; [h]$p < 0.05$: higher on coated (S2, S3 or S4) than control within groups, [i]$p < 0.05$: higher on S4 than S2 within groups; [j]$p < 0.05$: higher on S3 than S2 within groups. (Reprinted from *Biomaterials*, 26, Thian, E. S., Huang, J., Best, S. M., Barber, Z. H., and Bonfield, W., Magnetron co-sputtered silicon-containing hydroxyapatite thin films—An in vitro study, 2947–2956, Copyright 2005, with permission from Elsevier; *Biomaterials*, 27, Thian, E. S. et al., The response of osteoblasts to nanocrystalline silicon-substituted hydroxyapatite thin films, 2692–2698, Copyright 2006, with permission from Elsevier.)

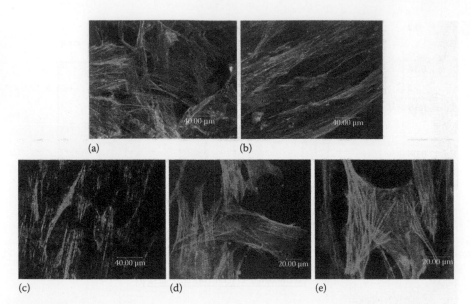

(a)   (b)

(c)   (d)   (e)

FIGURE 1.12
(See color insert.) Confocal fluorescence microscopy of nuclear DNA (stained blue), actin cytoskeleton (stained green), and vinculin plaque (stained red) in HOB cell as revealed with multiple labeling using TOTO-3, FITC conjugated phalloidin, and Texas red conjugated streptavidin on (a) S1, (b) S2, (c) uncoated Ti substrate (control), (d) S3, and (e) S4. (Reprinted from *Biomaterials*, 26, Thian, E. S., Huang, J., Best, S. M., Barber, Z. H., and Bonfield, W., Magnetron co-sputtered silicon-containing hydroxyapatite thin films—An in vitro study, 2947–2956, Copyright 2005, with permission from Elsevier; *Biomaterials*, 27, Thian, E. S. et al., The response of osteoblasts to nanocrystalline silicon-substituted hydroxyapatite thin films, 2692–2698, Copyright 2006, with permission from Elsevier.)

plaques (being stained red) throughout the cell membranes (Figure 1.12a and b). In contrast, very few adhesion plaques were expressed on the uncoated Ti substrate, and the cytoskeleton organization was poorly developed (Figure 1.12c).[85] The cytoskeleton organization revealed distinctive differences in its quality among SiHA films with varying Si levels.[86] Well-developed actin filaments aligning along the long axis of the cells were apparent within HOBs on both S3 and S4 samples (Figure 1.12d and e), but less distinct and diffuse within HOBs on S2 sample.

Cells seem to attach, spread, and grow on all samples, with filopodia extending from the cell edges.[86] Cell spreading was less pronounced, displaying underdeveloped lamellipodia on the uncoated substrate (Figure 1.13a). On the other hand, well-flattened cells were observed to cover the surfaces completely on all SiHA films (Figure 1.13b). Beyond day 42, sign of mineralization was observed on the surfaces of all samples. The formation of CaP mineral deposits on the uncoated Ti substrate was sparse (Figure 1.13c), while numerous and large mineral deposits were seen on all SiHA films, being the most significant for S4 (Figure 1.13d).

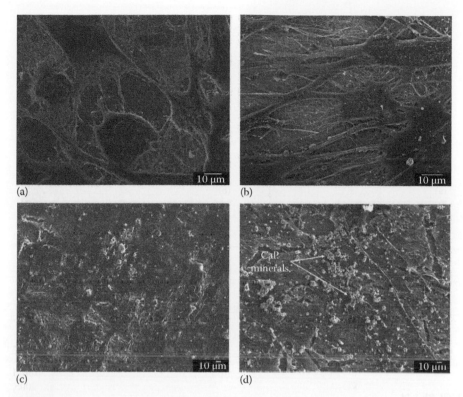

FIGURE 1.13
Cell morphology at different culture times. (a) Uncoated Ti substrate (control) at day 2, (b) S2 at day 2, (c) control at day 42, and (d) S4 at day 42. (With kind permission from Springer Science + Business Media: *J. Mater. Sci.*, Silicon-substituted hydroxyapatite (SiHA): A novel calcium phosphate coating for biomedical applications, 41, 2006, 709–717, Thian, E. et al.)

An apparent difference in the film behavior was revealed after subjecting to cell culture medium.[85,86] For S1 sample, the surface was covered by a lacy CaP network structure, with CaP spherulites seen precipitating and coalescing on the dissolving coating surface (Figure 1.14a). Nanocrystallites started to nucleate on the surfaces of S2, S3, and S4 samples (Figure 1.14b through d). Furthermore, surface dissolution was observed on S4, as evidenced by the appearance of dissolution pits. A porous CaP structure was obtained for S2 sample while a dense structure was achieved for both S3 and S4 samples. In contrast, a CaP layer was formed on the uncoated substrate only after 42 days in culture.

Silicon (Si) has shown to inhibit crystal growth and that this effect was more significant with increasing Si level.[68,69,81] As such, one might expect that with increased Si content, HA crystals tend to dissolve faster, thereby facilitating the rapid precipitation of a carbonated-apatite layer that provides favorable sites for the cells to attach and grow.[87] The released Si^{4+} ions could

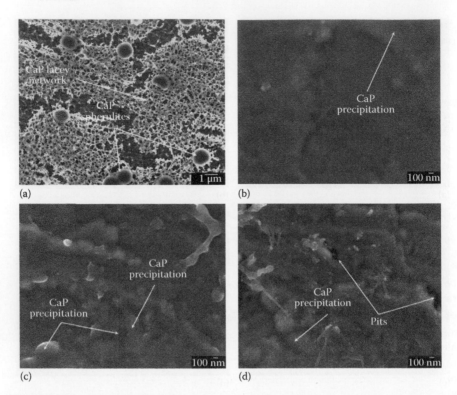

FIGURE 1.14
Coating morphology at different culture times. (a) S1 at day 4, (b) S2 at day 2, (c) S3 at day 2, and (d) S4 at day 2. (Reprinted from *Biomaterials*, 27, Thian, E. S. et al., The response of osteoblasts to nanocrystalline silicon-substituted hydroxyapatite thin films, 2692–2698, Copyright 2006, with permission from Elsevier.)

also have a positive protein adsorption. The formed silicate network structure has shown to be capable of adsorbing proteins;[88] thus, SiHA films tend to contain a higher protein concentration than that of either uncoated Ti substrates or HA-coated substrates. These adsorbed proteins are likely to trigger specific mRNA for gene expression, thereby carrying the code information to the ribosome sites of protein synthesis in the cell, thereby stimulating osteoblast outgrowth. Studies by Keeting et al.[89] and Reffitt et al.[90] supported the hypotheses since they showed that soluble Si simulated the proliferation and differentiation of HOB cells in vitro.

Cell growth was retarded at early culture period for films with high Si level (4.9 wt.%), due to the rapid dissolution of the films owing to their small crystallite size. However, when the surface was modified by the growth of a new carbonated apatite layer, cells began to migrate, adhere, and grow rapidly.

All these findings indicated that Si tends to control the dissolution rate of the film and, at the same time, plays an important role in the mineralization process. As such, of the three compositions studied, 2.2 wt.% SiHA is the preferred optimal one and merits further investigation.

1.7 Conclusions

The usage of magnetron sputtering process can be used to coat surfaces with metallic and inorganic materials. This process enables the coatings to be dense, highly uniform, and continuous. Unlike plasma-sprayed coatings, sputter coating produces a much higher coating–metal interfacial bonding strength.[33] Moreover, sputter coatings also produce nano-sized crystals, which are good for osteoblast activities.

In vitro studies showed that these films exhibited excellent bioactivity over uncoated substrates and HA-coated substrates. A high Si level not only resulted in the enhancement of biomineralization process but also led to a rapid coating dissolution rate due to different coating structures. The choice of coating depends on the final application, but these SiHA thin films offer a good substitute to the HA thin films. Even though sputter coating is seen as a better alternative in producing better bone strength and initial osseo-integration rate than plasma-spraying coating, there is still a need to carry out further studies on HA and CaP properties for better bone formation for HA coatings on dental and orthopedic implant surfaces with this method. Moreover, more clinical trials need to be carried out to further evaluate and optimize the sputter coating technique of HA and CaP for dental and orthopedic implants.

References

1. Jaffe, W. L. and Scott, D. F. Total hip arthroplasty with hydroxyapatite-coated prostheses. *J Bone Joint Surg Am.* **78**, 1918–1934 (1996).
2. Thomas, K. Hydroxyapatite coatings. *Orthopedics* **17**, 267–278 (1994).
3. Aoki, H., Kato, K., Ogiso, M., and Tabata, T. Studies on applications of apatite dental materials. *Shika Rikogaku Zasshi.* **18**, 86–89 (1977).
4. Bonfield, W. Designing porous scaffolds for tissue engineering. *Philos Transact A Math Phys Eng Sci.* **364**, 227–232 (2006).
5. Geesink, R. G., de Groot, K., and Klein, C. P. Bonding of bone to apatite-coated implants. *J Bone Joint Surg Br.* **70**, 17–22 (1988).
6. Geesink, R. G. T. Osteoconductive coatings for total joint arthroplasty. *Clin Orthop Relat Res.* **460**, 53–65 (2002).
7. Havelin, L. The Norwegian Arthroplasty Register: 11 years and 73,000 arthro-plasties. *Acta Orthop Scand 71.* **73**, 337–353 (2000).

8. Oh, S., Tobin, E., Yang, Y., Carnes, D. L., and Ong, J. L. In vivo evaluation of hydroxyapatite coatings of different crystallinities. *Int J Oral Maxillofac Implants.* **20**, 726–731 (2005).

9. Albrektsson, T., Johansson, C., and Spine, J. Osteoinduction, osteoconduction and osseointegration. *Eur Spine J.* **2**(10 Suppl.), S96–S101 (2001).

10. Kurashina, K., Kurita, H., Wu, Q., Ohtsuka, A., and Kobayashi, H. Ectopic osteo-genesis with biphasic ceramics of hydroxyapatite and tricalcium phosphate in rabbits. *Biomaterials* **23**, 407–412 (2002).

11. Surmenev, R. A review of plasma-assisted methods for calcium phosphate-based coatings fabrication. *Surf Coat Technol.* **206**, 2035–2056 (2012).

12. Abdel-Aal, E., Dietrich, D., Steinhaeuser, S., and Wielage, B. Electrocrystallization of nanocrystallite calcium phosphate coatings on titanium substrate at different current densities. *Saudi Med J.* **202**, 5895–5900 (2008).

13. Kamitakahara, M., Ohtsuki, C., and Miyazaki, T. Coating of bone-like apatite for development of bioactive materials for bone reconstruction. *Biomed Mater (Bristol, England).* **2**, R17–R23 (2007).

14. Eliaz, N. and. T. M. Sridhar. Electrocrystallization of hydroxyapatite and its dependence on solution conditions. *Cryst Growth Des* **8**, 3965–3977 (2008).

15. Yang, Y., Kim, K. H., and Ong, J. L. A review on calcium phosphate coatings produced using a sputtering process—An alternative to plasma spraying. *Biomaterials* **26**, 327–337 (2005).

16. Saha, N., Dubey, A. K., and Basu, B. Cellular proliferation, cellular viability, and biocompatibility of HA-ZnO composites. *J Biomed Mater Res B Appl Biomater.* **100**, 256–264 (2012).

17. Meleti, Z., Shapiro, I. M., and Adams, C. S. Inorganic phosphate induces apop-tosis of osteoblast-like cells in culture. *Bone* **27**, 359–366 (2000).

18. Adams, C. S., Mansfield, K., Perlot, R. L., and Shapiro, I. M. Matrix regulation of skeletal cell apoptosis. Role of calcium and phosphate ions. *J Biol Chem.* **276**, 20316–20322 (2001).

19. Oonishi, H. et al. Comparative bone growth behavior in granules of bioceramic materials of various sizes. *J Biomed Mater Res.* **44**, 31–43 (1999).

20. Driessens, F. The mineral in bone, dentin and tooth enamel. *Bull Soc Chim Belg.* **89**, 663–689 (1980).

21. Carlisle, E. M. Biochemical and morphological changes associated with long bone abnormalities in silicon deficiency. *J Nutr.* **110**, 1046–1056 (1980).

22. Carlisle, E. M. Silicon: An essential element for the chick. *Science* **178**, 619–621 (1972).

23. Carlisle, E. M. Silicon: A possible factor in bone calcification. *Science* **167**, 279–280 (1970).

24. Xynos, I. D., Edgar, A. J., Buttery, L. D., Hench, L. L., and Polak, J. M. Gene-expression profiling of human osteoblasts following treatment with the ionic products of Bioglass 45S5 dissolution. *J Biomed Mater Res.* **55**, 151–157 (2001).

25. Xynos, I. D. et al. Bioglass 45S5 stimulates osteoblast turnover and enhances bone formation in vitro: Implications and applications for bone tissue engineering. *Calcif Tissue Int.* **67**, 321–329 (2000).

26. Gough, J. E., Jones, J. R., and Hench, L. L. Nodule formation and mineralisa-tion of human primary osteoblasts cultured on a porous bioactive glass scaffold. *Biomaterials* **25**, 2039–2046 (2004).

27. Patel, N. et al. in vivo assessment of hydroxyapatite and silicate-substituted hydroxyapatite granules using an ovine defect model. *J Mater Sci Mater Med.* **16**, 429–440 (2005).
28. Porter, A. E., Patel, N., Skepper, J. N., Best, S. M., and Bonfield, W. Effect of sintered silicate-substituted hydroxyapatite on remodelling processes at the bone-implant interface. *Biomaterials* **25**, 3303–3314 (2004).
29. Patel, N. et al. A comparative study on the in vivo behavior of hydroxyapatite and silicon substituted hydroxyapatite granules. *J Mater Sci Mater Med.* **13**, 1199–1206 (2002).
30. Xu, J. L., Khor, K. A., Gu, Y. W., Kumar, R., and Cheang, P. Radio frequency (rf) plasma spheroidized HA powders: Powder characterization and spark plasma sintering behavior. *Biomaterials* **26**, 2197–2207 (2005).
31. Narayanan, R., Seshadri, S. K., Kwon, T. Y., and Kim, K. H. Calcium phosphate-based coatings on titanium and its alloys. *J Biomed Mater Res B Appl Biomater.* **85**, 279–299 (2008).
32. Sun, L., Berndt, C. C., Gross, K. A., and Kucuk, A. Material fundamentals and clinical performance of plasma-sprayed hydroxyapatite coatings: A review. *J Biomed Mater Res.* **58**, 570–592 (2001).
33. Massaro, C. et al. Surface and biological evaluation of hydroxyapatite-based coatings on titanium deposited by different techniques. *J Biomed Mater Res.* **58**, 651–657 (2001).
34. Kinbara, A., Kusano, E., and Kondo, I. Fundamentals of plasma and sputtering processes. *Biochem J.* **51**, 475–478 (1998).
35. Gambhir, S. K., Durgapal, M., and SiBanerjee, D. *Foundations of Physics, Vol. II*, New Age International, Darya Ganj, Delhi, India, 1993.
36. Kay, E. and Marton, L. Impact evaporation and thin film growth in a glow discharge. *Adv Electron Electron Phys.* **17**, 245–322 (1962).
37. Ding, S. J. Properties and immersion behavior of magnetron-sputtered multi-layered hydroxyapatite/titanium composite coatings. *Biomaterials* **24**, 4233–4238 (2003).
38. Nakamura, S., Hamagami, J. I., and Yamashita, K. Hydrothermal crystallization of carbonate-containing hydroxyapatite coatings prepared by radiofrequency-magnetron sputtering method. *J Biomed Mater Res B Appl Biomater.* **80**, 102–106 (2007).
39. Barshilia, H. and Rajam, K. Reactive sputtering of hard nitride coatings using asymmetric-bipolar pulsed DC generator. *Surf Coat Technol.* **201**, 1827–1835 (2006).
40. Long, J., Sim, L., Xu, S., and Ostrikov, K. Reactive plasma-aided deposition of hydroxyapatite bio-implant coatings. *Chem Vap Deposition* **13**, 299–306 (2007).
41. Wang, C. X., Chen, Z. Q., Wang, M., Liu, Z. Y., and Wang, P. L. Ion-beam-sputtering/mixing deposition of calcium phosphate coatings. I. Effects of ion-mixing beams. *J Biomed Mater Res.* **55**, 587–595 (2001).
42. Davidson, P. *An Introduction to Magnetohydrodynamics*, Cambridge University Press, Cambridge, U.K., 2001.
43. Thian, E. S., Huang, J., Best, S. M., Barber, Z. H., and Bonfield, W. A new way of incorporating silicon in hydroxyapatite (Si-HA) as thin films. *J Mater Sci Mater Med.* **16**, 411–415 (2005).
44. Thian, E. et al. Silicon-substituted hydroxyapatite (SiHA): A novel calcium phosphate coating for biomedical applications. *J Mater Sci.* **41**, 709–717 (2006).

45. Solla, E. Pulsed laser deposition of silicon substituted hydroxyapatite coatings from synthetical and biological sources. *Appl Surf Sci.* **254**, 1189–1193 (2007).
46. Solla, E. Pulsed laser deposition of silicon-substituted hydroxyapatite coatings. *Vacuum* **82**, 1383–1385 (2008).
47. López-Alvarez, M. et al. Silicon-hydroxyapatite bioactive coatings (Si-HA) from diatomaceous earth and silica. Study of adhesion and proliferation of osteoblast-like cells. *Anim Biotechnol.* **20**, 1131–1136 (2009).
48. Hijón, N., Victoria Cabañas, M., Peña, J., and Vallet-Regí, M. Dip coated silicon-substituted hydroxyapatite films. *Acta Biomater.* **2**, 567–574 (2006).
49. Balamurugan, A. et al. Suitability evaluation of sol-gel derived Si-substituted hydroxyapatite for dental and maxillofacial applications through in vitro osteoblasts response. *Dent Mater.* **24**, 1374–1380 (2008).
50. Xiao, X. F., Liu, R. F., and Tang, X. L. Electrophoretic deposition of silicon substituted hydroxyapatite coatings from n-butanol-chloroform mixture. *J Mater Sci Mater Med.* **19**, 175–182 (2008).
51. Zhang, B., Kobayashi, Y., Chiba, T., and Fujie, M. G. Robotic patch-stabilizer using wire driven mechanism for minimally invasive fetal surgery. *Conf Proc IEEE Eng Med Biol Soc.* **2009**, 5076–5079 (2009).
52. Zhang, E., Zou, C., and Zeng, S. Preparation and characterization of silicon-substituted hydroxyapatite coating by a biomimetic process on titanium substrate. *J Endocrinol.* **203**, 1075–1080 (2009).
53. Rao, W. R. and Boehm, R. F. A study of sintered apatites. *J Dent Res.* **53**, 1351–1354 (1974).
54. Yang, Y., Kim, K. H., Agrawal, C. M., and Ong, J. L. Influence of post-deposition heating time and the presence of water vapor on sputter-coated calcium phosphate crystallinity. *J Dent Res.* **82**, 833–837 (2003).
55. Yang, Y., Kim, K. H., Mauli Agrawal, C., and Ong, J. L. Effect of post-deposition heating temperature and the presence of water vapor during heat treatment on crystallinity of calcium phosphate coatings. *Biomaterials* **24**, 5131–5137 (2003).
56. Ong, J. et al. Surface characterization of ion-beam sputter-deposited Ca-P coatings after in vitro immersion. *Colloid Physicochem Eng Aspects* **87**, 151–162 (1994).
57. Zalm, P., Cuomo, J., Rossnagel, S., and Kaufman, H. Quantitative sputtering. In Cuomo J. J. (Ed.), *Handbook of Ion Beam Processing Technology: Principles, Deposition, Film Modification, and Synthesis*, Noyes Publications, NJ, pp. 78–111, 1989.
58. Li, P. et al. Apatite formation induced by silica gel in a simulated body fluid. *J Am Ceram Soc.* **75**, 2094–2097 (1992).
59. Ong, J. L., Lucas, L. C., Lacefield, W. R., and Rigney, E. D. Structure, solubility and bond strength of thin calcium phosphate coatings produced by ion beam sputter deposition. *Biomaterials* **13**, 249–254 (1992).
60. Golightly, F., Stott, F., and Wood, G. The influence of yttrium additions on the oxide-scale adhesion to an iron-chromium-aluminum alloy. *Oxid Met.* **10**, 163–187 (1976).
61. Ding, S., Lee, T., and Chu, Y. Environmental effect on bond strength of magnetron-sputtered hydroxyapatite/titanium coatings. *J Mater Sci Lett.* **22**, 479–482 (2003).
62. Klein, C. P., Wolke, J. G., de Blieck-Hogervorst, J. M., and de Groot, K. Features of calcium phosphate plasma-sprayed coatings: An in vitro study. *J Biomed Mater Res.* **28**, 961–967 (1994).

63. Bloebaum, R. D. and Dupont, J. A. Osteolysis from a press-fit hydroxyapatite-coated implant. A case study. *J Arthroplasty* **8**, 195–202 (1993).
64. Wolke, J. G., van Dijk, K., Schaeken, H. G., de Groot, K., and Jansen, J. A. Study of the surface characteristics of magnetron-sputter calcium phosphate coatings. *J Biomed Mater Res*. **28**, 1477–1484 (1994).
65. Morscher, E. W., Hefti, A., and Aebi, U. Severe osteolysis after third-body wear due to hydroxyapatite particles from acetabular cup coating. *J Bone Joint Surg Br*. **80**, 267–272 (1998).
66. de Bruijn, J. D., Bovell, Y. P., and van Blitterswijk, C. A. Structural arrangements at the interface between plasma sprayed calcium phosphates and bone. *Biomaterials* **15**, 543–550 (1994).
67. Driessens, F. C. Physiology of hard tissues in comparison with the solubility of synthetic calcium phosphates. *Ann NY Acad Sci*. **523**, 131–136 (1988).
68. Gibson, I., Best, S., and Bonfield, W. Effect of silicon substitution on the sintering and microstructure of hydroxyapatite. *J Am Ceram Soc*. **85**, 2771–2777 (2002).
69. Arcos, D., Carvajal, J., and Rodríguez- Vallet-Regí, M. Silicon incorporation in hydroxylapatite obtained by controlled crystallization. *Chem Mater*. **16**, 2300–2308 (2004).
70. Kim, S. R. et al. Synthesis of Si, Mg substituted hydroxyapatites and their sintering behaviors. *Biomaterials* **24**, 1389–1398 (2003).
71. Hankermeyer, C. R., Ohashi, K. L., Delaney, D. C., Ross, J., and Constantz, B. R. Dissolution rates of carbonated hydroxyapatite in hydrochloric acid. *Biomaterials* **23**, 743–750 (2002).
72. Wang, H., Lee, J. K., Moursi, A., and Lannutti, J. J. Ca/P ratio effects on the degradation of hydroxyapatite in vitro. *J Biomed Mater Res A*. **67**, 599–608 (2003).
73. Raemdonck, W. V., Ducheyne, P., Meester, P. D. Auger electron spectroscopic analysis of hydroxyapatite coating on titanium. *J Am Ceram Soc*. **63**, 381–384 (1984).
74. Boskey, A. L. and Posner, A. S. Bone structure, composition, and mineralization. *Orthop Clin North Am*. **15**, 597–612 (1984).
75. Maxian, S. H., Zawadsky, J. P., and Dunn, M. G. In vitro evaluation of amorphous calcium phosphate and poorly crystallized hydroxyapatite coatings on titanium implants. *J Biomed Mater Res*. **27**, 111–117 (1993).
76. LeGeros, R. Z. Biodegradation and bioresorption of calcium phosphate ceramics. *Clin Mater*. **14**, 65–88 (1993).
77. Sun, L., Berndt, C. C., Khor, K. A., Cheang, H. N., and Gross, K. A. Surface characteristics and dissolution behavior of plasma-sprayed hydroxyapatite coating. *J Biomed Mater Res*. **62**, 228–236 (2002).
78. Matsumoto, T. et al. Crystallinity and solubility characteristics of hydroxyapatite adsorbed amino acid. *Biomaterials* **23**, 2241–2247 (2002).
79. Serro, A. P., Fernandes, A. C., and de Jesus Vieira Saramago, B. Calcium phosphate deposition on titanium surfaces in the presence of fibronectin. *J Biomed Mater Res*. **49**, 345–352 (2000).
80. Kokubo, T., Kushitani, H., Sakka, S., Kitsugi, T., and Yamamuro, T. Solutions able to reproduce in vivo surface-structure changes in bioactive glass-ceramic A-W. *J Biomed Mater Res*. **24**, 721–734 (1990).
81. Thian, E. S., Huang, J., Best, S. M., Barber, Z. H., and Bonfield, W. Novel silicon-doped hydroxyapatite (Si-HA) for biomedical coatings: An in vitro study using acellular simulated body fluid. *J Biomed Mater Res B Appl Biomater*. **76**, 326–333 (2006).

82. Thian, E. S. *Ceramic Nanocoatings and Their Applications in the Life Sciences*, Wiley VCH, New York, 2011.
83. Rehman, I. and Bonfield, W. Characterization of hydroxyapatite and carbonated apatite by photo acoustic FTIR spectroscopy. *J Mater Sci Mater Med.* **8**, 1–4 (1997).
84. Botelho, C. M., Lopes, M. A., Gibson, I. R., Best, S. M., and Santos, J. D. Structural analysis of Si-substituted hydroxyapatite: Zeta potential and x-ray photoelectron spectroscopy. *J Mater Sci Mater Med.* **13**, 1123–1127 (2002).
85. Thian, E. S., Huang, J., Best, S. M., Barber, Z. H., and Bonfield, W. Magnetron co-sputtered silicon-containing hydroxyapatite thin films—An in vitro study. *Biomaterials* **26**, 2947–2956 (2005).
86. Thian, E. S. et al. The response of osteoblasts to nanocrystalline silicon-substituted hydroxyapatite thin films. *Biomaterials* **27**, 2692–2698 (2006).
87. Neo, M., Nakamura, T., Ohtsuki, C., Kokubo, T., and Yamamuro, T. Apatite formation on three kinds of bioactive material at an early stage in vivo: A comparative study by transmission electron microscopy. *J Biomed Mater Res.* **27**, 999–1006 (1993).
88. Schwarz, K. Proceedings: Recent dietary trace element research, exemplified by tin, fluorone, and silicon. *Fed Proc.* **33**, 1748–1757 (1974).
89. Keeting, P. E. et al. Zeolite A increases proliferation, differentiation, and transforming growth factor beta production in normal adult human osteoblast-like cells in vitro. *J Bone Miner Res.* **7**, 1281–1289 (1992).
90. Reffitt, D. M. et al. Orthosilicic acid stimulates collagen type 1 synthesis and osteoblastic differentiation in human osteoblast-like cells in vitro. *Bone* **32**, 127–135 (2003).

2

Electrochemical Deposition of Hydroxyapatite and Its Biomedical Applications

Toshiki Miyazaki and Masakazu Kawashita

CONTENTS

2.1 Introduction

Electrochemical deposition has been widely used for the preparation of thin films (TFs) and coatings [1]. It has been developed as an effective method of metal coating for the processing of electronic components, but electrochemical deposition of ceramic coatings onto metals has also been carried out for various applications, including the improvement of corrosion resistance of the metal or the creation of a more refractory surface for a high-temperature service. Although electrochemical deposition can be performed by cathodic or anodic methods, anodic deposition has limited utility in terms of the materials that can be deposited by this method and the types of substrates that can be used for deposition. Cathodic deposition, however, has important advantages for industrial applications. Two processes are commonly used to prepare ceramic coatings by cathodic electrochemical deposition: the electrophoretic process using suspensions of ceramic particles and the electrolytic process using solutions of metal salts, as shown in Figure 2.1 [2]. The electrophoretic deposition (EPD) process is often used

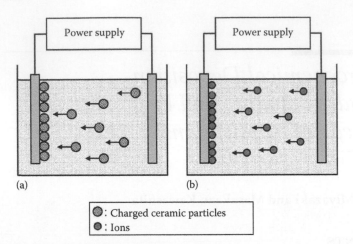

FIGURE 2.1
Schematic representations of (a) the EPD process and (b) ELD process.

for thick ceramic films, whereas the electrolytic deposition (ELD) process is often used for the formation of nanostructured thin ceramic films.

In the biomedical field, ceramic coatings have been used to modify the surfaces of orthopedic and dental metallic implants and, in some cases, to create an entirely new surface that gives the implant properties that are quite different from the uncoated device. Titanium (Ti) metal and its alloys are widely used for orthopedic and dental applications because they exhibit excellent mechanical properties and biocompatibility [3,4]. Ti metal has a passive layer of TiO_2 a few nanometers thick on its surface, which is responsible for its chemical stability and therefore its biocompatible characteristics. However, although Ti metal is biocompatible, it is not a bioactive material and hence cannot tightly bond to living bone. Here, a bioactive material is defined as a material that elicits a specific biological response at the interface of the material that results in the formation of a bond between the tissues and the material [5].

Calcium phosphate (CaP) ceramics such as hydroxyapatite (HA) are widely studied for orthopedic and dental applications due to their osteoconductive (i.e., bioactive) properties [5–10]. However, bulk CaP ceramics are intrinsically brittle and relatively weak compared to common implant metals and alloys and high-strength ceramics such as alumina and zirconia. Consequently, in order to combine the mechanical strength of metals with the bioactivity of CaP ceramics in load-bearing implant applications, many attempts to integrate CaP ceramic coatings with metallic implants, which are not intrinsically bioactive, have been made. As a coating method, the plasma-spraying technique [11,12] has been the most successful in applying CaP coatings to implants up to now because of its high deposition rate and the ability to coat large areas. The osteoconductivity of plasma-sprayed coatings has been confirmed by numerous studies, but there are some serious concerns about the process, such as the unpredictable phase change of CaP ceramics during the coating process, particle release, delamination, and so on [13]. In order to overcome the drawbacks of plasma spraying, various deposition methods such as hot isostatic pressing [14], sputter deposition [15], sol–gel deposition [16], biomimetic deposition [17], and others have been proposed. Here, we focus on EPD and ELD of CaP ceramics, including HA.

2.2 Electrochemical Deposition of CaP Ceramics

Numerous studies have been reported on EPD of CaP ceramics such as HA onto metallic biomaterials [18]. Early studies on EPD of CaP ceramics onto Ti metal and its alloys were done by Ducheyne and his colleagues from the late 1980s to the early 1990s [19–21]. They attempted EPD of various kinds of CaP ceramics, including commercially available HA particles, HA particles prepared by a precipitation method, β-calcium phosphate (β-TCP) particles, or HA–TCP particles. The CaP ceramic particles were suspended in isopropanol, and an electric field was applied between the lead anode and the cathode made of Ti metal or its alloys. After EPD, calcination was performed under various conditions. It was found that EPD coupled with calcination has a remarkable effect on the composition and crystalline phase of the deposited layer. In the late 1990s, Zhitomirsky and colleagues developed a stable suspension of stoichiometric HA nanoparticles 55 nm in size in isopropanol and achieved a high deposition rate as well as a low current density [22].

Over the last decade, most studies about EPD of HA nanoparticles have mainly focused on improving the uniformity and adhesion of the deposited layers. For example, it was reported that the ripening of HA nanoparticles [23], the stability of HA suspensions [24], and the surface finishing of substrates [25] affect the morphology and adhesion of the deposited HA. Recently, thin and uniform coatings of HA nanoparticles by the EPD process were found to have the potential to provide novel biosensors [26–29]. Figure 2.2 shows atomic force microscopic (AFM) images of Au/Cr-plated quartz substrates (a) before and (b) after EPD of HA nanoparticles, which indicate that the nanoparticles were coated onto the substrates [29].

Studies on ELD of CaP ceramics onto metallic substrates have been active since the 1990s. For example, Redepenning et al. [30,31] and Shirkhanzadeh [32,33] reported (ELDs) of dicalcium phosphate dihydrate (DCPD: $CaHPO_4 \cdot 2H_2O$) and HA on orthopedic alloys from acidic CaP solutions. Royer and Rey reported that amorphous CaP-containing carbonate was coated onto a Ti–6Al–4V alloy via electrochemical deposition using an aqueous solution saturated with DCPD and that it was transformed into carbonate apatite via heat treatment at 650°C under vacuum [34]. Monma investigated the influences of electrolyte concentration, temperature, and additives on

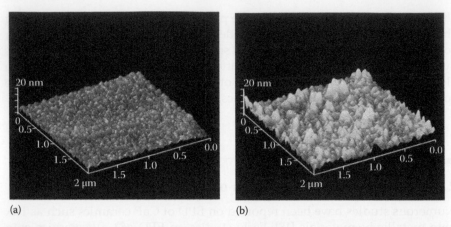

(a)　　　　　　　　　　　　　　　　　　　　(b)

FIGURE 2.2
AFM images of substrates. (a) Original Au/Cr-plated quartz substrates before EPD and (b) after EPD of HA nanoparticles. (With kind permission from Springer Science+Business Media: *J. Mater. Sci. Mater. Med.*, Preparation of low-crystalline apatite nanoparticles and their coating onto quartz substrates, 23, 2012, 1355–1362, Kawashita, M., Taninai, K., Li, Z., Ishikawa, K., and Yoshida, Y.)

the ELD of CaPs onto stainless steel substrates [35,36]. His studies revealed that apatite deposits were composed of ellipsoidal grains, needles, or fine granules, depending on the temperature and current, and that apatite deposited from fluorine-containing solutions resulted in fluoridated calcium-deficient apatites having a composition similar to $(Ca, Na)_9(HPO_4)(PO_4)_5(F, H_2O)_2 \cdot nH_2O$ with Ca/P molar ratios near 1.50.

The early investigations introduced earlier mainly focused on the electrochemical deposition of a CaP coating in a CaP solution having a relatively simple composition. However, in order to obtain CaP coatings with composition and structure similar to that of apatite in bone, simulated body fluid (SBF) [37,38] is appropriate to use as an electrolyte because the composition of SBF is almost identical to that of human blood plasma, and SBF has been used for biomimetic apatite coating on substrates [39]. Ban et al. first used modified simulated body fluid (m-SBF) as an electrolyte in the electrochemical deposition of a CaP coating on titanium (Ti) substrates [40,41]. This SBF had the composition of well-known SBF but without $MgCl_2 \cdot 6H_2O$, KCl, or $NaHCO_3$. They extensively investigated the morphology and microstructure of CaP coatings and found that needle-like apatite crystals were deposited on the Ti substrates in m-SBF at an electrolyte temperature higher than 52°C and at a current density higher than 12.9 mA/cm² [40]. They also proposed a hydrothermal–electrochemical method in an autoclave with two electrodes and showed that at 150°C–160°C, the HA rod grew very homogeneously perpendicular to the substrate. Their study also showed that the edge of the needle had a flat hexagonal plane [42,43], and thus the HA-coated titanium mesh showed excellent biocompatibility [44].

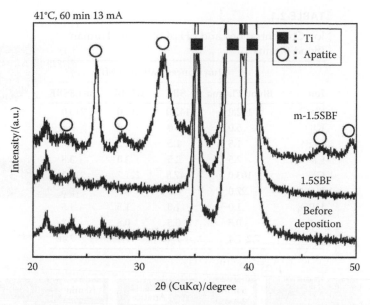

41°C, 60 min 13 mA

FIGURE 2.3
Thin-film (TF)-XRD patterns of titanium substrates after electrochemical deposition in 1.5SBF and m-1.5SBF at 41°C for 60 min at 13 mA. (With kind permission from Springer Science+Business Media: *J. Mater. Sci. Mater. Med.*, Apatite formation on titanium substrates by electrochemical deposition in metastable calcium phosphate solution, 19, 2008, 137–142, Kawashita, M., Itoh, S., Miyamoto, K., and Takaoka, G.H.)

Recently, apatite deposition on Ti substrates was attempted in the electrolyte of a metastable CaP solution (1.5SBF) [45] that had 1.5 times the ion concentrations of normal SBF. It was expected that apatite might form rapidly on the titanium by electrochemical deposition in 1.5SBF at a lower temperature and lower current density because the degree of supersaturation of 1.5SBF with respect to apatite is much higher than that of normal SBF. Indeed, as shown in Figure 2.3, apatite can be formed in modified 1.5SBF (m-1.5SBF), which has 1.5 times the ion concentrations of a normal SBF but does not contain $MgCl_2 \cdot 6H_2O$, at 41°C and 8.7 mA/cm^2, although it is not formed in 1.5SBF. Table 2.1 gives the ion concentrations of SBF, 1.5SBF, m-1.5SBF, and human plasma. These results indicate that magnesium ions may play an important role in the apatite formation in the electrochemical deposition of CaP in SBFs. Actually it was reported that magnesium ions enter the structure of forming HA nuclei by replacing calcium, resulting in a distorted atomic structure that slows the subsequent growth of HA [46]. In addition, apatite was not formed after 1 min deposition, but it started to precipitate after 5 min deposition. The area of apatite increased after 10 min deposition, and a dense apatite layer was partially formed after 20 min deposition. After 40 and 60 min deposition, a dense and uniform apatite layer covered whole surface of the titanium

TABLE 2.1

Ion Concentrations and pH of SBFs and Human
Blood Plasma

Ion	Ion Concentration (mM)			
	Blood Plasma	SBF	1.5SBF	m-1.5SBF
Na^+	142.0	142.0	213.0	213.0
K^+	5.0	5.0	7.5	7.5
Mg^{2+}	1.5	1.5	2.3	0
Ca^{2+}	2.5	2.5	3.8	3.8
Cl^-	103.0	147.8	221.7	217.2
HCO_3^-	27.0	4.2	6.3	6.3
HPO_4^{2-}	1.0	1.0	1.5	1.5
SO_4^{2-}	0.5	0.5	0.8	0.8
pH	7.2–7.4	7.40	7.40	7.40

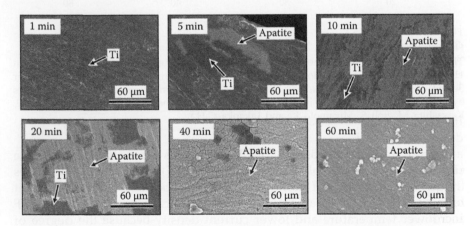

FIGURE 2.4
Field emission-SEM photographs of titanium substrates after electrochemical deposition in
m-1.5SBF at 41°C for different deposition times at 13 mA. (With kind permission from Springer
Science+Business Media: *J. Mater. Sci. Mater. Med.*, Apatite formation on titanium substrates
by electrochemical deposition in metastable calcium phosphate solution, 19, 2008, 137–142,
Kawashita, M., Itoh, S., Miyamoto, K., and Takaoka, G.H.)

substrates, as shown in Figure 2.4. The thickness of the deposited apatite
increased with increasing current, but the current did not have a significant
effect on the crystallite size of apatite, as shown in Figure 2.5. In the previous
study using m-SBF without $MgCl_2\cdot6H_2O$, KCl, or $NaHCO_3$ [40], CaP,
including HA and octacalcium phosphate (OCP: $Ca_8H_2(PO_4)_6\cdot5H_2O$), was
formed on the Ti substrates. However, only apatite formed on the titanium
substrates when the electrochemical deposition was performed in m-1.5SBF
that did not contain $MgCl_2\cdot6H_2O$ but did contain both KCl and $NaHCO_3$.

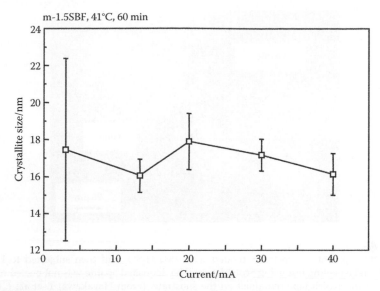

FIGURE 2.5
Change in crystallite size of apatite as a function of current. (With kind permission from Springer Science + Business Media: *J. Mater. Sci. Mater. Med.*, Apatite formation on titanium substrates by electrochemical deposition in metastable calcium phosphate solution, 19, 2008, 137–142, Kawashita, M., Itoh, S., Miyamoto, K., and Takaoka, G.H.)

It is therefore speculated that the apatite formed by electrochemical deposition in m-1.5SBF is more similar to bone apatite in its composition and structure than the apatite that was reported in the previous study using m-SBF [40].

The electrochemical deposition in m-1.5SBF makes it possible to coat apatite onto metallic substrates, but it still has the following drawbacks. This method increases the pH at the interface between Ti and the electrolyte due to electron incorporation, forming OH⁻ ions and H_2 through water reduction [47]. The H_2 gas evolution at the interface leads to a heterogeneous coating [48]. Furthermore, the adhesive strength between apatite and the substrate is often low, and the total amount of apatite crystals deposited on the Ti substrate is much smaller than that theoretically calculated from the total electric charge [49]. To solve these problems, electrochemical deposition of apatite under pulse current has been attempted. The electrodeposition under a long pulse (15 s) was more effective for crystal growth and apatite formation than that under direct current [50]. The studies also revealed that when the Ti metals were previously etched in 75% sulfuric acid (H_2SO_4), followed by electrodeposition and heat treatment, a dense and uniform apatite layer with good adhesive properties was formed onto the Ti substrates, as shown in Figure 2.6.

The pulse current has an effect on the structure and adhesion of apatite electrochemically deposited onto Ti substrates [51]. In this study, current is applied as a square-wave pulse, and the current-on time (T_{on}) is equal to the

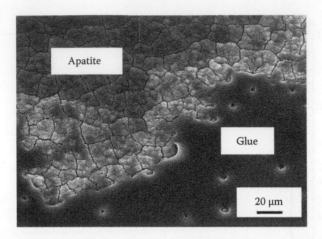

FIGURE 2.6
SEM photographs of Ti substrates treated with 75% H_2SO_4 and then subjected to ELD in m-1.5SBF and a peeling test using Scotch® tape. The deposited apatite was not peeled off, and the glue of the Scotch tape remained on the substrate. (From Hayakawa, T. et al., *J. Ceram. Soc. Jpn.*, 116, 68, 2008. With permission from The Ceramic Society of Japan.)

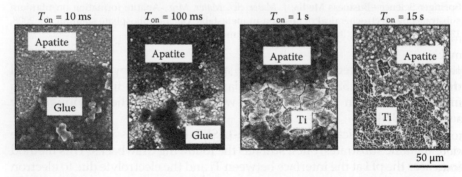

FIGURE 2.7
SEM photographs of Ti substrates subjected to ELD in m-1.5SBF at various T_{on} followed by a peeling test using Scotch® tape. (a) T_{on}=10 ms, (b) T_{on}=100 ms, (c) T_{on}=1 s, and (d) T_{on}=15 s. (From Hayakawa, T. et al., *J. Mater. Res.*, 23, 3176, 2008. With permission from Cambridge University Press.)

current-off time (Figure 2.7). As shown in Figure 2.7, the samples prepared using T_{on} of 100 and 10 ms exhibit better adhesive strength between the apatite and the Ti substrates than do the samples prepared using T_{on} of 1 and 15 s. The glue remains in the samples with T_{on} of 100 and 10 ms, but bare Ti is seen in the samples with T_{on} of 1 and 15 s. It is considered that small CaP crystals with low crystallinity were deposited on the Ti surface without reacting with other CaP crystals, H_2O, or HCO_3^- in the surrounding environment. This resulted in the relaxation of the lattice mismatch and improvement in the adhesive strength between the apatite crystals and Ti substrates.

Chitosan is a biodegradable and biocompatible polymer that has found widespread use in the cosmetics and food industries. It also has been widely studied for applications in medicine because it is known to be osteoconductive. Studies have shown that chitosan is a hemostatic material and that it has a beneficial influence on wound-healing and bone reformation. Recently, ELD of HA/chitosan composite coatings was proposed [52]. Composite coatings containing $CaHPO_4 \cdot 2H_2O$ and chitosan were prepared by electrochemical deposition. The $CaHPO_4 \cdot 2H_2O$/chitosan composites were converted to HA/chitosan composites in aqueous solutions of sodium hydroxide. The coatings ranged from approximately 1% to 15% chitosan by weight. Qualitative assessment of the coatings showed that adhesion significantly improved over that observed for electrochemical deposition of pure HA coatings.

2.3 Mechanism of Electrochemical Deposition of CaP Ceramics

Numerous studies have been done on EPD of CaP ceramics, but the exact mechanism is still not entirely clear. In a review paper on EPD [53], four mechanisms are introduced: (1) flocculation by particle accumulation mechanism, (2) particle charge neutralization mechanism, (3) electrochemical particle coagulation mechanism, and (4) electrical double-layer distortion and thinning mechanism [54]. Among them, mechanism (4) is the most plausible for the EPD of CaP ceramics such as HA because it considers the movement of a positively charged oxide particle, similar to HA particles in most solvents, toward the cathode without an increase in electrolyte concentration near the electrode. As shown in Figure 2.8, when positively charged particles move toward the cathode, the double-layer envelope is distorted to become thinner in front of and wider behind the particle (Figure 2.8a). The counter anions behind the positively charged particle will react with the cations around them to reduce the double-layer envelope behind the particle (Figure 2.8b). As a result, the next incoming particle can approach the particle to induce coagulation and deposition (Figure 2.8c).

The mechanism of ELD of a CaP coating in a CaP electrolyte system has also been discussed in some studies [47,55]. Figure 2.9 shows a schematic representation of the reactions occurring during the electrochemical deposition of a CaP coating [47]. The composition of the resulting CaP coating depends on the reaction pH, temperature, and solution composition [55]. Cathodic polarization of metallic substrates leads to an increase in pH at the interface between the alloy and electrolyte due to the formation of OH⁻ ions. The sudden increase in pH triggers crystal nucleation and initiates crystal growth of the desired CaP phase directly on the substrate surface.

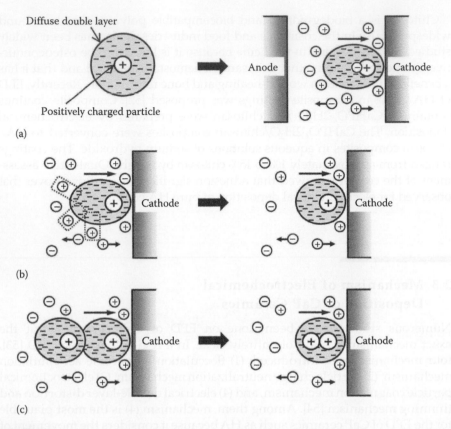

FIGURE 2.8
Schematic representation of mechanisms in EPD process. (a) Distortion of diffuse double layer by EPD, (b) local thinning of diffuse double layer, and (c) coagulation and deposition. (From *Prog. Mater. Sci.*, 52, Besra, L. and Liu, M., A review on fundamentals and applications of electrophoretic deposition (EPD), 1, Copyright 2007, with permission from Elsevier.)

Reactions at cathode surface are

$$2H_2O + 2e^- \rightarrow H_2 \uparrow + 2OH^- \tag{2.1}$$

$$2H_2O + O_2 + 4e^- \rightarrow 4OH^- \tag{2.2}$$

$$10Ca^{2+} + 6PO_4^{3-} + OH^- \rightarrow Ca_{10}(PO_4)_6(OH)_2 \downarrow (HA) \tag{2.3}$$

$$8Ca^{2+} + 2HPO_4^{2-} + 4PO_4^{3-} \rightarrow Ca_8H_2(PO_4)_6 \downarrow (OCP) \tag{2.4}$$

$$Ca^{2+} + HPO_4^{2-} + 2H_2O \rightarrow CaHPO_4 \cdot 2H_2O \downarrow (DCPD) \tag{2.5}$$

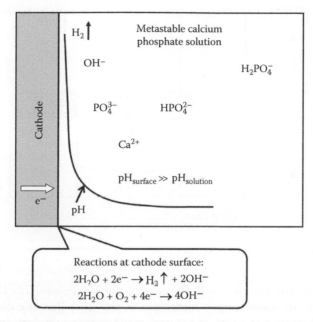

FIGURE 2.9
Schematic representation of reactions occurring during ELD process in metastable CaP solution. (From Lin, S et al., *J. Biomed. Mater. Res.*, 66A, 819, 2003. With permission.)

As can be seen in reactions (2.3) through (2.5), depending on the reaction pH, temperature, and solution composition, different crystalline phases of CaP coating are deposited on the substrates. Actually, HA is preferentially deposited when the pH of the electrolyte adjacent to cathode surface is higher than around 4.5 because the solubility of HA is lower than those of other CaPs such as DCPD and OCP [56], but by using an acidic CaP solution as the electrolyte, DCPD or OCP is likely to be deposited.

2.4 Mechanical Properties of the HA Films Fabricated by Electrodeposition

In order for the HA coating to exhibit bioactivity over a long implantation period, the coating should tightly adhere to the substrates. Several techniques have been used to quantitatively measure the adhesive strength of the electrochemically deposited HA layer. Among them, the tensile, shear, and scratch tests are most popular.

In the tensile test, the specimen is attached to metal jig with epoxy glue, and then tensile stress is applied perpendicular to the interface between the HA and the substrate until fracture occurs. ASTM standards of C633 and

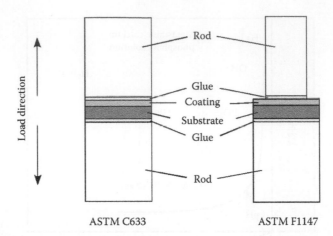

FIGURE 2.10
Tensile test designated by ASTM C633 and F1147.

F1147 are generally referred to in the case of the HA coating on Ti, as shown in Figure 2.10. In these tests, penetration of the glue into the coating layer should be taken into account. In general, most conventional plasma-sprayed HA coatings are highly dense due to complete melting by the plasma. Therefore, the effect of the glue penetration into the plasma-sprayed coating is minimal. On the other hand, most electrically deposited HA layers exhibit a porous structure composed of plate-like and rod-like nanocrystals. Therefore, it should be noted that the measured strength is easily affected by glue penetration.

In order to precisely analyze fracture behavior, the interface between the HA coating and the substrate after the test is observed by using an optical or electron microscope. A schematic representation of the fracture mode is shown in Figure 2.11. If the fracture occurs inside the coating or at the coating/substrate interface, the measured strength closely reflects the actual adhesion strength of the coating. On the other hand, if the fracture occurs inside the glue or in the glue/coating interface, the actual adhesion strength is higher than the measured value.

In shear strength measurements, jigs are attached to the specimens, as shown in Figure 2.12. Shear stress is applied to the specimens parallel to the interface between the coating and the substrate. In most research papers, shear strength was measured according to the ASTM F1044 standard. In a scratch test, a needle of diamond and sapphire is attached to the specimen and then the surface is scratched, as shown in Figure 2.13. During the test, the applied load is continuously increased. The bonding strength is then calculated from the critical load at fracture.

A significant amount of research on the tensile adhesion strength of electrochemically deposited HA coatings has been reported. LeGeros et al. measured the adhesive strength of the OCP layer deposited by ELD in an

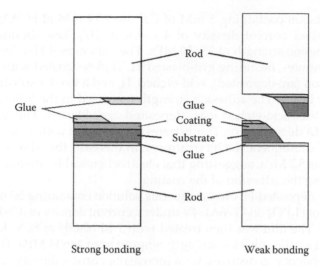

FIGURE 2.11
Schematic representation of the fracture mode.

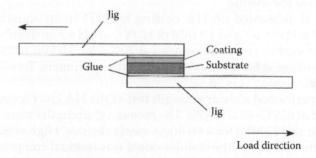

FIGURE 2.12
Schematic representation of shear strength measurement.

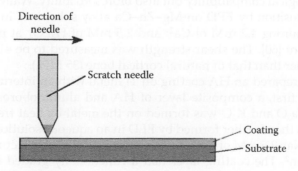

FIGURE 2.13
Schematic representation of scratch test.

aqueous solution containing 5 mM of Ca^{2+} and 3.4 mM of $H_2PO_4^-$ at pH 5.0 under a pulsed current density of 4 mA/cm² [47]. The obtained coating showed adhesion strength of 35–50 MPa. They attempted HA deposition on various substrates, including grit-blasted Ti, Ti plates coated with Ti wire by arc deposition (arc-deposited), acid-etched Ti, and a Co–Cr substrate coated with Co–Cr beads. The adhesion strength increased in the following order: Co–Cr beaded < acid-etched ≈ arc-deposited < grit-blasted. The authors also examined HA deposition on Ti substrates by treatment with a CaP solution at pH 2.67 and subsequent NaOH treatment. In this case, the adhesion strength was as low as 5.2 MPa, suggesting that electrochemical treatment is effective for improving the adhesion of the coating.

Han et al. deposited HA in an aqueous solution containing 50 mM of Ca^{2+} and 25 mM of $H_2PO_4^-$ on Ti–6Al–4V under a current density of 0.2–15 mA/cm² at 60°C [57]. The film was then treated with 1 M NaOH at 80°C for 2 h. The film showed tensile adhesion strength ranging from 4 to 14 MPa. The strength showed a tendency to decrease with increasing current density, and a dense coating was obtained at low-current density. In this case, fracture occurred at the interface between the coating and the substrate. On the other hand, a porous coating was formed at high-current density. In this case, fracture occurred inside the coating.

Huang et al. fabricated an HA coating by ELD in an aqueous solution containing 2 mM of Ca^{2+} and 1.2 mM of $H_2PO_4^-$ at pH 4.7 on Ti substrates [58]. Then, the HA-coated Ti was autoclaved at 200°C for 3 h and heated at 700°C for 2 h. DC voltage at 30 mV was applied to the specimens. Tensile adhesion strength of the coating was 38 MPa.

Wei et al. performed a shear strength test of the HA layer formed by EPD and sintered at 875°C–1000°C [59]. The measured strengths were 12 MPa for a Ti substrate and 22 MPa for a stainless steel substrate. High shear adhesion strength of the coating on the stainless steel was residual compressive stress in the coating caused by a higher thermal expansion coefficient for the stainless steel than for the HA.

Magnesium alloys are expected for novel metallic biomaterials exhibiting not only biological compatibility but also bioresorbability. Wang et al. examined HA deposition by EPD on Mg–Zn–Ca alloy substrates in an aqueous solution containing 4.2 mM of Ca^{2+} and 2.5 mM of $H_2PO_4^-$ at pH 5.0 under a pulse current [60]. The shear strength was measured to be 41.8 ± 2.7 MPa, which is higher than that of natural cortical bone (35 MPa).

Ban et al. prepared an HA coating on Ti metal with an intermediate glass layer [61]. At first, a composite layer of HA and aluminoborosilicate glass containing Na_2O and K_2O was formed on the metal by heat treatment. The HA layer was then further formed by ELD in an aqueous solution containing 137.8 mM of Na^+, 2.5 mM of Ca^{2+}, and 1.6 mM of HPO_4^{2-} with a current density of 160 mA/cm². The coating layer had a functionally graded structure, as shown in Figure 2.14. The authors compared adhesion after implantation in rabbit tibia with and without composite coating. After 9 weeks of implantation,

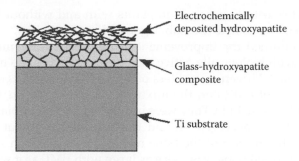

Electrochemically
deposited hydroxyapatite

Glass-hydroxyapatite
composite

Ti substrate

FIGURE 2.14
HA coating on Ti intermediated with glass–HA composite.

the HA layer with an intermediate glass–HA composite layer showed a pullout strength of 29.8 MPa, while the electrochemical HA coating with directly deposited Ti showed a pullout strength of 20.6 MPa. This means that the intermediate composite coating is effective in improving in vivo bone adhesion.

The scratch test has also been applied for the mechanical property evaluation of HA coatings formed by ELD. Kuo and Yen fabricated an HA coating on Ti substrates in an aqueous solution containing 42 mM of Ca^{2+} and 25 mM of $H_2PO_4^-$ at pH 4.1 [62]. The current density ranged from 1 to 20 mA/cm^2, and the critical scratch load for the fracture of the coating was 20 N. The corresponding shear stress was 106.3 MPa.

Plasma-sprayed HA coatings are reported to show bonding strength ranging from 10 to 80 MPa [63]. The bonding strength of most electrochemically deposited HA coatings is lower than the maximum value of plasma-sprayed HA coatings. NaOH- and heat-treated Ti alloys show high bioactivity and are clinically used for a novel artificial hip joint in orthopedic surgery [64]. The apatite layer formed on the modified Ti in a simulated body environment shows a tensile bonding strength of 30 MPa [65]. This means that most of the electrochemically deposited HA coatings match the bonding strength of commercially available bioactive materials.

2.5 Biomedical Applications

Research showing good biological performance has been reported for in vitro and in vivo tests on electrochemically coated Ti metals. Many electrochemically deposited HA coatings have been proven to be highly biocompatible, similar to clinically utilized biomaterials.

Huang et al. performed cell and animal experiments of HA-coated Ti [58]. They showed that HA-coated Ti implant materials have little cytotoxicity in vitro and little inflammatory reaction in vivo. In addition, there has been no statistical disparity between Ti implants with and without HA coating. This means that HA-coated Ti exhibits good biocompatibility.

Ban et al. examined the improvement in the biological compatibility by combinations with biologically active molecules such as bone morphogenic protein (BMP) [66]. Fifteen microliters of 10% gelatin aqueous solution was mixed with 5 mg of BMP, and the mixture was deposited on Ti substrates coated with HA TFs by ELD. They were implanted in rabbit tibia for 3 weeks. Bone formation was observed around the substrates without BMP, while it extended to the interior of the bone marrow with BMP. In addition, the amount of the formed bone was twice as large with BMP as it was without.

Generally, HA ceramics and HA-coated materials are well known to induce bone formation along the material surfaces after implantation in bony effects. This phenomenon is called osteoconduction and is recognized as an important trigger for bone integration.

In addition, it is known that some specific kinds of materials can induce bone formation even in non-bony tissues, meaning that the implanted materials trigger ectopic bone formation. This specific biological phenomenon is called osteoinduction and is quite attractive for the design of novel biomaterials that facilitate bone regeneration. Various artificial materials such as HA [67], tricalcium phosphate (TCP) [68], and NaOH- and heat-treated Ti metals [69] have been known to be osteoinductive. In addition, it is considered that a porous structure is essential for achieving osteoinduction [70]. The difference between osteoconduction and osteoinduction is schematically illustrated in Figure 2.15.

Chai et al. investigated osteoinduction of three-dimensional (3D) porous Ti scaffolds coated with HA [71]. The HA layer was deposited on Ti by ELD with current densities of 5 and 10 A/m². The specimens were seeded with human periosteum-derived cells and implanted subcutaneously on the backs of mice in the cervical region. Larger bone formation was observed for the HA coating deposited at a current density of 10 A/m² than for that at 5 A/m². The former showed a greater amount of dissolution than the latter. The authors assume that osteoclastic activity was vigorously occurred to enhance the osteoinduction in the former.

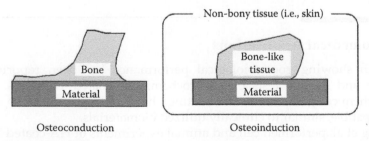

FIGURE 2.15
Schematic representation of osteoconduction and osteoinduction.

2.6 Electrodeposition of HA-Based Composite Films

Not only pure HA but also various kinds of HA-based composites can be coated on metals by electrochemical deposition. These composite coatings can control various mechanical and chemical properties, as well as improve biological performance.

Electrochemical composite coatings can enhance the mechanical properties of the coating. Xiao et al. produced an HA–TiO_2 composite coating in an aqueous solution containing various concentrations of TiO_2 powder (particles 1 μm in diameter), as well as 10 mM of Ca^{2+} and 6 mM of $H_2PO_4^-$ at pH 4.6 [72]. The current density ranged from 0.1 to 10 mA/cm^2. The coating was then heat treated at temperatures ranging from 500°C to 1200°C. The coating was composed of HA and anatase TiO_2 after heat treatment at 500°C, while rutile TiO_2 and α-TCP were newly observed at 1200°C. The bonding strength increased with increasing TiO_2 content. The improvement of the bonding strength was attributed to dispersion strengthening by the homogeneous distribution of TiO_2 particles in the HA coatings and the increase in density of the coating. The increase in the bonding strength was attributed to graded structure construction on the surface. An HA–ZrO_2 composite coating is also available by electrochemical processing [73].

Biological performance is also enhanced by construction of a composite layer. Ou et al. compared the in vitro differentiation behavior of mesenchymal stem cells (MSc) on nanoporous HA films with and without the incorporation of reconstructed collagen [74]. As a result, cells exhibited a fibroblast-like morphology without collagen, whereas they exhibited more spread-out filopods with collagen. These results mean that an HA–collagen composite enhances osteogenic differentiation of MSc.

Antibacterial properties can be provided on the HA coating by EPD processing. Pang and Zhitomirsky attempted an HA–chitosan composite coating on stainless steel substrates [75,76]. The composite film was deposited in chitosan solutions containing needle-like HA nanoparticles in a mixed ethanol–water solvent. The electrodeposition was performed at a constant current density of 0.1 mA/cm^2. The XRD patterns of the composite coatings showed an increased intensity of the 300 diffraction and a suppression of the 002 diffraction of HA. These results indicate that a preferred orientation of the c axes of the HA crystals parallel to the coating surface was constructed. The authors also evaluated the addition of Ag into the HA–chitosan composite coating to provide it with antibacterial properties. The coating on stainless 3×6 cm^2 in size released 30 ppm of Ag within 200 h into 20 mL of pure water.

The chemical durability of the surface HA film can be controlled by the composite coating. HA–hyaluronic acid composite deposition was examined by Sun and Zhitomirsky [77]. Composite formation was performed in a mixed solution of sodium hyaluronate and HA. The composite film formation of different thicknesses in the range of 0.1–100 μm was possible with this method. The composite films showed corrosion protection of stainless steel substrates in Ringer's physiological solutions.

Hu et al. produced an HA–poly(vinyl acetate) composite coating [78]. The electrolyte was an aqueous solution containing 1 mM of $Ca(NO_3)_2$, 1.67 mM of $NH_4H_2PO_4$, with or without 0.0125 vol% of vinyl acetate at pH 6 and at 80°C. The deposition was performed under constant current density ranging from 1 to 2 mA/cm^2. The cell adhesion number was significantly improved by the addition of the poly(vinyl acetate). The authors attributed the enhancement of cell affinity to the formation of a highly oriented structure by the addition of the poly(vinyl acetate) and secondary HA nucleation on the electrochemically deposited HA in a simulated body environment.

Biological molecules such as proteins can be incorporated by electrochemical deposition. Cheng et al. examined the preparation of HA/bovine serum albumin (BSA) composites [79]. The electrolyte was an aqueous solution containing 42 mM of $Ca(NO_3)_2$ and 25 mM of $NH_4H_2PO_4$ with 2 mg/mL of BSA at 0.09 mA/cm^2. Release of the BSA was measured in a phosphate buffer solution. The substrates released 15% of incorporated BSA after 70 h. These results indicate that novel biomaterials with a potential for drug delivery can be designed by electrodeposition.

Composites of carbon nanotubes (CNTs) and HA have been fabricated by ELD on CNT thin films (CNT TFs) [80] and by an EPD process in chitosan solution dispersed with CNT and HA [81]. This type of coating is expected to improve mechanical properties such as adhesion strength and hardness of the surface layer.

In addition to the research described previously, preparations of various kinds of HA-based composites by electrochemical deposition have been reported. They are summarized in Table 2.2.

TABLE 2.2

Various HA-Based Composites
Prepared by Electrodeposition

Counterpart for HA	Reference
Inorganic material	
TiO_2	[72]
ZrO_2	[73]
Carbon nanotubes (CNTs)	[80,81]
Organic material	
Collagen	[74]
Chitosan	[75,76]
Hyaluronic acid	[77]
Poly(vinyl acetate)	[78]
Alginate	[82]
Bovine serum albumin (BSA)	[79]

2.7 Summary

In this chapter, electrochemical deposition of CaP ceramics represented by HA was introduced. The electrochemical deposition method can be performed by simple apparatus compared with vacuum processes such as the plasma-spray method, and hence is environment friendly. This method can be also used to control the composition, structure, and adhesion of the deposited layer with comparative ease. It is believed that the electrochemical deposition of CaP ceramics such as HA will provide novel biomedical devices with novel surface functions in future.

References

1. Paunovic, M. and Schlesinger, M. 2006. *Fundamentals of Electrochemical Deposition*, 2nd edn. Hoboken, NJ: John Wiley & Sons, Inc.
2. Zhitomirsky, I. 2002. Cathodic electrodeposition of ceramic and organoceramic materials. Fundamental aspects. *Adv. Colloid Interface Sci.* 97:279–317.
3. Niinomi, M. 1998. Mechanical properties of biomedical titanium alloys. *Mater. Sci. Eng. A* 243:231–236.
4. Niinomi, M. 2002. Recent metallic materials for biomedical applications. *Metall. Mater. Trans. A* 33:477–486.
5. Hench, L. L. and Wilson, J. 1993. *An Introduction to Bioceramics*, eds. L. L. Hench and J. Wilson, pp. 7–8. Singapore: World Scientific.
6. Liu, X. Y., Chu, P. K., and Ding, C. X. 2004. Surface modification of titanium, titanium alloys, and related materials for biomedical applications. *Mater. Sci. Eng. R* 47:49–121.
7. Nery, E. B. and Lynch, K. L. 1978. Preliminary clinical studies of bioceramic in periodontal osseous defects. *J. Periodontol.* 49:523–527.
8. Daculsi, G., Passuti, N., Martin, S., Deudon, C., LeGeros, R. Z., and Raher, S. 1990. Macroporous calcium-phosphate ceramic for long-bone surgery in humans and dogs: Clinical and histological study. *J. Biomed. Mater. Res.* 24:379–396.
9. Dorozhkin, S. V. and Epple, M. 2002. Biological and medical significance of calcium phosphates. *Angew. Chem. Int. Ed.* 41:3130–3146.
10. LeGeros, R. Z., Lin, S., Rohanizadeh, R., Mijares, D., and LeGeros, J. P. 2003. Biphasic calcium phosphate bioceramics: Preparation, properties and applications. *J. Mater. Sci. Mater. Med.* 14:201–209.
11. Cook, S. D., Thomas, K. A., Kay, J. F., and Jarcho, M. 1988. Hydroxyapatite-coated titanium for orthopedic implant applications. *Clin. Orthop. Relat. Res.* 232:225–243.
12. Klein, C. P. A. T., Patka, P., van der Lubbe, H. B. M., Wolke, J. G. C., and de Groot, K. 1991. Plasma-sprayed coatings of tetracalciumphosphate, hydroxyl-apatite, and α-TCP on titanium-alloy: An interface study. *J. Biomed. Mater. Res.* 25:53–65.

13. Lusquinos, F., De Carlos, A., Pou, J. et al. 2003. Calcium phosphate coatings obtained by Nd: YAG laser cladding: Physicochemical and biologic properties. *J. Biomed. Mater. Res.* 64A:630–637.

14. Hero, H., Wie, H., Jorgensen, R. B., and Ruyter, I. E. 1994. Hydroxyapatite coatings on Ti produced by hot isostatic pressing. *J. Biomed. Mater. Res.* 28:343–348.

15. Ong, J. L., Lucas, L. C., Lacefield, W. R., and Rigney, E. D. 1992. Structure, solubility and bond strength of thin calcium-phosphate coatings produced by ion-beam sputter deposition. *Biomaterials* 13:249–254.

16. Gross, K. A., Chai, C. S., Kannangara, G. S. K., Ben-Nissan, B., and Hanley, L. 1998. Thin hydroxyapatite coatings via sol–gel synthesis. *J. Mater. Sci. Mater. Med.* 9:839–843.

17. Habibovic, P., Barrere, F., van Blitterswijk, C. A., de Groot, K., and Layrolle, P. 2002. Biomimetic hydroxyapatite coating on metal implants. *J. Am. Ceram. Soc.* 85:517–522.

18. Boccaccini, A. R., Kleim, S., Ma, R., Li, Y., and Zhitomirsky, I. 2010. Electrophoretic deposition of biomaterials. *J. R. Soc. Interface* 7:S581–S613.

19. Ducheyne, P., Van Raemdonck, W., Heughebaert, J. C., and Heughebaert, M. 1986. Structural analysis of hydroxyapatite coatings on titanium. *Biomaterials* 7:97–103.

20. Ducheyne, P., Radin, S., Heughebaert, M., and Heughebaert, J. C. 1990. Calcium-phosphate ceramic coatings on porous titanium: Effect of structure and composition on electrophoretic deposition, vacuum sintering and in vitro dissolution. *Biomaterials* 11:244–254.

21. Kim, C. S. and Ducheyne, P. 1991. Compositional variations in the surface and interface of calcium phosphate ceramic coatings on Ti and Ti–6Al–4V due to sintering and immersion. *Biomaterials* 12:461–469.

22. Zhitomirsky, I. and Gal-Or, L. 1997. Electrophoretic deposition of hydroxyapatite. *J. Mater. Sci. Mater. Med.* 8:213–219.

23. Wei, M., Ruys, A. J., Mithorpe, B. K., and Sorrell, C. C. 1999. Solution ripening of hydroxyapatite nanoparticles: Effects on electrophoretic deposition. *J. Biomed. Mater. Res.* 45:11–19.

24. Xiao, X. F. and Liu, R. F. 2006. Effect of suspension stability on electrophoretic deposition of hydroxyapatite coatings. *Mater. Lett.* 60:2627–2632.

25. de Sena, L. A., de Andrade, M. C., Rossi, A. M., and Soares, G. A. 2002. Hydroxyapatite deposition by electrophoresis on titanium sheets with different surface finishing. *J. Biomed. Mater. Res.* 60:1–7.

26. Monkawa, A., Ikoma, T., Yunoki, S. et al. 2006. Fabrication of hydroxyapatite ultra-thin layer on gold surface and its application for quartz crystal microbalance technique. *Biomaterials* 27:5748–5754.

27. Ikoma, T., Tagaya, M., Hanagata, N. et al. 2009. Protein adsorption on hydroxyapatite nanosensors with different crystal sizes studied in situ by a quartz crystal microbalance with the dissipation method. *J. Am. Ceram. Soc.* 92:1125–1128.

28. Tagaya, M., Ikoma, T., Hanagata, N., Chakarov, D., Kasemo, B., and Tanaka, J. 2010. Reusable hydroxyapatite nanocrystal sensors for protein adsorption. *Sci. Technol. Adv. Mater.* 11:045002.

29. Kawashita, M., Taninai, K., Li, Z., Ishikawa, K., and Yoshida, Y. 2012. Preparation of low-crystalline apatite nanoparticles and their coating onto quartz substrates. *J. Mater. Sci. Mater. Med.* 23:1355–1362.

30. Redepenning, J. and McIsaac, J. P. 1990. Electrocrystallization of brushite coatings on prosthetic alloys. *Chem. Mater.* 2:625–627.
31. Redepenning, J., Schlessinger, T., Burnham, S., Lippiello, L., and Miyano, J. 1996. Characterization of electrolytically prepared brushite and hydroxyapatite coatings on orthopedic alloys. *J. Biomed. Mater. Res.* 30:287–294.
32. Shirkhanzadeh, M. 1991. Bioactive calcium phosphate coatings prepared by electrodeposition. *J. Mater. Sci. Lett.* 10:1415–1417.
33. Shirkhanzadeh, M. 1994. Calcium phosphate coatings prepared by electrocrystallization from aqueous electrolytes. *J. Mater. Sci. Mater. Med.* 5:91–93.
34. Royer, P. and Rey, C. 1991. Calcium phosphate coatings for orthopaedic prosthesis. *Surf. Coat. Technol.* 45:171–177.
35. Monma, H. 1993. Electrochemical deposition of calcium deficient apatite on stainless steel substrate. *J. Ceram. Soc. Jpn.* 101:737–739.
36. Monma, H. 1994. Electrolytic depositions of calcium phosphates on substrate. *J. Mater. Sci.* 29:949–953.
37. Kokubo, T., Kushitani, H., Sakka, S., Kitsugi, T., and Yamamuro, T. 1990. Solutions able to reproduce in vivo surface-structure changes in bioactive glass–ceramic A-W. *J. Biomed. Mater. Res.* 24:721–734.
38. Kokubo, T. and Takadama, H. 2006. How useful is SBF in predicting in vivo bone bioactivity? *Biomaterials* 27:2907–2915.
39. Tanahashi, M., Yao, T., Kokubo, T. et al. 1994. Apatite coating on organic polymers by a biomimetic process. *J. Am. Ceram. Soc.* 77:2805–2808.
40. Ban, S. and Maruno, S. 1995. Effect of temperature on electrochemical deposition of calcium-phosphate coatings in a simulated body fluid. *Biomaterials* 16:977–981.
41. Ban, S. and Maruno, S. 1998. Morphology and microstructure of electrochemically deposited calcium phosphates in a modified simulated body fluid. *Biomaterials* 19:1245–1253.
42. Ban, S. and Maruno, S. 1998. Hydrothermal–electrochemical deposition of hydroxyapatite. *J. Biomed. Mater. Res.* 42:387–395.
43. Ban, S. and Hasegawa, J. 2002. Morphological regulation and crystal growth of hydrothermal-electrochemically deposited apatite. *Biomaterials* 23:2965–2972.
44. Yuda, A., Ban, S., and Izumi, Y. 2005. Biocompatibility of apatite-coated titanium mesh prepared by hydrothermal–electrochemical method. *Dent. Mater.* 24:588–595.
45. Kawashita, M., Itoh, S., Miyamoto, K., and Takaoka, G. H. 2008. Apatite formation on titanium substrates by electrochemical deposition in metastable calcium phosphate solution. *J. Mater. Sci. Mater. Med.* 19:137–142.
46. Blumenthal, N. C. 1989. Mechanisms of inhibition of calcification. *Clin. Orthop. Relat. Res.* 247:279–289.
47. Lin, S., LeGeros, R. Z., and LeGeros, J. P. 2003. Adherent octacalciumphosphate coating on titanium alloy using modulated electrochemical deposition method. *J. Biomed. Mater. Res.* 66A:819–828.
48. Manso, M., Jimenez, C., Morant, C., Herrero, P., and Martinez-Duart, J. M. 2000. Electrodeposition of hydroxyapatite coatings in basic conditions. *Biomaterials* 21:1775–1761.
49. Ban, S. and Maruno, S. 1993. Effect of pH buffer on electrochemical deposition of calcium phosphate. *Jpn. J. Appl. Phys.* 32:L1545–L1548.

50. Hayakawa, T., Kawashita, M., and Takaoka, G. H. 2008. Coating of hydroxyapatite films on titanium substrates by electrodeposition under pulse current. *J. Ceram. Soc. Jpn.* 116:68–73.
51. Hayakawa, T., Kawashita, M., Takaoka, G. H., and Miyazaki, T. 2008. Effect of pulse current on structure and adhesion of apatite electrochemically deposited onto titanium substrates. *J. Mater. Res.* 23:3176–3183.
52. Redepenning, J., Venkataraman, G., Chen, J., and Stafford, N. 2003. Electrochemical preparation of chitosan/hydroxyapatite composite coatings on titanium substrates. *J. Biomed. Mater. Res.* 66A:411–416.
53. Besra, L. and Liu, M. 2007. A review on fundamentals and applications of electrophoretic deposition (EPD). *Prog. Mater. Sci.* 52:1–61.
54. Sarkar, P. and Nicholson, P. S. 1996. Electrophoretic deposition (EPD): Mechanisms, kinetics and application to ceramics. *J. Am. Ceram. Soc.* 79:1987–2002.
55. Hu, R., Lin, C., Shi, H., and Wang, H. 2009. Electrochemical deposition mechanism of calcium phosphate coating in dilute Ca–P electrolyte system. *Mater. Chem. Phys.* 115:718–723.
56. Elliott, J. C. 1994. *Structure and Chemistry of the Apatites and Other Calcium Orthophosphates*, p. 4. Amsterdam, the Netherlands: Elsevier.
57. Han, Y., Fu, T., Lu, J., and Xu, K. 2001. Characterization and stability of hydroxyapatite coatings prepared by an electrodeposition and alkaline-treatment process. *J. Biomed. Mater. Res.* 54:96–101.
58. Huang, S., Zhou, K., Huang, B., Li, Z., Zhu, S., and Wang, G. 2008. Preparation of an electrodeposited hydroxyapatite coating on titanium substrate suitable for in-vivo applications. *J. Mater. Sci. Mater. Med.* 19:437–442.
59. Wei, M., Ruys, A. J., Swain, M. V., Kim, S. H., Milthorpe, B. K., and Sorrell, C. C. 1999. Interfacial bond strength of electrophoretically deposited hydroxyapatite coatings on metals. *J. Mater. Sci. Mater. Med.* 10:401–409.
60. Wang, H. X., Guan, S. K., Wang, X., Ren, C. X., and Wang, L. G. 2010. In vitro degradation and mechanical integrity of Mg–Zn–Ca alloy coated with Ca-deficient hydroxyapatite by the pulse electrodeposition process. *Acta Biomater.* 6:1743–1748.
61. Ban, S., Maruno, S., Arimoto, N., Harada, A., and Hasegawa, J. 1997. Effect of electrochemically deposited apatite coating on bonding of bone to the HA-G-Ti composite and titanium. *J. Biomed. Mater. Res.* 36:9–15.
62. Kuo, M. C. and Yen, S. K. 2002. The process of electrochemical deposited hydroxyapatite coatings on biomedical titanium at room temperature. *Mater. Sci. Eng.* C 20:153–160.
63. Lacefield, W. R. 1993. Hydroxyapatite coatings. In *An Introduction to Bioceramics*, eds. L. L Hench and J. Wilson, pp. 223–238. Singapore: World Scientific.
64. Kim, H.-M., Miyaji, F., Kokubo, T., and Nakamura, T. 1996. Preparation of bioactive Ti and its alloys via simple chemical surface treatment. *J. Biomed. Mater. Res.* 32:409–417.
65. Kim, H.-M., Miyaji, F., Kokubo, T., and Nakamura, T. 1997. Bonding strength of bonelike apatite layer to Ti metal substrate. *J. Biomed. Mater. Res. B* 38:121–127.
66. Ban, S., Kawai, T., Arimoto, N. et al. 1997. Osteoinductive activity of electrochemical surface modified Ti and HA-G-Ti composite with BMP in hard tissue. *J. Jpn. Soc. Dent. Mater. Dev.* 16:374–381 (in Japanese).
67. Yamasaki, H. and Sakai, H. 1992. Osteogenic response to porous hydroxyapatite ceramics under the skin of dogs. *Biomaterials* 13:308–312.

68. Yuan, H., deBruijn, J. D., Li, Y., Feng, J., Yang, Z., de Groot, K., and Zhang, X. 2001. Bone formation induced by calcium phosphate ceramics in soft tissue of dogs: A comparative study between porous alpha-TCP and beta-TCP. *J. Mater. Sci. Mater. Med.* 12:7–13.
69. Fujibayashi, S., Neo, M., Kim, H.-M., Kokubo, T., and Nakamura, T. 2004. Osteoinduction of porous bioactive titanium metal. *Biomaterials* 25:443–450.
70. deBrujin, J. D., Shankar, K., Yuan, H., and Habibovic, P. 2008. Osteoinduction and its evaluation. In *Bioceramics and Their Clinical Applications*, ed. T. Kokubo, pp. 199–219. Cambridge, U.K.: Woodhead Publishing Ltd.
71. Chai, Y. C., Kerckhofs, G., Roberts, S. J. et al. 2012. Ectopic bone formation by 3D porous calcium phosphate-Ti6Al4V hybrids produced by perfusion electrodeposition. *Biomaterials* 33:4044–4058.
72. Xiao, X. F., Liu, R. F., and Zheng, Y. Z. 2006. Characterization of hydroxyapatite/titania composite coatings codeposited by a hydrothermal–electrochemical method on titanium. *Surf. Coat. Technol.* 200:4406–4413.
73. Qiu, D., Wang, A., and Yin, Y. 2010. Characterization and corrosion behavior of hydroxyapatite/zirconia composite coating on NiTi fabricated by electrochemical deposition. *Appl. Surf. Sci.* 257:1774–1778.
74. Ou, K.-L., Wu, J., Lai, W.-F, Yang, C.-B. et al. 2010. Effects of the nanostructure and nanoporosity on bioactive nanohydroxyapatite/reconstituted collagen by electrodeposition. *J. Biomed. Mater. Res.* 92A:906–912.
75. Pang, X. and Zhitomirsky, I. 2007. Electrophoretic deposition of composite hydroxyapatite–chitosan coatings. *Mater. Charact.* 58:339–348.
76. Pang, X. and Zhitomirsky, I. 2008. Electrodeposition of hydroxyapatite–silver chitosan nanocomposite coatings. *Surf. Coat. Technol.* 202:3815–3821.
77. Sun, F. and Zhitomirsky, I. 2009. Electrodeposition of hyaluronic acid and composite films. *Surf. Eng.* 25:621–627.
78. Hu, H. B., Lin, C. J., Hu, R., and Leng, Y. 2002. A study on hybrid bioceramic coatings of HA/poly(vinyl acetate) co-deposited electrochemically on Ti–6Al–4V alloy surface. *Mater. Sci. Eng. C* 20:209–214.
79. Cheng, X., Filiaggi, M., and Roscoe, S. G. 2004. Electrochemically assisted co-precipitation of protein with calcium phosphate coatings on titanium alloy. *Biomaterials* 25:5395–5403.
80. Lobo, A. O., Corat, M. A. F., Ramos, S. C. et al. 2010. Fast preparation of hydroxyapatite/superhydrophilic vertically aligned multiwalled carbon nanotube composites for bioactive application. *Langmuir* 26:18308–18314.
81. Rath, P. C., Bimal, S., Laxmidhar, B., and Bhattacharjee, S. 2012. Multiwalled carbon nanotubes reinforced hydroxyapatite–chitosan composite coating on Ti metal: Corrosion and mechanical properties. *J. Am. Ceram. Soc.* 95:2725–2731. doi/10.1111/j.1551-2916.2012.05195.x
82. Cheong, M. and Zhitomirsky, I. 2008. Electrodeposition of alginic acid and composite films. *Colloids Surf. A* 328:73–78.

3

Simultaneous Incorporation of Magnesium and Fluorine Ions in Hydroxyapatite Coatings on Metallic Implant for Osseointegration and Stability

Yanli Cai, Sam Zhang, Soon-Eng Ong, Xianting Zeng, and Wilson Wang

CONTENTS

3.1 Introduction

In orthopedic medical engineering, there are two main concerns regarding the application of implants: short-term osseointegration and long-term stability. Osseointegration is the stable anchorage of an implant achieved by direct bone-to-implant contact [1]. Stability refers to a long-term service life of the implant with the ability to carry and sustain the static and dynamic loads [2]. It is obviously very important to achieve both of them within the shortest possible healing time and with a very small failure rate.

Metallic, ceramic, and polymeric biomaterials have been widely used as bone replacements. Metals like titanium and titanium alloys are often employed as basic materials for load-bearing implants because of their excellent mechanical properties. Even though these materials are mostly bioinert, they are not bioactive, i.e., they do not form a mechanically stable link between the implant and bone tissue. Thus, it is difficult to integrate into the body. As such, the metal surface has to be properly engineered. Many ways have been sought to improve the mechanical contact at the interface. The most effective way is to coat the metal with calcium phosphate ceramics that can promote bone bonding between the metal and the bone. Hydroxyapatite ($Ca_{10}(PO_4)_6(OH)_2$; HA) has similar composition and molecular structure as that of the natural bone; thus, it has been widely used as a coating on metallic implants to promote cellular activities and stimulate formation of new bone. It has been suggested that the more similar HA is to bone mineral in composition, crystallinity, crystal structure, crystal size, and morphology, the better osteoconductivity would be achieved [3].

However, bulk HA or HA coatings vary extensively from natural bone mineral in composition. Natural bone is a non-stoichiometric apatite containing many trace elements, such as Na^+ (*ca.* 0.7 wt%), K^+ (*ca.* 0.03 wt%), Mg^{2+} (*ca.* 0.55 wt%), Cl^- (*ca.* 0.10 wt%), and F^- (*ca.* 0.02 wt%), which play important roles in the physiological systems [4]. Pure HA does not have this composition; therefore, its bioactivity is inferior to the bone mineral [5], which adversely affects the short-term osseointegration. Pure HA coating does not have strong adhesion strength on metallic substrate either. Moreover, it dissolves rapidly in body fluids, which in turn creates a long-term stability problem.

To resemble bone mineral in composition, various cations and anions are incorporated into HA or HA coatings. In brief, incorporation of cations like Mg^{2+} into HA decreases its crystallinity and improves the biological performances, i.e., good bioactivity and better cell behaviors [6]. The incorporation of anions (F^-) into HA crystal structure decreases the solubility of the HA coating [3,7] and enhances the adhesion strength between the coating and the substrate [8]. Therefore, it is interesting and desirable to incorporate both Mg and F ions into HA coatings on Ti6A14V substrates to have a combination of their individual functions in one coating/substrate system to achieve both short-term osseointegration and long-term stability.

3.2 Cation- and Anion-Incorporated Hydroxyapatite

Bone has an apatite structure, and the general chemical formula is expressed as $M_{10}(XO_4)_6Y_2$, where M is a bivalent ion, XO_4 a trivalent ion, and Y a monovalent ion. In general, it is represented as $Ca_{10}(PO_4)_6(OH)_2$, and many trace elements are contained in the bone mineral. They do not only play pivotal roles in the behavior of biological apatite but also contribute to various biochemical reactions in the human body. For example, sodium (Na^+) is a monovalent ion available in abundance next to calcium and phosphorus, playing a significant role in bone metabolism and osteoporosis [9,10]. Magnesium (Mg^{2+}) has its own significance in cell proliferation, calcification process, and bone fragility. Moreover, it indirectly influences mineral metabolism. Mg deficiency may directly result in a decrease in osteoblastic bone formation and delay the onset of mineralization of newly formed cartilage matrix and bones [4,11,12]. Potassium (K^+) is active in mineralization and biochemical processes [13]. Fluorine (F⁻) is well recognized for its potential behavior of decreasing the solubility of the apatite and its prevention role in dental caries in a bacteria-containing acidic environment. Also F⁻ promotes the mineralization and crystallization of calcium phosphate in the bone-forming process [7,14,15]. The significance of chloride (Cl⁻) ions resides in their ability to develop an acidic environment on the surface of bone that activates osteoblasts in the bone resorption process [16,17].

With the presence of trace elements, partial/total substitutions may occur without significant changes in the crystal structure. This is due to the apatite structure readily accepting a wide variety of anionic and cationic dopants [18]. Numerous attempts of substituted apatite have been reported: the substitution of Ca^{2+} with Mg^{2+} [19–21], Zn^{2+} [19,22–24], Sr^{2+} [25,26], Pb^{2+} [27], etc., the replacement of PO_4^{3-} by HPO_4^{2-} [28], CO_3^{2-} [19,29–33] or SiO_4^{4-} [34–36], and OH^- by F^- [19,37–40] or Cl^- [41].

3.2.1 Incorporation of Cation into HA

The chemical composition of hexagonal HA (within the ideal $P6_3/m$ space group) is $Ca(I)_4Ca(II)_6(PO_4)_6(OH)_2$. The local atomic configurations are displayed in Figure 3.1. The Ca(I) site is surrounded by six PO_4^{3-} tetrahedral and coordinated by nine oxygen ions. The Ca(II) site is seven-coordinated with six oxygen ions from PO_4^{3-} and one oxygen ion from OH^- [42,43]. It has been reported that the larger cations usually occupy the Ca(II) sites and the smaller ones prefer the Ca(I) sites, because the nearest distance among Ca(I) sites is smaller than that among Ca(II) sites [44]. The substitution of cations into HA tends to cause the HA structure disordered and influence the phase, its mechanical and biological properties.

Among the incorporated cations, Sr^{2+} and Pb^{2+} ions could replace Ca^{2+} ions in the whole range of composition [45,46], while the incorporation of Mg^{2+} ions into the HA crystal structure is still under debate. Generally,

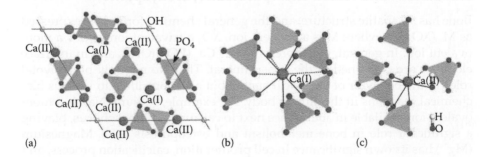

(a) (b) (c)

FIGURE 3.1

Crystal structure of HA viewed along the *c*-axis (a) and the local atomic structures of Ca(I) (b) and Ca(II) (c). (From Matsunaga, K. et al., Theoretical calculations of the thermodynamic stability of ionic substitutions in hydroxyapatite under an aqueous solution environment, *J. Phys. Cond. Matter*, 22, 384210, 2010. With permission from Institute of Physics.)

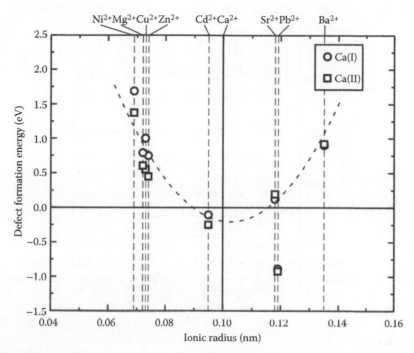

FIGURE 3.2
Formation energies of substitutional divalent cations against experimental ionic radii (Ionic substitution in HA under an aqueous solution at pH 7). (From Matsunaga, K. et al., Theoretical calculations of the thermodynamic stability of ionic substitutions in hydroxyapatite under an aqueous solution environment, *J. Phys. Cond. Matter*, 22, 384210, 2010. With permission from Institute of Physics.)

it seems that cations having larger atomic numbers tend to be more easily exchanged with Ca^{2+} in HA. However, ionic size and/or electronegativity of cations also affect the substitution of Ca^{2+} in HA. The ionic size difference between Ca^{2+} and M^{2+} (substitutional divalent cation) causes variations in the interatomic distances between substitutional divalent cations and surrounding oxygen atoms; thus, the formation energy varies. The relevance of the formation energies to the ionic sizes was investigated and plotted in Figure 3.2. It was found that the formation energies increased more with increasing the ionic size difference between Ca^{2+} and M^{2+}. This is because the smaller- or larger-sized cations allow for larger lattice relaxations of the surrounding oxygen by substitution; as such, they may be suffering from more significant energy expenses [42].

3.2.1.1 Phase Evolution and Substitution of Cation into HA

The substitution of Ca^{2+} with M^{2+} leads to the changes of as-formed phases. Figure 3.3 shows the phase evolution of Mg-substituted HA powders

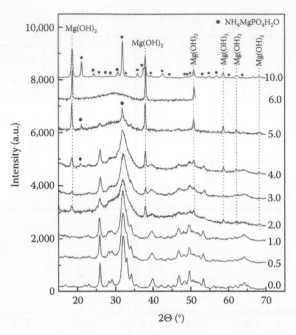

FIGURE 3.3
XRD patterns of as-prepared Mg-HA powders. (From *Biomaterials*, 25, Suchanek, W.L., Byrappa, K., Shuk, P., Riman, R.E., Janas, V.F., and Tenhuisen, K.S., Preparation of magnesium-substituted hydroxyapatite powders by the mechanochemical-hydrothermal method, 4647, Copyright 2004, with permission from Elsevier.)

(MgHA) $(Ca_{10-x}Mg_x(PO_4)_6(OH)_2$, x varies from 0 to 10) prepared by the mechanochemical–hydrothermal method. With increasing x (lower than 6.0), the HA diffraction peaks broaden and weaken, indicating the gradual decrease in crystallinity of the apatite phase. At $x = 6.0$, only an amorphous phase (Mg-containing amorphous calcium phosphate, Mg-ACP) is detected; at $x = 10.0$, only the $NH_4MgPO_4·H_2O$ phase is produced instead of HA or ACP. These results reveal that Mg has limited solubility in HA, and it is difficult to obtain a pure HA phase with a complete Mg substitution ($x = 10.0$) [20]. The decrease in crystallinity with increasing Mg content was also proved by other researchers [44,47]. The gradual decrease in crystallinity is one of the indications that the incorporation of Mg into the HA crystal structure leads to the increased lattice disorder [20]. Heat treatment of as-prepared MgHA powders at 900°C for 1 h in air shows that MgHA phase becomes thermally unstable with the increasing Mg content. Besides HA peaks at low Mg contents ($x = 0.5 - 1.0$), weak peaks of whitlockite [$(Ca, Mg)_3(PO_4)_2$] are also observed; at $x = 2.0 - 4.0$, HA-derived peaks disappear, and other phases such as $(Ca, Mg)_3(PO_4)_2$, $Ca_4Mg_5(PO_4)_6$, and β-$Ca_2P_2O_7$ form [20].

When the synthesis method is different, the substitution amount of Mg ions in the HA lattice changes. Ren et al. [48] synthesized Mg-substituted

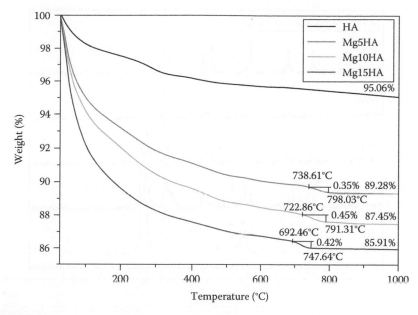

FIGURE 3.4

TGA curves for Mg-substituted apatite. (From *Acta Biomater.*, 6, Ren, F., Leng, Y., Xin, R., and Ge, X., Synthesis, characterization and ab initio simulation of magnesium substituted hydroxyapatite, 2787, Copyright 2010, with permission from Elsevier.)

apatite ($Ca_{10-x}Mg_x(PO_4)_6(OH)_2$, x varies from 0 to 10) by the wet-chemical precipitation method. Phase identification revealed that the apatite phase could be preserved only in the Mg concentration range of 0–1.5. However, the Mg/(Ca + Mg) molar ratio in the final apatite was significantly lower than that in the initial solutions. In their research, the upper limit of Mg substitution in the HA lattice was between 5 and 7 mol%. The thermal stability of Mg-substituted apatite was studied by thermogravimetric analyzer (TGA) (Figure 3.4). The decomposition temperature of HA to β-TCP in the region of 690°C–800°C decreases with Mg content as indicated in the TGA curves. Since Mg substitution in HA lattice could destabilize the crystal structure, the decrease in the decomposition temperature confirms that Mg enters the HA lattice because the surface adsorbed Mg on the apatite surface would not affect the decomposition temperature of HA to β-TCP.

Mg-incorporated apatite coating ($Ca_{10-x}Mg_x(PO_4)_6(OH)_2$, x varies from 0 to 2) has been deposited on Ti6Al4V substrates by the sol–gel method [49]. With the incorporation of Mg, Mg-substituted β-TCP (whitlockite, β-TCMP) forms in the coating and dominates at $x = 1.0$–2.0 (Figure 3.5). These results are consistent with Mg-substituted HA powders as discussed earlier: Mg-substituted HA is thermally unstable, and Mg stabilizes whitlockite with increasing Mg contents.

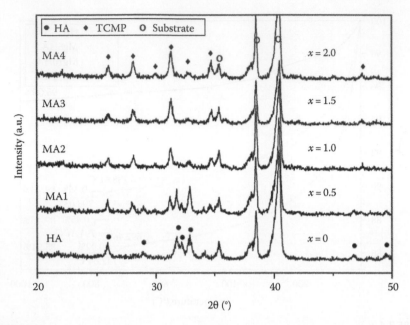

FIGURE 3.5
XRD patterns of the coatings with different Mg contents. (From *Appl. Surf. Sci.*, 255, Qi, G., Zhang, S., Khor, K.A., Lye, S.W., Zeng, X., Weng, W., Liu, C., Venkatraman, S.S., and Ma, L.L., Osteoblastic cell response on magnesium-incorporated apatite coatings, 304, Copyright 2008, with permission from Elsevier.)

The substitution of cation into HA lattice can be reflected by tracing the change of the lattice parameters. The incorporation of larger or smaller cations into the HA lattice will cause the expansion or contraction of the unit cell. In MgHA powder, the unit cell dimensions a and c decrease almost linearly with increasing x_{Mg} ($x_{Mg} = Mg/(Mg+Ca)$) (shown in Figure 3.6), induced by the substitution of Ca^{2+} ions (0.99 Å) by smaller Mg^{2+} (0.69 Å) ions [44]. The same results were also found by other researchers [50]. When the larger cations such as Sr^{2+} ions (1.13 Å) are incorporated into the apatite, the lattice parameters of both a and c linearly increase with the addition of Sr (Figure 3.7) [46], indicating structural incorporation of Sr^{2+} into HA lattice. Similar results were also reported by Li et al. [51].

3.2.1.2 Mechanical Properties

Currently, most researchers are focusing on the preparation of cation-incorporated HA bulk, and the mechanical properties of bulk materials depend on the structures processed. For example, when Mg ions were added into the apatite reactant solution to be introduced into a cement system, it resulted in a reduction in compressive strength of the cement [21]. When nanocrystalline Mg- and Zn-incorporated HA bulks were synthesized, the

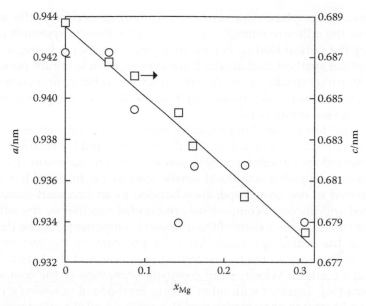

FIGURE 3.6
Unit cell dimensions *vs.* x_{Mg}: a (○); c (□). (From Yasukawa, A., Ouchi, S., Kandori, K., and Ishikawa, T., Preparation and characterization of magnesium-calcium hydroxyapatites, *J. Mater. Chem.*, 6, 1401, 1996. Reproduced by permission of The Royal Society of Chemistry.)

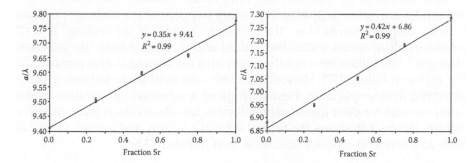

FIGURE 3.7
Variation in lattice parameters, a and c for $(Sr_xCa_{1-x})_5(PO_4)_3OH$. (From *Acta Biomater.*, 4, O'Donnell, M.D., Fredholm, Y., de Rouffignac, A., and Hill, R.G., Structural analysis of a series of strontium-substituted apatites, 1455, Copyright 2008, with permission from Elsevier.)

addition of 1.0 wt% of MgO enhanced the compressive strength of HA by 33.8%, and the addition of 1.0 wt% of ZnO increased it by 6.2% [52].

Adhesion strength between the coating and the substrate is a critical factor for the successful implantation and long-term stability of the coated implant. Normally, scratch and pull-off tests are widely used. Scratch testing is a simple method to determine the coating adhesion and failure modes [53–55]. While a sample is displaced at a constant speed, elastic and/or plastic deformation takes place. At a certain load, damage occurs in the scratching region: the adhesive strength of the coating–substrate system is characterized by the critical load L_c. For the increasing load scratch test, it can be defined as the smallest load at which some recognizable failure occurs and the load-bearing capacity is lost. As the scratch test is basically a comparison test, the critical load L_c determined by this method is the representative of coating adhesion behavior [55].

The pull-off test is useful in providing information on the adhesion properties at the interface of the coating and the substrate and is a relatively quantitative method for adhesion strength assessment. The adhesion strength is determined by applying a uniaxial tensile load to a cylindrical test assembly composed of one coated specimen bonded to an uncoated component. The coated and uncoated components are bonded together by the adhesive. The materials and dimensions of the uncoated component shall be the same as those of the coated specimen. After the preparation, the test assembly is placed in the grips of the tensile testing machine, and the tensile load is applied at a constant velocity until complete separation of the components is achieved [56]. Together with other testing methods of adhesion strength, the detailed bonding characteristics of the coating and the substrate can be obtained.

ASTM C633 standard governs the determination of the degree of adhesion of a coating to a substrate or the cohesion strength of the coating in a tension normal to the surface. According to the standard, two kinds of failure modes are defined: (1) adhesive failure happens at the coating–substrate interface and is referred to as "the adhesion strength of the coating" and (2) cohesive failure occurs within the coating and is referred to as "the cohesion strength." The failure between the epoxy and the coating is also regarded as the cohesive failure [57]. However, a combination of these failures is often observed in one specimen. Figure 3.8 gives a schematic illustration of the failure modes for clear identification. Figure 3.8a shows the typical adhesive failure, Figure 3.8b shows the typical cohesive failure, and a combination of both adhesive and cohesive failures is seen in Figure 3.8c.

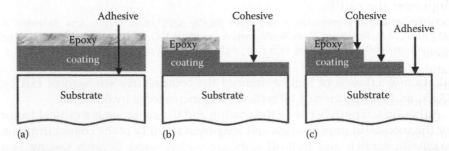

FIGURE 3.8
Schematic illustration of the failure modes: (a) adhesive failure, (b) cohesive failure, and (c) combination of all the failures.

The adhesion strength between Mg-incorporated apatite coating and Ti6Al4V substrate was evaluated by the scratch test [58]. The adhesion strength increases almost linearly as more Mg ions are incorporated in the apatite coatings. The chemical bonds Ca-P-H-O-Ti-Mg at the interface of the coating and substrate contribute to the enhancement of adhesion strength. More Mg-Ti-O chemical bonds are formed as more Mg ions are incorporated in the coating, giving rise to higher adhesion strength in Mg-containing apatite coatings.

3.2.1.3 Biological Performances

Understanding the performance of both natural and synthetic biomaterials is important for the development of new and more biocompatible implant materials. The evaluation involves both in vitro and in vivo assessments. In vitro assessment uses cell and/or tissue derived from both animals and humans and maintained in tissue-culturing conditions. In vivo test uses animal models to evaluate the biological responses to biomaterials in living animals. By obtaining the clinical data during the trial period and histological data from the examination of postmortem tissues, we are able to assess the tissue responses to the implanted materials.

The prerequisite for the biomaterials to bond with living bone is the formation of a layer of biologically active hydroxylcarbonate apatite (HCA) on the biomaterial surface when it is exposed to physiological fluids. The behavior of HCA formation on biomaterials in a simulated body fluid (SBF) is known as the in vitro bioactivity, which is used to evaluate the in vivo bioactivity of bone-forming biomaterials [59]. Zn-incorporated HA with Zn concentrations (Zn/Zn+Ca) at 0.5, 1, 2, and 5 at% was immersed into biological fluids to study the physicochemical reactions at the material's periphery [60]. A Ca-P-Mg layer was formed at the periphery of the HA incorporated with 5 at% Zn, representing the bioactive properties of Zn-incorporated HA. Biomimetic Mg-substituted HA (~1.0 wt% of Mg) prepared via wet-chemical synthesis made the apatite more bone-like, enhanced the solubility of the material at the physiological pH value, and improved the cell behaviors of MSCs and MG63 in terms of adhesion, proliferation, and metabolic activation compared with stoichiometric HA [61]. Sr-containing HA (Sr-HA) cement prepared for primary hip replacement showed good bioactivity and bone-bonding ability after the implantation using a rabbit model, and the direct contact was confirmed between the Sr-HA cement and cancellous and cortical bone under load-bearing conditions [62,63].

Cation-incorporated HA coatings were also evaluated by in vitro and in vivo tests. Sr-HA coatings on Ti6Al4V substrates were prepared by plasma spraying, aiming at the applications of load-bearing implants such as artificial hip replacement. The coating exhibited strong bonding with substrate, its good bioactivity was demonstrated by forming a bone-like apatite layer on the surface of the coating when immersed in the SBF solution, and its excellent biocompatibility was indicated by cell culture tests with human osteoblasts [64]. Also Sr-substituted apatite coating could be grown on Ti6Al4V substrate through biomimetic synthesis. The presence of Sr ions in the solution inhibited the coating formation and resulted in the decrease in coating thickness. The as-formed Sr-substituted biomimetic coatings presented a bone-like structure similar to human bone, which might have a positive effect in enhancing bone formation and contributing to better osseointegration [25]. Mg-containing apatite coatings prepared by the sol–gel method promoted the formation of apatite layer in the SBF solution and showed good bioresponses with MG63 cells just like the HA coatings [65,66].

3.2.2 Incorporation of Anion into HA

There are two kinds of anions in HA crystal structure, phosphate group (PO_4^{3-}) and hydroxyl group (OH^-). PO_4^{3-} can be replaced by HPO_4^{2-}, CO_3^{2-}, or SiO_4^{4-}, and OH^- by F^- or Cl^-. In this review, the emphasis is on the substitution of OH^- groups. In turn, the effect of anion incorporation into HA will be discussed through phase evolution, solubility, and mechanical and biological properties.

3.2.2.1 Phase Evolution and Substitution of Anion into HA

When OH^- groups in HA are partially substituted by F^- ions, fluoride-substituted or fluoridated hydroxyapatite (FHA) forms with a composition of $Ca_{10}(PO_4)_6F_x(OH)_{2-x}$ ($0 < x < 2$). If all the OH^- groups are substituted by F^- ions, it is called fluorapatite (FA, $Ca_{10}(PO_4)_6F_2$). FA has a very similar crystal structure to HA and belongs to the same space group as HA. The arrangement of OH^- groups and F^- ions in HA, FHA, and FA is shown in Figure 3.9. Theoretically, F^- ions easily replace OH^- groups in HA on thermodynamic ground ($\Delta E_x = -0.4–0.6 \text{ kJ mol}^{-1}$). The lowest energy of the crystal structure is reached at $x = 1.0$ with a configuration that HA crystal consists of alternating planes of F^- ions and OH^- groups, where they are alternating in the c-axis but colocated within the a/b-plane [67].

The incorporation of F ions in HA improved the thermal and chemical stability of HA. Figure 3.10 is the TGA data of HA, FHA, and FA powders heated from 25°C to 1400°C in dry air, which shows different degrees of weight loss of samples with different F contents [68]. For pure HA powder, the weight loss from 25°C to 200°C is caused by the evaporation of absorbed water in the powder, that from 800°C to 850°C is due to the release of OH^- groups from the HA crystals, and the decomposition of HA and further loss of OH^- groups happen above 1000°C. However, the decomposition stage is not observed with F incorporation, suggesting the retarded decomposition and improved thermal stability of HA. Moreover, the dilatometer analysis

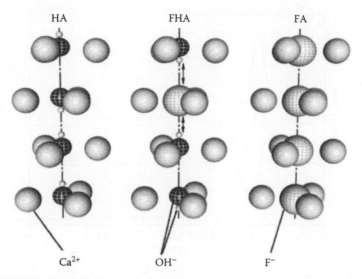

FIGURE 3.9
Arrangement of OH⁻ groups and F⁻ ions in HA, FHA, and FA crystal structures. (From *Biomaterials*, 26, Chen, Y., and Miao, X., Thermal and chemical stability of fluorohydroxyapatite ceramics with different fluorine contents, 1205, Copyright 2005, with permission from Elsevier.)

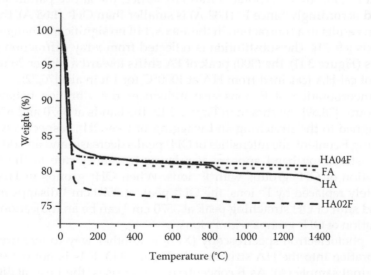

FIGURE 3.10
TGA data of HA, FHA, and FA powders heated to 1400°C in dry air. (From *Biomaterials*, 26, Chen, Y., and Miao, X., Thermal and chemical stability of fluorohydroxyapatite ceramics with different fluorine contents, 1205, Copyright 2005, with permission from Elsevier.)

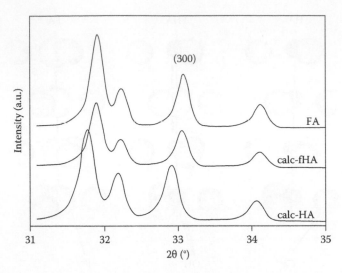

FIGURE 3.11
Short-range XRD patterns of calc-HA, calc-FHA, and FA. (With kind permission from Springer Science + Business Media: *J. Mater. Sci. Mater. Med.*, Synthesis and characterization of hydroxy-apatite, fluoride-substituted hydroxyapatite and fluorapatite, 14, 2003, 311, Wei, M., Evans, J.H., Bostrom, T., and Grøndahl, L.)

showed that the onset sintering temperature of FHA powder was increased to 1150°C compared with that of HA powder around 1000°C.

When F^- ions are incorporated into HA lattice, the lattice parameters are changed accordingly. Since F^- (1.32 Å) is smaller than OH^- (1.68 Å), the substitution results in a contraction in the a-axis, but no significant change along the c-axis [69–71]. The substitution is reflected from x-ray diffraction (XRD) patterns (Figure 3.11): the (300) peak of FA shifts toward a higher 2θ relative to that of cal-HA (calcined from HA at 1000°C for 1 h in air) [70,72].

The incorporation of F^- ions also influences the vibration behavior of OH^- groups [38,69]. As shown in Figure 3.12, the bands at 3570 and 630 cm^{-1} are assigned to the stretching and wagging of free OH^-, respectively. With increasing F content, the intensities of OH^- peaks decrease, and an additional OH^-–F^- stretching band appears at a lower frequency due to the partial substitution of OH^- groups with F^- ions. When OH^- groups in HA were completely replaced by F^- ions, the OH^- peak at 3570 cm^{-1} disappears. The split and shift of OH stretching peak at 3570 cm^{-1} can be an indication of the substitution of OH^- groups with F^- ions.

X-ray photoelectron spectroscopy (XPS) is another way to identify the F incorporation into the HA structure. In Figure 3.13, F 1s is not detected in the original sample (A). As F concentration increases, the peak at 684.3 eV intensifies. While for CaF_2, F 1s peak locates at 685.0 eV. Moreover, the binding energy of Ca 2p shifts to 347.4 eV, which is different from that in the original sample (at 347.1 eV) and that for CaF_2 (at 348.3 eV). This is due to

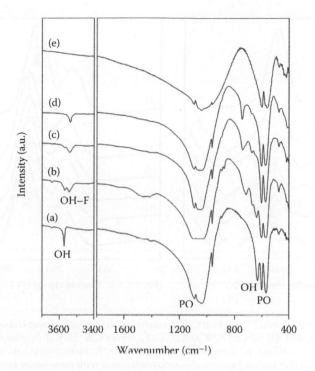

FIGURE 3.12
FTIR spectra of the coatings: (a) HA, (b) 25 FHA, (c) 50 FHA, (d) 75 FHA, and (e) FA. (From *Biomaterials*, 26, Lee, E.-J., Lee, S.-H., Kim, H.-W., Kong, Y.-M., and Kim, H.-E., Fluoridated apatite coatings on titanium obtained by electron-beam deposition, 3843, Copyright 2005, with permission from Elsevier.)

the chemical shift relative to the incorporation of F$^-$ ions in the apatite [73]. The fingerprint of F 1s in FHA or FA at 684.3 eV was also reported by other researchers [74,75].

When OH$^-$ groups in HA are partially replaced by Cl$^-$ ions, chlorine-substituted HA (chlorapatite, Clap) is obtained [41]. The hexagonal apatite structure retains for all the chlorine substituted samples with no major structural distortion. The calculated lattice parameters of the samples provide the evidence for the incorporation of Cl$^-$ ions in the apatite structure. A significant expansion of all the parameters of *a*-axis and *c*-axis was observed for Clap with respect to that of HA due to the substitution of bigger Cl$^-$ ions (1.81 Å) for OH$^-$ groups (1.68 Å).

3.2.2.2 Solubility

Solubility is an important factor influencing both the short-term osseo-integration and the long-term stability. Immediately after the implantation, a certain extent of dissolution of the coating is necessary to increase

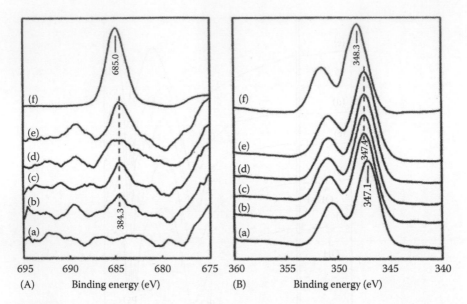

FIGURE 3.13
XPS spectra of F 1s (A) and Ca 2p (B) of CaHA treated with various F concentrations (mol/dm^3): (a) 0, (b) 0.01, (c) 0.03, (d) 0.05, (e) 0.10, and (f) CaF_2. (From *Coll. Surf. A Physicochem. Eng. Asp.*, 204, Tanaka, H., Yasukawa, A., Kandori, K., and Ishikawa, T., Surface structure and properties of fluoridated calcium hydroxyapatite, 251, Copyright 2002, with permission from Elsevier.)

the local ion concentrations of calcium and phosphate to stimulate the formation of new bone. From the viewpoint of the long-term stability of the implant, the dissolution resistance of the coating should be strong enough to avoid the detachment of the coating from the substrate. Therefore, it is vital to investigate the solubility of the coating. Factors affecting the solubility of a material include the preparation method, the resultant crystallinity and density, and the extent of ion substitutions into the apatite lattice [76].

It is well known that fluoride can prevent caries. The mechanism of reduced solubility of HA by addition of fluoride has been investigated intensively [77–80]. Moreno et al. [80] discussed the solubility of FHA $(Ca_5(PO_4)_3(OH)_{1-x}F_x)$ theoretically and experimentally. A series of solid solutions with different fluoridation degrees x, from 0 to 1, were prepared. The solubility of these compounds was obtained by equilibrating them for more than 1 month with a series of dilute phosphoric acid solutions at 37°C. The solubility product constant K_x at various substitution concentrations was calculated according to the experimental data shown in Figure 3.14. From Figure 3.14, a minimum in K_x is seen at a substitution degree of 0.56. The decreased solubility can be explained on the basis of the disordered structure of HA and the hydrogen bonding between OH⁻ groups and

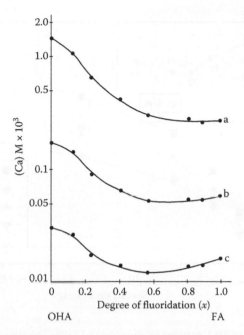

FIGURE 3.14
Solubility of the solid solutions as a function of the degree of fluoride substitution $(Ca_5(PO_4)_3(OH)_{1-x}F_x$, [a] pH 5; [b] pH 6; [c] pH 7). (Reprinted by permission from Macmillan Publishers Ltd., *Nature*, Moreno, E.C., Kresak, M., and Zahradnik, R.T., Fluoridated hydroxyapatite solubility and caries formation, 247, 64, Copyright 1974.)

F^- ions. Also the solubility of heterogeneous FHA was studied by Okazaki et al. [81,82]. The apparent solubility of FHA decreased with the increased fluoridation degree no matter if the fluoride was supplied in the initial or final half of the experimental period.

Similar dissolution behaviors were also found in FHA coatings regardless of preparation methods (such as thermal spraying [83], electron-beam deposition [38], and sol–gel deposition [7,15]). For example, Figure 3.15 shows that the incorporation of F^- ions decreased the solubility of HA coatings prepared by sol–gel deposition. Besides the experimental studies of the effect of F incorporation on the solubility of HA, a computer modeling study was also done using molecular dynamic simulations [67,84]. In pure HA material, the onset of dissolution is induced through the leaching of surface hydroxyl groups from the material into solution in the presence of water. When F^- ions partially replace OH^- groups, OH^- group may dissolve, but the F^- ion at the surface does not dissolve as readily but remains in the crystal. Moreover, the surface Ca^{2+} ions persist more closely coordinated to the material surface in the vicinity of the substituted F^- ions. Thus, the incorporation of F^- ions into the lattice makes

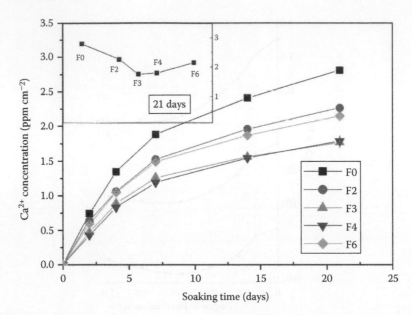

FIGURE 3.15
Dissolution behavior of FHA coatings in physiological saline solution: Ca^{2+} concentration as a function of soaking time. (From *Acta Biomater.*, 3, Wang, Y.S., Zhang, S., Zeng, X.T., Ma, L.L., Weng, W.J., Yan, W.Q., and Qian, M., Osteoblastic cell response on fluoridated hydroxyapatite coatings, 191, Copyright 2007, with permission from Elsevier.)

the materials more dissolution resistant through two aspects: the resistance toward dissolution into the solvent of F^- ions themselves and strong anchoring of the surface Ca^{2+} ions.

3.2.2.3 Mechanical Properties

Fluorine incorporation into HA also changes the mechanical properties of HA. Bulk FHA has been widely synthesized, and their mechanical properties were measured using indentation, including hardness, elastic modulus, fracture toughness, and brittleness [85,86]. Both the fluoridation degree and the sintering temperature have an effect on the mechanical properties of bulk HA, which are not reviewed here. However, the dependence of F incorporation on the mechanical properties of FHA coatings is reviewed later in detail.

Wang et al. [8,87,88] have systematically investigated the mechanical properties of sol–gel derived FHA ($Ca_{10}(PO_4)_6(OH)_{2-x}F_x$, x varies from 0 to 2) coatings on Ti6Al4V substrates. The adhesion strength between the coating and the substrate was evaluated through both scratch and pull-out tensile tests. Scratch test revealed that the adhesion strength was up to 35% better as the F concentration increased in the coating, and this adhesion enhancement was

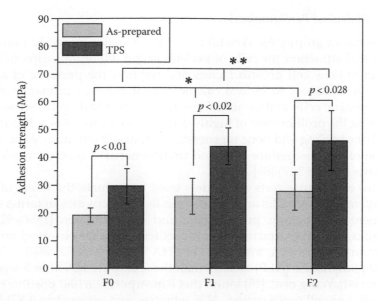

FIGURE 3.16
Adhesion strength of FHA coating before and after soaking in TPS solutions. (From *Thin Solid Films*, 516, Zhang, S., Wang, Y.S., Zeng, X.T., Khor, K.A., Weng, W., and Sun, D.E., Evaluation of adhesion strength and toughness of fluoridated hydroxyapatite coatings, 5162, 2008, with permission from Elsevier.)

much more obvious at higher firing temperatures. The pull-out tensile test showed that the adhesion strength increased up to about 40% from ~19 MPa for HA to ~26 MPa for FHA ($x = 1$) as shown in Figure 3.16. After soaking in tris-buffered physiological saline (TPS) solutions for 21 days, the enhancement became more significant because the rougher surface after the dissolution resulted in stronger bonding between the epoxy and the coating and in turn reduced the area fraction of the "adhesive failure."

The toughness, interfacial shear strength, and residual stress of FHA coatings have also been evaluated [8,87,88]. At the fluoridation degree of $x = 1$, the toughness of the coating doubled as compared with HA coating. The interfacial shear strength was evaluated by the shear strain lag method, and the "wafer curvature method" was employed to measure the residual stresses. The resultant interfacial shear strength increased from ~393 MPa (pure HA) to ~459 MPa (1.96 at% fluorine) and further to ~572 MPa (3.29 at% fluorine). The residual stress in the coating also decreased from ~273 to ~190 MPa and further to ~137 MPa as F concentration in the coating increased. Other researchers also studied the mechanical properties of FHA coatings on metal substrates. For instance, Kim et al. [69] deposited an HA/FHA double-layered film on titanium by a sol–gel method, which was highly dense and uniform and adhered to the substrate strongly with an adhesion strength of about 40 MPa.

3.2.2.4 Biological Performances

Fluoride has an affinity for skeletal tissue and also has a specific mitogenic activity and attraction for cells of skeletal origin. Fluoride seems to be an enhancer of bone cell growth factors and requires the presence of appropriate growth factors to induce calcification. It may act primarily on the osteoprogenitor cells and/or undifferentiated osteoblasts cells rather than stimulating the proliferation of highly differentiated osteoblasts. It is important in bone healing and bone regeneration because it will induce the differentiation of osteoprogenitor cells and undifferentiated precursor cells into osteoblasts (Figure 3.17) [89].

In vitro cell culture tests are widely used to evaluate the effect of fluorine content in FHA bulks and coatings on the cell activities in terms of cell attachment, morphology, proliferation, and differentiation [15,90–92]. For FHA bulks, more cell attachment and proliferation were observed on FHA bulks than on pure HA, and cells on FHA bulks demonstrated a higher alkaline phosphatase (ALP) activity than those on pure HA after 3 weeks of culturing [92]. Wang et al. [15] found that F incorporation had positive stimulating effect on cell proliferation, ALP activities, and osteocalcin (OC) levels,

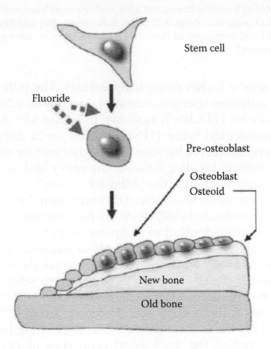

Stem cell

Fluoride

Pre-osteoblast

Osteoblast
Osteoid

New bone

Old bone

FIGURE 3.17
Effect of fluoride in cell differentiation and new bone formation. (From Ellingsen, J.E., Thomsen, P., and Lyngstadaas, S.P.: Advances in dental implant materials and tissue regeneration. *Periodontology*. 2000, 2006. 41. 136. Copyright Wiley-VCH Verlag GmbH & Co. KGaA. Reproduced with permission.)

especially at the fluoridation degree of 0.8–1.1. However, Kim et al. [7] found that the osteoblast-like MG63 and HOS cells grew and proliferated favorably on all the HA and FHA coatings as well as on pure Ti. The osteoblast cells proliferated on FHA coatings to a lower extent than on pure HA coating obtained by electron-beam deposition, and the ALP activity of the cells on the FHA coatings was not significantly different from that on the HA coating [38]. The positive or negative effect of fluorine ions in FHA coatings on the cell behaviors is still under argument.

In addition, several phenomena were observed during in vivo studies of FHA coatings. The degradation of plasma-sprayed HA, FA, and FHA coatings on dental implants evaluated in dog jaws till 12 months indicated good integration between FHA coating and bone tissue. HA and FA coatings (even at 100 μm thickness) were almost completely degraded within the implantation period. In contrast, FHA coatings did not show significant degradation during the same period [93]. Savarino et al. [94] have investigated the in vivo compatibility and degradation of plasma-sprayed FHA, HA coatings, and titanium oxide (TiO_x). The bone at the interface was not mineralized in all samples after 2 weeks of implantation. After 12 weeks, the newly formed bone tissue at the interface with both FHA and HA coating was shown to be fully mineralized. Also the in vivo tests done by Caulier et al. [95] proved that no significant differences were found in bone reaction among uncoated, plasma-sprayed HA and FA coatings on titanium on the initial bone healing of oral implants.

3.2.3 Problems Involved in Cation- or Anion-Incorporated HA

To make the implant more acceptable to bone cells and stimulate rapid integration with bone rather than fibrous encapsulation, the surface coating should have suitable biochemical and structural characteristics to elicit wanted cell activity and finally result in full osseointegration. Moreover, the resultant bone–implant interface must have the internal strength and bonding capacity to the underlying metal substrate so as to endure the forces produced by the implant loading. Cation or anion substitution into HA improves the mechanical and biological properties in comparison with pure HA. For example, the cations like Mg^{2+} incorporation is beneficial to the biological performance of HA, while the anions like F^- incorporation improves the adhesion strength between the coating and the substrate. However, it is found that the short-term osseointegration and the long-term stability are difficult to be addressed at the same time if only a cation or an anion is incorporated. As such, simultaneous incorporation of cation and anion into HA is proposed to exert their special advantages in biological and mechanical performances and achieve the short-term osseointegration and the long-term stability of the implant, which will be reviewed in the next section.

3.2.4 Simultaneous Incorporation of Cations and Anions into HA

The apatite structure has the ability to accept a wide number of possible substitutions in its lattice, and trace elements are important to bone mineral, so the simultaneous incorporation of several elements into apatite structure is expected to improve the performances of apatite. So far, many attempts have been made to synthesize the apatite substituted with different cations and anions. The simultaneous incorporation of cations and anions into HA is summarized in Table 3.1. As shown in Table 3.1, aqueous precipitation is the most commonly used method for the synthesis of cation- and anion-incorporated bulk HA, while sol–gel method is preferred to deposit HA and derivative coatings because the method allows precise control of the coatings' composition.

When cations and anions are incorporated into HA simultaneously, crystallographic properties, thermal behaviors, mechanical and biological properties, etc., differ from individual substitution of cations or anions into HA. For example, cations themselves (Al^{3+}, Mg^{2+}, Zn^{2+}, or Sr^{2+}) at low levels in the solution had negative effects on the properties of apatite, e.g., lower crystallinity, lower Ca/P ratios, higher HPO_4/PO_4, and higher dissolution in an acid buffer. When only F^- ions were present in the precipitation solution, positive effects on the apatite were observed, e.g., greater crystallinity, higher Ca/P, lower HPO_4/PO_4, and lower dissolution. However, the simultaneous presence of F^- ions with the cations (Al^{3+}, Mg^{2+}, Zn^{2+}, or Sr^{2+}) minimized the negative effects caused by the cations, suppressed the formation

TABLE 3.1

Summary of Simultaneous Incorporation of Cations and Anions into HA

	Cation	Anion	Method	References
Bulk	K^+	Cl^-	Melt salt method	[142]
	Al^{3+}	F^-	Aqueous precipitation method	[96]
	Sr^{2+}	F^-	Aqueous precipitation method	[96]
	Zn^{2+}	F^-	Aqueous precipitation method	[96]
	Mg^{2+}	F^-	Aqueous precipitation method	[96,98,100,101]
	Mg^{2+}	F^-	Heterogeneous synthesis with two-step supply system	[99]
	Mg^{2+}	CO_3^{2-}	Aqueous precipitation method	[97,143]
	Mg^{2+}	CO_3^{2-}	Precipitation method with gradient Mg supply system	[144]
	Mg^{2+}	CO_3^{2-}	Neutralization method	[61]
	Mg^{2+}	CO_3^{2-}, F^-	Aqueous precipitation method	[145]
	Na^+, Mg^{2+}	F^-	Aqueous precipitation method	[14]
	Na^+, K^+, Mg^{2+}	F^-, Cl^-	Aqueous precipitation method	[146]
Coating	Zn^{2+}	F^-	Sol–gel method	[23,24,147,148]

of Mg-substituted β-TCP (β-TCMP), and allowed a greater incorporation of cations in the apatite [96].

Among different cations and anions, Mg and F ions have their own advantages. As the fourth abundant element in the human body, Mg could reduce the risks of cardiovascular disease, promote the catalytic reactions, and control the biological functions. Mg content in the natural bone is affected by a great variability depending on aging: it is contained in high concentrations in cartilage and natural bone tissue during the initial stages of osteogenesis, while it tends to disappear when the bone is mature. For example, 1.0 wt% Mg was found in the new-born rats, and this level decreased to 0.5 wt% after 1 year [97]. So it is vital to introduce Mg ions during the stage of new bone formation. Furthermore, F ions as well are recognized for their potential behavior on the long-term stability of the apatite coating, such as the improvement of mechanical properties and dissolution resistance.

Okazaki [98,99] has systematically investigated the crystallographic properties of Mg-free and Mg-containing fluoridated apatite (Mg-FHAp$_I$ and Mg-FHAp$_{II}$ with different Mg concentration) synthesized by aqueous precipitation and two-step supply system. The inverse of the half-value breadth calculated from (300) and (002) reflections is represented as the crystallinity along the a-axis and c-axis, as shown in Figure 3.18. The crystallinity varies in each series with increasing F content. A minimum crystallinity of FHAp is observed at the low F content, and then the crystallinity increases with the increasing F content (Figure 3.18a). At lower Mg content, Mg-FHAp$_I$, this minimum crystallinity is invisible, and an overall increase in crystallinity is found (Figure 3.18b). At higher Mg content, Mg-FHAp$_{II}$, the crystallinity steadily reaches a maximum at about half of the maximum F content (Figure 3.18c), and the whole crystallinity of Mg-FHAp$_{II}$ is lower than that of Mg-FHAp$_I$. The minimum crystallinity seen at lower F content in FHAp might be due to the imbalance between F$^-$ ions and OH$^-$ groups, which is relieved by the increase in F content. When Mg^{2+} ions are added, especially at higher concentrations, the positive effect of F$^-$ ions on the crystallinity is weakened by the addition of Mg^{2+} ions.

The influence of Mg and F ions on the crystallinity of Mg-containing fluoridated apatite was also reported by Hidouri et al. [100]. As shown in Figure 3.19, a slight broadening of the reflections is observed as the increased Mg concentration (Figure 3.19b through d), indicating the lower crystallinity of Mg-containing fluoridated apatite compared with pure fluoridated apatite (Figure 3.19a).

The lattice dimensions of three series of FHAp were calculated from (300) and (002) reflections using silicon as standard, as shown in Figure 3.20. The decrease in a-axis of Mg-free and Mg-containing FHAp with the increased fluoridation degree suggests the substitution of F$^-$ ions into the apatite crystal structures. The c-axis also decreases with the increased Mg content

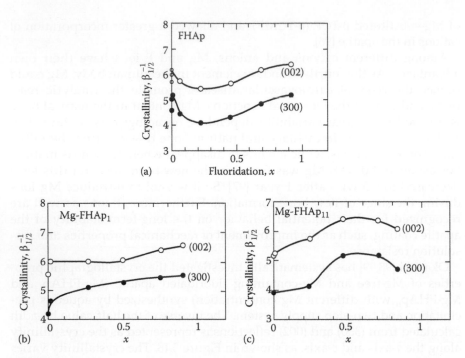

FIGURE 3.18
Crystallinity of Mg-free (FHAp [a]) and Mg-containing fluoridated hydroxyapatite (Mg-FHAp$_I$ [b] and Mg-FHAp$_{II}$ [c]). (From *J. Fluor. Chem.*, 41, Okazaki, M., Magnesium-containing fluoridated apatites, 45, Copyright 1988, with permission from Elsevier.)

over the whole range of F content due to the substitution of Ca^{2+} with Mg^{2+} ions. This substitution can also be reflected from the shifting of (002) peaks (Figure 3.21). With the increasing Mg content, (002) peak shifts toward higher degree. In addition, the decreased a- and c-axes with the increase in Mg content further confirmed the substitution [100].

The thermal behavior of Mg-containing FHA was investigated by Hidouri [100]. As-prepared powders were calcined at different temperatures to determine possible changes of sintered phases. Take 1.00 MgFap as an example (Figure 3.22). A secondary-phase $Mg_2F(PO_4)$ is detected in the range of 650°C–1120°C, and the amount of this phase increases with increasing Mg concentration. $Mg_2F(PO_4)$ phase disappears at 1300°C due to its dissolution in the liquid phase or incorporation into the apatite structure. Since no decomposition of the material is observed at 1300°C, it is concluded that the incorporation of Mg into apatite does not affect its thermal stability.

Figure 3.23 shows the solubility of Mg-free and Mg-containing FHAp as the function of dissolved Ca^{2+} concentration in 0.5 mol/L acetate buffer solutions (pH 4.0, 37°C). The solubility of Mg-free FHAp is lower than that of Mg-containing FHAp at lower F content, but the difference between them is negligible at higher F content [98]. Even if the inclusion of Mg^{2+} ions increases

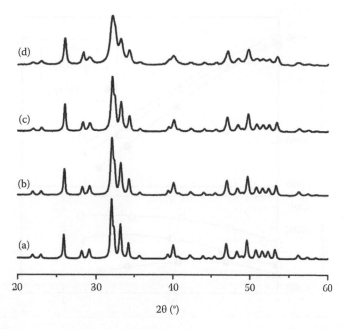

FIGURE 3.19

XRD patterns of as-prepared powders with different Mg concentrations: (a) Fap, (b) 0.25 MgFap, (c) 0.50 MgFap, and (d) 1.00 MgFap. (From *Mater. Chem. Phys.*, 80, Hidouri, M., Bouzouita, K., Kooli, F., and Khattech, I., Thermal behaviour of magnesium containing fluorapatite, 496, Copyright 2003, with permission from Elsevier.)

the solubility of FHA, the strong bonding of F^- ions with neighboring ions may be maintained to stand against the acid attack, so that the solubility of different series at higher F content does not vary extensively.

Most recently, Sun et al. [101] synthesized Mg^{2+}/F^--doped nanophase HA powder by a precipitation method followed by sintering at 1100°C for 1 h. The highest microhardness was achieved in 1% Mg- and 2.5% F-doped HA. Cytocompatibility testing shown in Figure 3.24 demonstrates no statistical difference in osteoblast cell densities after 4 h between HA, HA1Mg, HA2.5Mg, HA7.5Mg, HA1Mg2.5F, and HA2.5Mg2.5F. Significant increases in osteoblast cell densities are found on HA7.5Mg2.5F and HA2.5Mg7.5F (where 7.5Mg and 2.5F refer to the mole percent of Mg and F in HA, respectively). However, the reason for the greater osteoblast adhesion on these two samples is still unknown.

The preliminary investigation shows that the simultaneous incorporation of cations and anions results in different crystallinity, thermal stability, solubility, and mechanical and biological performances from those of cation- or anion-incorporated HA. However, up to date, the exploration of simultaneous incorporation of cations and anions (especially Mg and F ions) into HA is not systematical and needs further investigation, especially when they are employed as coatings on metal substrates.

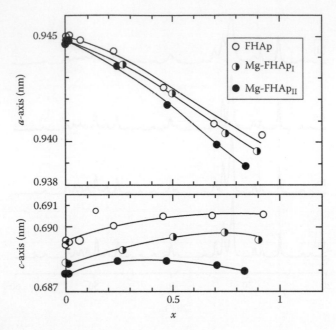

FIGURE 3.20
Dimensions of *a*- and *c*-axes for Mg-free (FHAp) and Mg-containing fluoridated hydroxyapatite (Mg-FHAp$_I$ and Mg-FHAp$_{II}$). (From *J. Fluor. Chem.*, 41, Okazaki, M., Magnesium-containing fluoridated apatites, 45, Copyright 1988, with permission from Elsevier.)

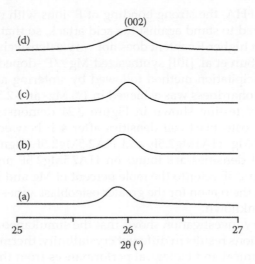

FIGURE 3.21
(002) peak reflections of as-prepared powders: (a) Fap, (b) 0.25 MgFap, (c) 0.50 MgFap, and (d) 1.00 MgFap. (From *Mater. Chem. Phys.*, 80, Hidouri, M., Bouzouita, K., Kooli, F., and Khattech, I., Thermal behaviour of magnesium-containing fluorapatite, 496, Copyright 2003, with permission from Elsevier.)

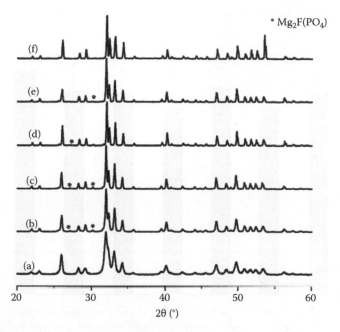

FIGURE 3.22
XRD patterns of as-prepared 1.00 MgFap powder sintered at different temperatures.
(a) 1300°C, (b) 1120°C, (c) 1050°C, (d) 750°C, (e) 650°C, and (f) 500°C. (From *Mater. Chem. Phys.*,
80, Hidouri, M., Bouzouita, K., Kooli, F., and Khattech, I., Thermal behaviour of magnesium-
containing fluorapatite, 496, Copyright 2003, with permission from Elsevier.)

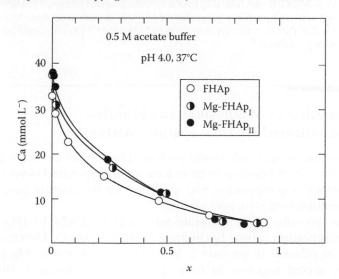

FIGURE 3.23
Solubility of Mg-free (FHAp) and Mg-containing fluoridated hydroxyapatite (Mg-FHAp$_I$ and
Mg-FHAp$_{II}$) as dissolved Ca^{2+} concentration. (From *J. Fluor. Chem.*, 41, Okazaki, M., Magnesium-
containing fluoridated apatites, 45, Copyright 1988, with permission from Elsevier.)

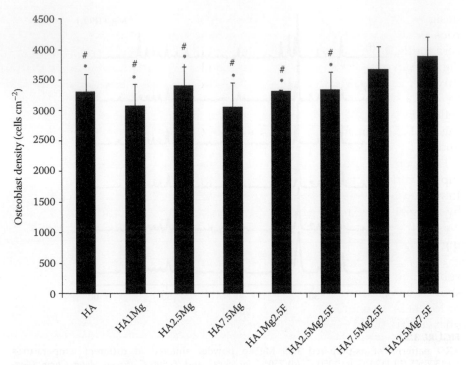

FIGURE 3.24

Four-hour osteoblast adhesion test (*$p < 0.05$ with respect to HA7.5Mg2.5F, #$p < 0.05$ with respect to HA2.5Mg7.5F). (From Sun, Z.P., Ercan, B., Evis, Z., and Webster, T.J.: Microstructural, mechanical, and osteocompatibility properties of Mg^{2+}/F^--doped nanophase hydroxyapatite. *J. Biomed. Mater. Res. Part A.* 2010. 94A, 806. Copyright Wiley-VCH Verlag GmbH & Co. KGaA. Reproduced with permission.)

3.3 Preparation of Magnesium and Fluorine Co-Substituted Hydroxyapatite Coatings*

Sol–gel dip-coating method is most preferred to deposit ion-substituted HA coating, especially when two or more ions are incorporated simultaneously. It can control the composition and thickness of the coating precisely and realize a multilayered structure easily.

To make the sols, calcium nitrate tetrahydrate ($Ca(NO_3)_2 \cdot 4H_2O$, Merck, AR) and magnesium nitrate hexahydrate ($Mg(NO_3)_2 \cdot 6H_2O$, Merck, AR) were dissolved in ethanol to prepare 2 M of Ca-precursor and Mg-precursor. They were mixed together to form a Ca–Mg mixture according to the

---

* From *Thin Solid Films*, 517, Cai, Y., Zhang, S., Zeng, X., Wang, Y., Qian, M., and Weng, W., Improvement of bioactivity with magnesium and fluorine ions incorporated hydroxyapatite coatings via sol-gel deposition on Ti6Al4V alloys, 5347, 2009, with permission from Elsevier.

designed amount. 2 M of P-precursor was prepared by dissolving phosphorous pentoxide (P_2O_5, Merck, GR) in ethanol followed by a refluxing process for 24 h. Then F-precursor (diluting hexafluorophosphoric acid (HPF_6, Sigma-Aldrich, GR) in ethanol) was added to P-precursor to obtain a P–F mixture. The Ca–Mg mixture was added drop-wise into the P–F mixture to form a solution with a (Ca, Mg)/P ratio of 1.67. This mixture was further refluxed for 24 h to obtain the sol. The coating composition is controlled during this process by adjusting the ratios of Ca-precursor and Mg-precursor, and P-precursor and F precursor.

Fine-polished Ti6Al4V ($20 \times 30 \times 2$ mm^3) were used as substrates. It was dipped vertically into the sol and withdrawn at a speed of 3 cm min^{-1} for the first layer. The as-dipped coating was then dried at 150°C for 15 min followed by firing at 600°C for 15 min. This dipping–drawing–drying–firing process was further repeated four times at a speed of 4.5 cm min^{-1} for a thicker coating with a thickness of 1.5–2 μm.

This work aims at developing a bilayer structured Mg and F ion cosubstituted HA coating on Ti6Al4V substrate to achieve an integration of short-term osseointegration and long-term stability. $Ca_{10-x}Mg_x(PO_4)_6F_1(OH)_1$ or Mg_xFHA in short, $Ca_9Mg_1(PO_4)_6F_y(OH)_{(2-y)}$ or MgF_yHA in short, and the bilayer structured HA (BHA in brief) coatings were deposited by a sol–gel dip-coating method. The compositional dependence of co-substitution of Mg and F ions was systematically studied on material structure, in vitro bioactivity, cell response, dissolution rate, and adhesion strength.

3.4 Effect of Magnesium Incorporation in Fluoridated Hydroxyapatite Coatings*

Magnesium ions are incorporated into FHA coatings to improve the biological performances. The as-deposited coatings are denoted by Mg_xFHA ($Ca_{(10-x)}Mg_x(PO_4)_6F(OH)$, x varies from 0 to 2). The corresponding samples were labeled as FMA0, FMA1, FMA2, FMA3, and FMA4, or the corresponding coatings as Mg_0FHA, $Mg_{0.5}FHA$, Mg_1FHA, $Mg_{1.5}FHA$, and Mg_2FHA for further discussion.

* From *Thin Solid Films*, 519, Cai, Y., Zhang, S., Zeng, X., Qian, M., Sun, D., and Weng, W., Interfacial study of magnesium-containing fluoridated hydroxyapatite coatings, 4629, 2011, with permission from Elsevier.

From *Thin Solid Films*, 517, Cai, Y., Zhang, S., Zeng, X., Wang, Y., Qian, M., and Weng, W., Improvement of bioactivity with magnesium and fluorine ions incorporated hydroxyapatite coatings via sol–gel deposition on Ti6Al4V alloys, 5347, 2009, with permission from Elsevier.

From Cai, Y., Zhang, J., Zhang, S., Mondal, D., Venkatraman, S.S., and Zeng, X., Osteoblastic cell response on fluoridated hydroxyapatite coatings: The effect of magnesium incorporation, *Biomed. Mater.*, 5, 054114, 2010. With permission from Institute of Physics.

3.4.1 Co-Substitution of Mg and F Ions in the Mg$_x$FHA Coatings

3.4.1.1 Phase Evolution

Grazing incidence x-ray diffraction (GIXRD) patterns of Mg$_x$FHA coatings on Ti6Al4V substrates are shown in Figure 3.25. The diffraction peaks of all the samples are assigned to apatite phase and Ti6Al4V substrate. With reference to pure HA (PDF#09-0432), there is no other calcium phosphate found in the coating. All the main diffraction peaks in FMA0 sample have a slight shift toward higher degree, especially the (300) peak shifting from 32.90° of pure HA to 33.04°. This is because some of the hydroxyl groups (OH$^-$) are substituted by the fluorine ions (F$^-$) to form FHA [91]. There is not much difference in peak positions when the concentration of Mg is low (see FMA1 and FMA2) except slight peak shift toward higher degree. It is expected as the smaller Mg^{2+} (0.69 Å) substituted the larger Ca^{2+} (0.99 Å) ion. As Mg concentration increases further, i.e., FMA3 and FMA4, a new phase of Mg-substituted β-tricalcium phosphate (β-TCMP) is detected, signaling it beyond the maximum "solubility" of Mg in HA crystal structure.

In our previous study [65], β-TCMP is the main phase when Mg is incorporated into HA coatings with the same concentration as FMA2 in the sol. This is because the substitution of Mg^{2+} into Ca^{2+} positions makes the apatite structure resemble β-TCP. Moreover, Mg favors the conversion of HA into β-TCMP during the heat treatment [47,102]. However, when Mg and F ions are incorporated into HA coatings at the same time, only a small amount

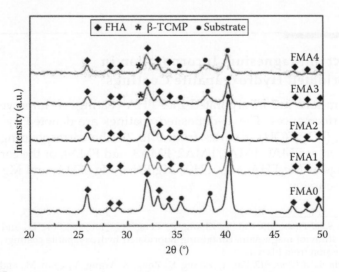

FIGURE 3.25
GIXRD patterns of as-deposited Mg$_x$FHA coatings on Ti6Al4V substrates. (From *Thin Solid Films*, 517, Cai, Y., Zhang, S., Zeng, X., Wang, Y., Qian, M., and Weng, W., Improvement of bioactivity with magnesium and fluorine ions incorporated hydroxyapatite coatings via sol-gel deposition on Ti6Al4V alloys, 5347, 2009, with permission from Elsevier.)

of β-TCMP is found in FMA3 and FMA4, meaning F-incorporation in HA increases the "solubility" of Mg in the HA structure. Similar results are also reported by other researchers who prepared apatite powders in the presence of Mg and F ions [96]. The shrinkage along c-direction caused by the substitution of OH⁻ groups by F⁻ ions would lead to larger "rattling" space. This makes the lattice structure less stable, and as a result, Mg ions can have more chances to enter the HA crystal structure to occupy the Ca position.

3.4.1.2 Co-Substitution of Mg and F Ions in the Mg$_x$FHA Coatings

Theoretically, Mg^{2+} (0.69 Å) substitutes Ca^{2+} (0.99 Å), and F⁻ (1.32 Å) replaces OH⁻ (1.68 Å) in HA crystal structure when Mg or F ions are incorporated into HA powders or coatings. The incorporation of foreign ions into HA crystal structure influences the lattice parameters, chemical states, and mechanical and biological properties of HA correspondingly. Therefore, it is vital to confirm the substitution of Mg and F ions into HA crystal structure, which is studied from lattice parameters, chemical states, and chemical groups.

The lattice parameters are modified when fluorine ions are incorporated into HA crystal structure. There is a significant decrease in the a-axis from 9.418 Å for pure HA down to 9.3710 Å for pure FA, but the c-axis shows no obvious change [69–71]. In this case, the shift of (002) peak and the change of c-axis can reflect the substitution of Ca^{2+} with Mg^{2+} when Mg and F ions are simultaneously incorporated into HA coating. Table 3.2 lists the two-theta of (002) peak and calculated c-axis of Mg$_x$FHA coatings. As Mg concentration increases from 0 to 1.0, the (002) peak shifts toward higher degree due to the substitution of larger Ca^{2+} (0.99 Å) with smaller Mg^{2+} (0.69 Å). However, when Mg concentration increases further to 1.5 and 2.0, (002) peak shifts to lower degree as compared to when $x = 1.0$, but still higher than that when $x = 0.5$. The same trend is also observed from the change of c-axis. The change of c-axis is related to the phase analysis stated earlier. At higher Mg concentrations (FMA3 and FMA4), β-TCMP is detected as well as FHA, implying that Mg^{2+} cannot completely replace Ca^{2+}. The formation of β-TCMP competes with the Ca^{2+} substitution on the assumption of Mg^{2+} in the sol. Therefore, there is a little increase in c-axis at higher Mg concentrations. Although the change of c-axis is not linear with the increasing Mg concentration, the decrease in c-axis indicates the replacement of Ca^{2+} by Mg^{2+} in the HA crystal structure.

TABLE 3.2

Two-Theta of (002) Peak and Calculated c-Axis of Mg$_x$FHA Coatings

Sample	FMA0	FMA1	FMA2	FMA3	FMA4
Two-theta (°)	25.84	25.86	25.94	25.90	25.90
c-axis (Å)	6.89	6.88	6.86	6.87	6.87

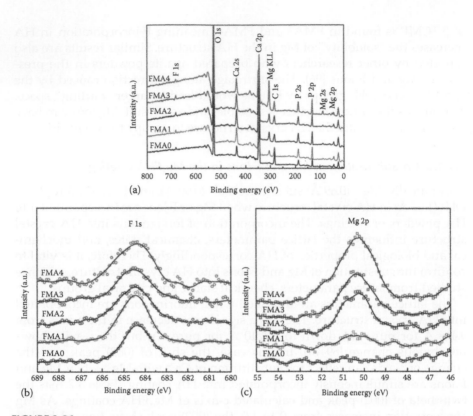

FIGURE 3.26
XPS profiles of Mg$_x$FHA coatings on Ti6Al4V substrates: (a) survey scan, (b) F 1s narrow scan, and (c) Mg 2p narrow scan. (From *Thin Solid Films*, 517, Cai, Y., Zhang, S., Zeng, X., Wang, Y., Qian, M., and Weng, W., Improvement of bioactivity with magnesium and fluorine ions incorporated hydroxyapatite coatings via sol-gel deposition on Ti6Al4V alloys, 5347, 2009, with permission from Elsevier.)

The chemical compositions of Mg$_x$FHA coatings are confirmed by XPS, as shown in Figure 3.26. The survey scans (Figure 3.26a) demonstrate that all the elements in Mg$_x$FHA coatings are detected (the presence of C 1s in all the samples is due to inevitable surface contamination even from air). The narrow scans of F 1s are shown in Figure 3.26b, and the binding energy of F 1s at 684.3 ± 0.1 eV reveals that F ions have been incorporated into the HA crystal structure (F 1s at 684.3 ± 0.1 eV is the fingerprint of F in FHA or FA, while the peak of F 1s in CaF$_2$ is at 686.7 eV [8,75]). This is also supported by the GIXRD analysis. Figure 3.26c shows the narrow scans of Mg 2p. With increasing Mg in the sols, Mg 2p peak gradually intensifies, indicating more Mg ions are included in the HA coating. Moreover, only one symmetrical peak is located at 50.3 eV belonging to Mg 2p, which means only one chemical state of Mg is existing in the coating. This peak is the fingerprint

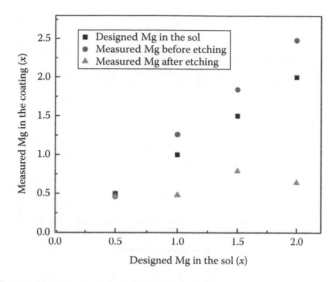

FIGURE 3.27
Comparison of the measured Mg concentration in the Mg_xFHA coatings before and after etching with the designed Mg concentration in the sols. (From *Thin Solid Films*, 519, Cai, Y., Zhang, S., Zeng, X., Qian, M., Sun, D., and Weng, W., Interfacial study of magnesium-containing fluoridated hydroxyapatite coatings, 4629, 2011, with permission from Elsevier.)

of Mg^{2+} in apatite structure [65]. The result confirms that Mg ions have been incorporated into the HA lattice. Mg bonds with phosphate groups in both Mg-containing HA and Mg-substituted β-TCP, thus the binding energy of Mg 2p between them has no difference. As a result, only one symmetrical peak of Mg 2p is detected for FMA3 and FMA4 samples, though a second phase of β-TCMP is discovered by GIXRD.

Element concentration in the coating can be determined by the ratio of area under the respective elemental peak in an XPS narrow scan. Figure 3.27 shows the measured Mg concentration in the coating (indicated by capital X) in comparison with the designed value before and after Ar^+ etching. The squares show the ideal situation when all the designed Mg are completely incorporated into the HA crystal structure. The circles represent the measured Mg concentrations on the surface of the coatings before the Ar^+ etching, and triangles for that after the etching. Although the measured Mg concentration (X value) increases with the increase in designed x value, a discrepancy is observed between the designed and measured concentrations, especially after $x > 0.5$. After the etching, the measured Mg concentration has shown to decrease significantly with contents much lower than that on the surface of the coating. The difference in the Mg concentration before and after the etching indicates that only a portion of Mg is incorporated into the HA crystal structure ($x > 0.5$), and the remaining ions are abundant on the surface of the coating. After the removal of the Mg on the surface of the coating, the measured Mg determines the concentration of Mg incorporated into HA crystal

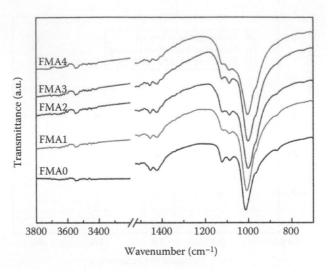

FIGURE 3.28

FTIR spectra of Mg_xFHA coatings on Ti6Al4V substrates.

structure, and the corresponding formula is $Ca_{9.2-9.5}Mg_{0.8-0.5}(PO_4)_6F_1(OH)_1$, i.e., about 1.2–1.9 wt% of Mg replaces Ca in the HA crystal structure.

Fourier transform infrared spectroscopy (FTIR) identifies chemical groups in different coatings, as shown in Figure 3.28. There are three peaks at 1002–1015, 1086–1088, and 1120–1122 cm^{-1} for each coating, due to the triply asymmetric $v3$ stretching vibration of the phosphate group (PO_4^{3-}). A weak peak at 964–965 cm^{-1} represents the $v1$ symmetric stretching vibration of PO_4^{3-}. The presence of typical peaks of PO_4^{3-} indicates the formation of apatite on the substrate [103]. Also there are two weak peaks between 1400 and 1500 cm^{-1}, ascribing to the $v3$ stretching vibration of carbonate (CO_3^{2-}) [104]. The existing of carbonate groups in the coating is from the coating preparation process.

Furthermore, there is a weak peak around 3540 cm^{-1} relating to the F incorporation into the apatite crystal structure. For pure HA coating, a peak at about 3570 cm^{-1} is detected due to the stretching of O–H in an infinite OH chain. While for FHA coatings, the new peak at about 3540 cm^{-1} is caused by the direct influences of incorporated F on the stretching vibration of OH groups through the hydrogen bonds along OH channel, i.e., –OH–F– [69]. The presence of the peak at 3540 cm^{-1} is another evidence of the replacement of OH⁻ groups by F⁻ ions.

3.4.1.3 Surface and Interface

Figure 3.29 shows the SEM surface and cross-sectional views of the coating on Ti6Al4V substrate. The coatings are dense and uniform with a thickness of about 1.5–2.0 µm. No delamination is seen at the interface even after

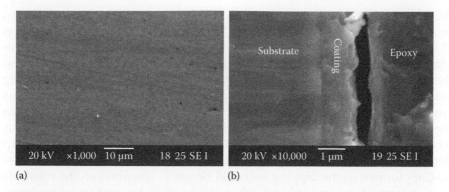

(a)　　　　　　　　　　　　　　　　　　(b)

FIGURE 3.29
Surface (a) and cross-sectional view (b) of FMA4 on Ti6Al4V substrates. (From *Thin Solid Films*, 519, Cai, Y., Zhang, S., Zeng, X., Qian, M., Sun, D., and Weng, W., Interfacial study of magnesium-containing fluoridated hydroxyapatite coatings, 4629, 2011, with permission from Elsevier.)

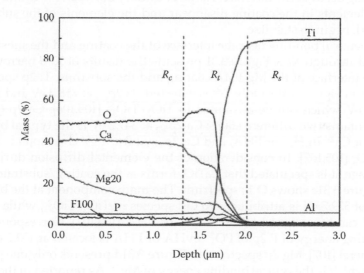

FIGURE 3.30
GDOES depth profile of FMA4 (Mg20: 20 times of Mg concentration; F100: 100 times of F concentration). (From *Thin Solid Films*, 519, Cai, Y., Zhang, S., Zeng, X., Qian, M., Sun, D., and Weng, W., Interfacial study of magnesium-containing fluoridated hydroxyapatite coatings, 4629, 2011, with permission from Elsevier.)

repeated abrading and polishing, suggesting good bonding between the coating and the substrate.

Glow discharge optical emission spectroscopy (GDOES) provides elemental information and is used to acquire elemental distribution from coating to substrate through depth profiling. GDOES depth profile of FMA4 is plotted in Figure 3.30. It shows the distribution of main elements of the coating and the substrate along the depth direction, including Ca, P, O, Mg, F, Ti, and Al.

Along the depth direction, the elemental distribution at cross section can be divided into three regions: the coating region (R_c), the transitional region (R_t), and the substrate region (R_s). The masses of Ca, P, O, and F in the coating are very stable in the coating region. The mass of Mg decreases from the coating to the substrate, in agreement with our previous study that Mg ions can only be partially incorporated into HA lattice, whereas extraneous ions are abundant on the surface of the coating. At the boundary between the R_c and the R_t, the concentrations of Ca, P, Mg, and F begin to decrease, whereas there is a slight increase in O concentration followed by a sharp decrease. The increase in O might owe to the presence of F ions, because nc-CaF$_2$ formed in the dipping sol could be easily absorbed on the substrate surface through hydrogen bond, which could attract more O near the interface [8]. As can be seen in Figure 3.30, the transitional region R_t is around 700 nm. As all the elements of the coating and substrate are seen in the R_t, diffusion of coating elements into the substrate and substrate elements into the coating occur at the interface. At the boundary between the R_t and the R_s, all the elements in the coating disappear and the elements of the substrate, Ti and Al, begin to stabilize.

The chemical bonding near the interface of the coating and the substrate is analyzed through XPS. Figure 3.31 presents the results of XPS narrow scan near the interface of the Mg$_2$FHA coating and the substrate. Ti 2p spectrum is shown in Figure 3.31a. Only Ti^{4+} is detected, Ti 2p$_{3/2}$ at 458.3 eV and Ti 2p$_{1/2}$ at 464.0 eV, which can be attributed to TiO$_2$/TiO$_3^{2-}$ [105,106]. Ca 2p (Figure 3.31b) contains two different states: Ca 2p$_{3/2}$ at 347.2 eV is the typical binding energy of Ca^{2+} in HA or FHA, and Ca 2p$_{3/2}$ at 346.4 eV corresponds to that in CaTiO$_3$ [105,107]. In consideration of the elemental diffusion during the firing stage, it is speculated that CaTiO$_3$ forms at the coating/substrate interface. Figure 3.31c shows O 1s spectrum. The main component at the binding energy of 529.7 eV is attributable to O^{2-} species in TiO$_2$/TiO$_3^{2-}$, while that at 531.3 eV can be assigned to O^{2-} in HA or FHA [105,107]. Correspondingly, the binding energy of P 2p of PO$_4^{3-}$ in HA or FHA is located at 133.2 eV (not shown here) [107]. Mg 2p spectrum in Figure 3.31d presents only one peak at 50.4 eV, which is the typical binding energy of Mg^{2+}. As reported in the literature [108], MgTiO$_3$ could be formed around 400°C. Also Mg is detected in the transitional region (R_t) (Figure 3.30). Therefore, it is speculated that MgTiO$_3$ also forms during the firing at 600°C. From the aforementioned GDOES and XPS analyses, it can be concluded that chemical bonds, especially Ca-O-Ti and Mg-O-Ti bonds, form at the coating/substrate interface.

3.4.2 Short-Term Osseointegration

It is a great concern that a stable anchorage achieves by direct bone-to-implant contact within the shortest time, so it is worthy of investigating the short-term osseointegration in vitro. Moreover, Mg incorporation has shown to result in better biological performances. Thus, it is expected to have positive

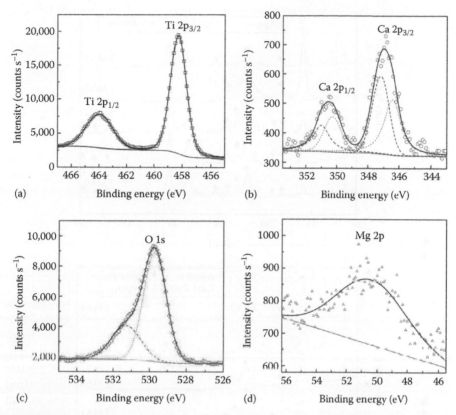

FIGURE 3.31
XPS spectra of Ti 2p (a), Ca 2p (b), O 1s (c), and Mg 2p (d) near the interface of the Mg₂FHA coating and the substrate. (From *Thin Solid Films*, 519, Cai, Y., Zhang, S., Zeng, X., Qian, M., Sun, D., and Weng, W., Interfacial study of magnesium-containing fluoridated hydroxyapatite coatings, 4629, 2011, with permission from Elsevier.)

effects on the in vitro bioactivity and cell response of the Mg$_x$FHA coatings, which are studied in the following text.

3.4.2.1 In Vitro Bioactivity

Figure 3.32 exhibits the GIXRD profiles of Mg$_x$FHA coatings after soaking in the SBF solutions for 7 and 28 days. After 7 days, as shown in Figure 3.32a, an obvious difference is observed around 32° for FMA3 and FMA4 as compared with that in Figure 3.25. For FMA3 and FMA4, the peaks of (211), (112), and (300) are combined into one broad peak, which implies the poorly crystallized apatite grown in the solution—evidence of precipitation of a new apatite layer on the primary coating. The diffraction peaks of substrate and calcium phosphate are found in all the samples. Figure 3.32b shows the GIXRD profiles after 28 days. By now, broad peaks around 32° are also obvious for FMA2,

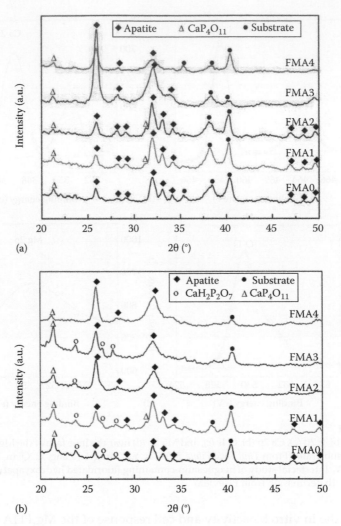

FIGURE 3.32
GIXRD profiles of Mg$_x$FHA coatings on Ti6Al4V substrates after soaking in the SBF solutions for (a) 7 days and (b) 28 days. (From *Thin Solid Films*, 517, Cai, Y., Zhang, S., Zeng, X., Wang, Y., Qian, M., and Weng, W., Improvement of bioactivity with magnesium and fluorine ions incorporated hydroxyapatite coatings via sol-gel deposition on Ti6Al4V alloys, 5347, 2009, with permission from Elsevier.)

which are assigned to the newly formed apatite layer. Some other intermediate products like $CaH_2P_2O_7$ and CaP_4O_{11} are also detected.

Figure 3.33 shows SEM images of Mg$_x$FHA coatings before and after soaking in the SBF solutions for 7 and 28 days. After soaking in the SBF solution for 7 days, the surface morphology of FMA3 shows a newly formed porous layer (Figure 3.33, FMA3a). Similar observations are found on FMA4 (Figure 3.33, FMA4a). Other surfaces exhibit some grooves due to the dissolution of the

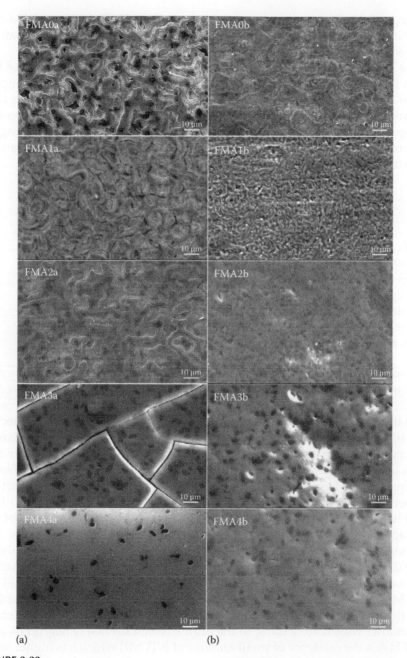

FIGURE 3.33
Surface morphology of Mg_xFHA coatings on Ti6Al4V substrates after soaking in the SBF solutions for (a) 7 days and (b) 28 days. (From *Thin Solid Films*, 517, Cai, Y., Zhang, S., Zeng, X., Wang, Y., Qian, M., and Weng, W., Improvement of bioactivity with magnesium and fluorine ions incorporated hydroxyapatite coatings via sol-gel deposition on Ti6Al4V alloys, 5347, 2009, with permission from Elsevier.)

coatings. After 28 days (the b-series images), FMA2b, FMA3b, and FMA4b exhibit similar porous layer as FMA3a and FMA4a (precipitation of apatite), but FMA0b and FMA1b still show the dissolution characteristics.

The dissolution and precipitation behavior of HA coating are two of the main factors governing the bioactivity of HA coating [109]. When samples are immersed in the SBF solution, both the dissolution of the coating and the precipitation of the new apatite layer occur at the same time. When the precipitation is dominant, the new apatite layer forms. At the beginning of the process, the dissolution dominates. The kinetics of this process can be described by the following empirical equation [109]:

$$\frac{dc}{dt} = k_1 s_1 (c_\infty - c)^n$$

where
  dc/dt is the dissolution rate
  k_1 is the rate constant for dissolution
  s_1 represents the specific area
  c_∞ stands for the equilibrium concentration
  c for the concentration of solution at different times
  n is the effective order of reaction

As soon as the coatings are immersed in the SBF solution, the difference of coatings in terms of $(c_\infty - c)$ can be ignored. Thus, the specific area of the coatings, s, has main influence on the dissolution rate. The dissolution increases with the increase in the contact surface area with the solution. The specific area of the coatings depends on the crystallite size of the coating: the larger the crystallite size, the less the specific area of the coating is. Scherrer formula $(D_{hkl} = K\lambda(\beta_{1/2}\cos\theta)^{-1})$ is used to calculate the crystallite size in different coatings. In the formula, D_{hkl} is the crystallite size measured in the direction perpendicular to the diffraction planes hkl; K is the Scherrer constant (0.91 is used here); λ is the wavelength of x-ray (1.5406 Å); $\beta_{1/2}$ is the broadening of the diffraction peak at half-maximum intensity; and θ is the half value of the diffraction angle 2θ for hkl. The crystallite size of Mg$_x$FHA coatings calculated at (002) is 187.8, 182.0, 173.9, 155.6, and 169.7 Å; therefore, Mg$_{1.5}$FHA coating has the highest dissolution rate, followed by Mg$_2$FHA, Mg$_1$FHA, Mg$_{0.5}$FHA, and Mg$_0$FHA coating, respectively. Namely, the higher the Mg concentration in the coatings (i.e., Mg$_{1.5}$FHA and Mg$_2$FHA), the faster the dissolution is. Another reason attributing to the high dissolution rate of Mg$_{1.5}$FHA and Mg$_2$FHA is the existence of small amount of β-TCMP, which has higher solubility than Mg-containing HA [110]. As a result, the concentrations of calcium ions and phosphate groups increase sharply, leading to the increase in local supersaturation, which is beneficial to the nucleation and growth of the new apatite crystals. The crystal growth (precipitation) can be described by a similar empirical equation [109]:

$$-\frac{dc}{dt} = k_2 s_2 (c - c_\infty)^n$$

where

    $-dc/dt$ is the crystal growth rate (precipitation)

    k_2 is the rate constant for crystal growth (precipitation)

    s_2 is a function of the total number of available nucleation sites

    c_∞ stands for the equilibrium concentration

    c for the concentration of solution at different times

    n is the effective order of reaction

The dissolution and precipitation are reversible, of which the direction is determined by the difference between the concentration of the solution (c) and the equilibrium concentration (c_∞). When $c_\infty > c$, the precipitation process predominates. Since the dissolution of $Mg_{1.5}FHA$ and Mg_2FHA proceeds much faster than other coatings, $Mg_{1.5}FHA$ and Mg_2FHA firstly reverse to the precipitation. This is evident in Figure 3.33a where a new apatite layer is observed only after 7 days. The new apatite layer is also observed on FMA2 after 28 days, while not found on FMA0 and FMA1 even after 28 days. The rate of precipitation is governed by the concentration difference ($c - c_\infty$) and the number of available nucleation sites (s_2). This again confirms that high Mg concentration favors the formation of new apatite layer. Both surface morphology observation and phase analysis show that the existence of Mg in the coating significantly affects the bone growth (formation of new apatite layer) in the SBF solution, i.e., the in vitro bioactivity of the coating. Higher concentration of Mg has better in vitro bioactivity.

The chemical groups are identified by FTIR after the in vitro bioactivity test, as shown in Figure 3.34. Compared with the FTIR spectra of the as-deposited coatings (Figure 3.28), some changes are seen. For all the coatings, the weak peak between 1120 and 1122 cm^{-1} vanishes. The two peaks located at around 603 and 565 cm^{-1} correspond to the phosphate group (PO_4^{3-}), ascribing to the deformation vibration of PO_4^{3-} [22]. In addition, in Mg_0FHA and $Mg_{0.5}FHA$ coatings, two peaks around 1041 and 1095 cm^{-1} belonging to (PO_4^{3-}) are visible. In Mg_1FHA, $Mg_{1.5}FHA$, and Mg_2FHA coatings, PO_4^{3-} peak around 1031 cm^{-1} broadens, indicating the poor crystallinity of apatite. Besides PO_4^{3-} peaks, carbonate group (CO_3^{2-}) peaks are also detected in Mg_1FHA, $Mg_{1.5}FHA$, and Mg_2FHA coatings between 1400 and 1500 cm^{-1} due to the $v3$ stretching vibration of carbonate, and another peak at about 872 cm^{-1} ascribes to the $v2$ mode of CO_3^{2-} [104]. Since the CO_3^{2-} peaks do not show in the Mg_0FHA and $Mg_{0.5}FHA$ coatings, CO_3^{2-} forms during the growth of new apatite layer. The changes in chemical groups before and after soaking in the SBF solutions support the phase analysis and surface morphology observation that a new apatite layer forms on the surface of FMA2, FMA3, and FMA4.

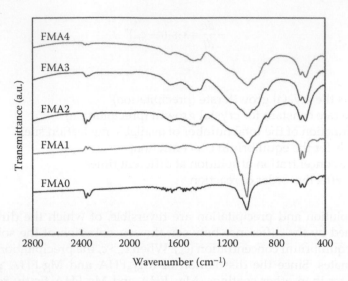

FIGURE 3.34
FTIR spectra of Mg$_x$FHA coatings on Ti6Al4V substrates after soaking in the SBF solutions for 28 days.

The chemical compositions of Mg$_x$FHA coatings after the bioactivity test are studied using the XPS technique. The survey scan profiles are shown in Figure 3.35. The main elements of the apatite coatings, Ca, P, and O, are detected in all the samples. F 1s is visible in FMA0 and FMA1 profiles, while not in FMA2, FMA3, and FMA4. It suggests a new apatite layer on the primary coating of FMA2, FMA3, and FMA4, while not on that of FMA0 and FMA1. A small amount of Mg is also detected after the in vitro bioactivity test. For FMA2, FMA3, and FMA4, Mg may be contained in the newly formed apatite layer during the biomimetic growth in the SBF solutions [111]. While for FMA0 and FMA1, the detected Mg might be due to the immersion in Mg-containing SBF solutions.

In summary, after soaking in the SBF solutions for 28 days, phase analysis shows the formation of poorly crystallized apatite layer on the surface of FMA2, FMA3, and FMA4. FTIR spectra further confirm the existence of CO$_3^{2-}$ as well as PO$_4^{3-}$ in the new layer. This newly formed layer is also observed from SEM images, and all the elements in the apatite coatings are also detected by XPS. On the other hand, no obvious change is found in the phases, chemical groups, and chemical compositions in FMA0 and FMA1, and only the dissolution features are observed. Therefore, it is concluded that HCA layer forms on the surfaces of FMA2, FMA3, and FMA4, i.e., Mg in the coating results in better in vitro bioactivity as revealed by faster formation of a new apatite layer on the surface of the coating as submerged in the SBF solution.

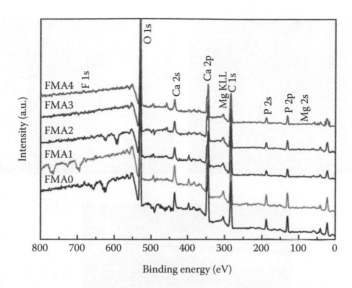

FIGURE 3.35

XPS survey scan profiles of Mg_xFHA coatings on Ti6Al4V substrates after soaking in the SBF solutions for 28 days.

3.4.2.2 Cell Response

Osteoblastic cells are responsible for the synthesis and mineralization of bone during both initial bone formation and later bone remodeling. Human osteosarcoma MG63 cells exhibit fundamental phenotypic characteristics of osteoblastic cells, thus they are cultured for the evaluation of cell response on the coatings. Cell response on the Mg_xFHA coatings is studied in terms of cell morphology, cell proliferation, and differentiation.

Figure 3.36 shows typical SEM micrographs of MG63 cells after 3 days of culture on Mg_xFHA coatings. The dark areas signify the spreading of cell while the rough base is due to the coating dissolution in the culture medium. No significant difference is observed in cell morphology on different coatings. All cells spread well and grow favorably on the coating surface. Under higher magnification (Figure 3.36c), the cell's filopodium and lamellipodium are seen to be flat and adhered tightly on the coating surface, suggesting good viability on the coating. The well-spread cells on all Mg_xFHA coatings are expected to be favorable for cell growth.

Figure 3.37 shows the quantification of cell numbers at each cell culture period to monitor the cell proliferation process. On the first day, there is no significant difference between the coatings. After incubation for 3 days, FMA1 and FMA3 have significantly higher cell numbers than that on FMA0. After 5 days, the cell numbers on FMA3 are significantly higher than those on all the other coatings. Compared with the initial seeded cell density, all the coatings show a significant increase in cell numbers after incubation for

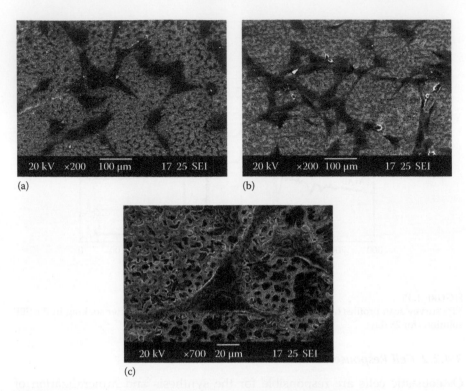

FIGURE 3.36
Cell morphology after incubation on Mg$_x$FHA coatings for 3 days: (a) FMA0, (b) FMA3, and (c) FMA2 (higher magnification of a single cell). (From Cai, Y., Zhang, J., Zhang, S., Mondal, D., Venkatraman, S.S., and Zeng, X., Osteoblastic cell response on fluoridated hydroxyapatite coatings: The effect of magnesium incorporation, *Biomed. Mater.*, 5, 054114, 2010. With permission from Institute of Physics.)

5 days (four to six times as many as the seeded cell density). It indicates that all the coatings, regardless of the degree of Mg incorporation, have favorable viability with MG63 cells. Meanwhile, the most cell numbers on FMA3 after 5 days predict that FMA3 coating (a relatively higher Mg concentration) stimulates cell proliferation more.

To evaluate the cell differentiation, ALP activity, an early differentiation marker, was first measured. Figure 3.38 reveals the intracellular ALP activity of MG63 cells cultured on Mg$_x$FHA coatings for different periods. A significant increase ($p < 0.05$) in ALP activity is observed for all the coatings except for FMA4 in the first 2 weeks. From day 14 to day 21, a downregulation is seen for all the coatings: a significant decrease for FMA0, FMA1, and FMA3 and an insignificant change for FMA2 and FMA4. The significant decrease indicates that more cells step into the next differentiation stage. However, there is no significant difference between Mg-incorporated coatings (from FMA1 to FMA4) and Mg-free one (FMA0).

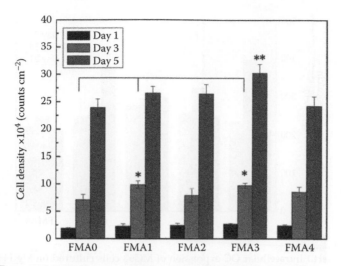

FIGURE 3.37
(See color insert.) Cell proliferation on Mg$_x$FHA coatings in terms of cell density: *At day 3, FMA1 and FMA3 have a significantly higher cell number than that on FMA0 ($p < 0.05$). **At day 5, cell numbers on FMA3 are significantly higher than all the other coatings ($p < 0.05$). (From Cai, Y., Zhang, J., Zhang, S., Mondal, D., Venkatraman, S.S., and Zeng, X., Osteoblastic cell response on fluoridated hydroxyapatite coatings: The effect of magnesium incorporation, *Biomed. Mater.*, 5, 054114, 2010. With permission from Institute of Physics.)

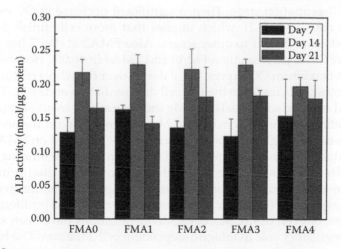

FIGURE 3.38
(See color insert.) Intracellular ALP activities of MG63 cells cultured on Mg$_x$FHA coatings over the culture period: no significant difference between Mg-incorporated coatings (from FMA1 to FMA4) and Mg-free one (FMA0). (From Cai, Y., Zhang, J., Zhang, S., Mondal, D., Venkatraman, S.S., and Zeng, X., Osteoblastic cell response on fluoridated hydroxyapatite coatings: The effect of magnesium incorporation, *Biomed. Mater.*, 5, 054114, 2010. With permission from Institute of Physics.)

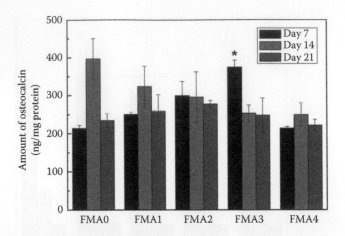

FIGURE 3.39

(See color insert.) Intracellular OC expression of MG63 cells cultured on Mg_xFHA coatings over the culture period: *FMA3 has a significantly higher OC expression than all the other coatings ($p < 0.05$). (From Cai, Y., Zhang, J., Zhang, S., Mondal, D., Venkatraman, S.S., and Zeng, X., Osteoblastic cell response on fluoridated hydroxyapatite coatings: The effect of magnesium incorporation, *Biomed. Mater.*, 5, 054114, 2010. With permission from Institute of Physics.)

As a late marker of cell differentiation, OC expression is also measured as shown in Figure 3.39. At day 7, FMA3 exhibits a significantly higher OC expression than all the other coatings ($p < 0.05$), showing the fastest entry to the late differentiation stage. Then, a significant decrease in OC expression follows at days 14 and 21, which implies that more cells finish the differentiation stage and shift to other stages. Also FMA2 at day 7 has a significant higher OC expression than FMA0 and FMA4 ($p < 0.05$). However, FMA0 reaches its maximum OC expression at day 14 over the whole culture period, meaning most cells undergo the late cell differentiation at day 14. The aforementioned comparison reveals that Mg incorporation can accelerate late cell differentiation. However, considering Mg concentration on the accelerating effect, it is not the higher the better, since it is not observed on FMA4.

The healing of damaged bone should originate from the binding or adhesion of cells to biomaterials, which is thought to be mediated primarily by membrane-associated adhesion receptors belonging to the integrin family [112]. The expression of specific integrin will determine cell proliferation, differentiation, or even death. Moreover, divalent cations have been suggested to play a regulatory role in integrin-dependent cell adhesion. The binding of integrins to specific ligands depends on the presence of divalent cations, with most but not all cells preferring Mg^{2+} to Ca^{2+} ions. It was also reported that Mg^{2+} enhanced cell adhesion to collagen type I and to laminin in a concentration-dependent manner [113]. Zreiqat et al. also proved that the adhesion of human bone-derived cell to Mg^{2+} ions modified bioceramic (Al_2O_3-Mg^{2+}) was increased compared to that on Mg^{2+}-free Al_2O_3. This cation-promoted cell adhesion depends on integrin-associated signal transduction pathways involving signaling protein [114].

Since the F concentration is fixed for all the coatings, the stimulating effect on cell responses should come from incorporation of Mg^{2+} ions. Mg^{2+} ions seem to contribute to cell adhesion on the surface of the coating through the action of the integrin family existing on the cell membrane. Mg^{2+} ions bind cells to the coating surface to accelerate cell adhesion, proliferation, differentiation, and finally bone formation. Considering that FMA3 shows a significant difference in cell proliferation (the most cell numbers) and late cell differentiation (the highest OC expression), it is probable that Mg^{2+} ions both in the apatite lattice and in the surface of apatite crystals act directly or indirectly to promote cell responses. However, regarding the Mg^{2+} concentration in the coating, it is not the higher the better. In the present study, Mg_xFHA coating exhibits optimum cell responses at $x = 1.5$. Furthermore, the dissolved Mg^{2+} ions from the coating can contribute to becoming a source for the remodeling of new bone.

3.4.3 Long-Term Stability

The long-term stability study of surface-coated medical implants is to relieve mainly two concerns: dissolution rate of coating and adhesion strength between coating and substrate. Therefore, in this part, the long-term stability of the Mg_xFHA coatings is investigated based on both the dissolution rate of coating and the adhesion strength between the coating and the substrate.

3.4.3.1 Dissolution Rate

The dissolution behaviors of Mg_xFHA coatings in the tris-buffered saline (TBS) solution are shown in Figure 3.40. The concentration of Ca^{2+} released from the coating into the solution increases (Figure 3.40), while the dissolution rate of the coating decreases (Figure 3.41) with the soaking time. The dissolution curves divide into two distinct stages: the initial sharp increase in Ca^{2+} concentration and the sharp decrease in the dissolution rate during the first 7 days, followed by a slow increase in Ca^{2+} concentration and finally a nearly constant dissolution rate. The different dissolution behaviors in the two stages are due to the different driving force to the dissolution. In the first stage, the dissolution should be surface dissolution limited. In the prepared coatings, other phases, such as amorphous phase, must be existent in the coatings. The solubilities of some calcium phosphates are as follows [115]:

$$ACP \gg TTCP \approx \alpha\text{-}TCP > DCPD > DCP > OCP > \beta\text{-}TCP \gg HA > FHA$$

Therefore, more soluble phases on or near the coating surface are dissolved into the solution very quickly, leading to the sharp increase in Ca^{2+} concentration during the first 7 days. As the driving forces for all the coatings are

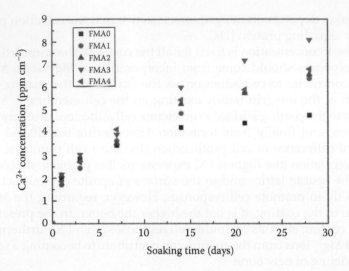

FIGURE 3.40
Dissolution behaviors of Mg_xFHA coatings in the TBS solution: Ca^{2+} concentration as a function of both soaking time and Mg concentration. (From Cai, Y., Zhang, J., Zhang, S., Mondal, D., Venkatraman, S.S., and Zeng, X., Osteoblastic cell response on fluoridated hydroxyapatite coatings: The effect of magnesium incorporation, *Biomed. Mater.*, 5, 054114, 2010. With permission from Institute of Physics.)

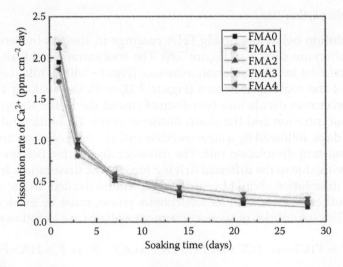

FIGURE 3.41
Dissolution rates of Mg_xFHA coatings in the TBS solution as a function of both soaking time and Mg concentration.

same, there is no significant difference on the dissolution behavior among different coatings. After the removal of these more soluble phases near the surface of the coating, the dissolution of the coating becomes diffusion limited, and the effective dissolution rate is approaching to that of the phases in the coating. As discussed in Section 3.4.1.1, FHA phases are the main phases in all the coatings, and a small amount of β-TCMP exists in FMA3 and FMA4. Because β-TCMP is more soluble than FHA, FMA3 and FMA4 releases more Ca^{2+} into the solution after the first 7 days, and FMA3 shows the fastest dissolution.

On the other hand, the dissolution behavior shows that Mg incorporation into the coating has a negative effect on the dissolution resistance of the coating, which was also reported by other researchers [98]. The possible reason is that the substitution of Ca atom with Mg atom in the apatite structure destabilizes the crystal structure. When the central Ca atom is replaced by Mg atom, the structure becomes asymmetrical. Hydroxyl groups locate on one side of the central atom, and the six external calcium atoms displace contralaterally. Also the replacement of the central Ca atom by Mg atom shortens the bond lengths between the central atom and surrounding oxygen atoms and shrinks the structure [116]. When Mg ions are simultaneously incorporated into the HA coating with F ions, this deformation may not only destabilize the apatite structure, but also influence the ordered structure formed by the F incorporation in the apatite structure. It is well known that the solubility of HA coating can be decreased by the substitution of OH^- with F^- in the HA lattice [7,15,83]. In the HA crystal structure, H^+ in a OH^- can bond with O^{2-} from different directions along the c-axis; once an OH^- is replaced by a F^-, a stronger hydrogen bond forms between the F^- and the nearby OH^-, forming a more ordered structure, especially at a 50% substitution [15]. Thus, the replacement of Ca^{2+} by Mg^{2+} results in the increased solubility of Mg_xFHA coating ($x > 0$) in comparison with FHA coating.

The release of Mg^{2+} ions from the coating into the solutions is also measured. Taking FMA3 as an example, the Mg^{2+} concentration increases with the soaking time, shown in its typical dissolution behavior in Figure 3.42. As shown in the inset of Figure 3.42, the Mg^{2+} concentration of different coatings exhibits the same trend as that of the Ca^{2+} concentration after soaking in the TBS solution for 28 days. The dissolution behavior of Mg^{2+} can be explained in terms of coating composition, which is measured by XPS (Figure 3.27). Excessive Mg distributes on the surface of the coatings, and the measured Mg concentration increases with the increase in the designed Mg in the sol. Therefore, the coating with higher Mg concentration releases more Mg^{2+} ions into the solution (FMA1 < FMA2 < FMA3); however, FMA4 (the highest Mg concentration) releases less Mg^{2+} than FMA3. This might be due to the extra Mg ions on the surface of the coating, which protect the coatings from dissolution because the vacancies or

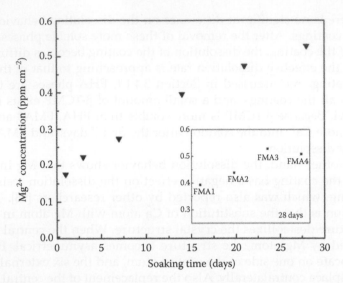

FIGURE 3.42
Dissolution behavior of FMA3 in the TBS solution: Mg^{2+} concentration as a function of soaking time (inset: total Mg^{2+} concentration after 28 days). (From Cai, Y., Zhang, J., Zhang, S., Mondal, D., Venkatraman, S.S., and Zeng, X., Osteoblastic cell response on fluorided hydroxyapatite coatings: The effect of magnesium incorporation, *Biomed. Mater.*, 5, 054114, 2010. With permission from Institute of Physics.)

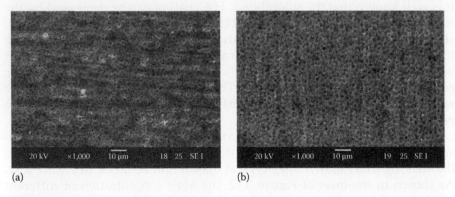

(a)   (b)

FIGURE 3.43
Surface morphology of Mg_xFHA coatings on Ti6Al4V substrates after soaking in the TBS solutions for 28 days: (a) FMA3 and (b) FMA4.

tangling bonds that are active in dissolution are filled with Mg ions, thus leading to slower dissolution.

After the dissolution test, the surface morphologies of the dissolved Mg_xFHA coatings have no significant difference, as shown in Figure 3.43. A rough and porous surface is observed on both FMA3 and FMA4, and the pore diameter is 2–3 μm. As a result of different solubility of different phases in the coating, the rapid dissolution of the more soluble phases leaves behind a porous morphology with the dissolution-resistant phases (HA or FHA) within the coating.

Since the initial dissolution of the coating is beneficial to the healing process as well as the osseointegration [117], the current dissolution behaviors of different coatings are promising. When a synthetic HA ceramic was exposed in vivo, a small amount of the ceramic was phagocytized, but the major remaining part acted as a secondary nucleator as evidenced by the appearance of a newly formed mineral [118]. The initial sharp dissolution of the coating increases the concentrations of calcium and phosphate ions to create a locally supersaturated environment, causing the precipitation and growth of biological apatite crystals from the surface of the initial synthetic crystals with simultaneous incorporation of various ions presented in the biologic fluid. This is favorable to speed up the formation of links between the implant and the bone. On the other hand, it is easier for osteoblast cells to attach and grow on a rough and porous surface and then form new bone.

3.4.3.2 Adhesion Strength

The adhesion strength between the coating and the substrate is measured by the pull-off test. The typical morphology of rupture surfaces of the Mg_2FHA coating after the pull-off test is shown in Figure 3.44. It can be seen that both the adhesive and cohesive failures are present. Figure 3.44a shows the typical adhesive failure where a piece of the coating is detached from the substrate. Cohesive failure can be observed in Figure 3.44b. It happens within the coating and is spot flaking around the surface of the coating. "Pull-off adhesion strength" test is conducted on the coatings (data shown in Figure 3.45), but it is not able to distinguish the failure modes. Without Mg incorporation (FMA0), the pull-off adhesion

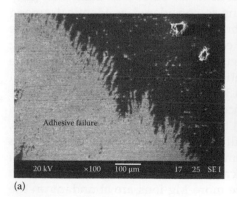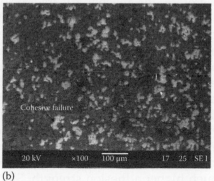

(a)   (b)

FIGURE 3.44
Typical morphology of rupture surfaces of the Mg_2FHA coating: (a) adhesive failure and (b) cohesive failure.

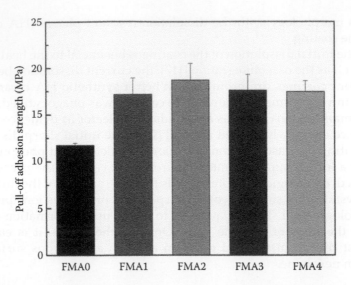

FIGURE 3.45

Pull-off adhesion strength between Mg_xFHA coatings and Ti6Al4V substrates: Mg incorporation enhances the pull-off adhesion strength significantly ($p < 0.05$), but no significant difference between different Mg concentrations.

strength is about 12 MPa. When Mg is incorporated, it increases significantly ($p < 0.05$) to about 18 MPa. However, there is no significant difference between different Mg concentrations.

Both the mechanical interlocking and the chemical bonding contribute to the adhesion between the coating and the substrate. In this work, all the substrates were polished to grade #1200 of SiC sandpapers to achieve the same finish. Moreover, Blackwood et al. proved that the adhesion became independent of surface finish at levels smoother than #600 [119]. Therefore, the mechanical interlocking is considered as identical among all the samples, thus the main contribution to the adhesion strength is the chemical bonds at the interface of the coating and substrate formed during the coating deposition process, especially at the firing stage. During the firing process, the elements of the coating and the substrate diffuse toward each other. As can be seen in Figure 3.30, all the elements are present in the transitional region (R_t). Without Mg incorporation, chemical bonds between Ti, O, Ca, P, and F atoms are formed at the interface, such as Ca-O-Ti bonds, which could contribute to higher adhesion strength [8]. When Mg ions are incorporated, Mg-O-Ti bonds form at the interface as well as Ca-O-Ti bonds, which would make the chemical bonds even stronger and result in much higher adhesion strength. Since more Mg ions are abundant on the surface of the coating, Mg concentration near the interface might be not significantly different among Mg_xFHA coatings, thus there is no significant difference among different Mg concentrations.

3.5 Effect of Fluorine Substitution in Magnesium-Containing Hydroxyapatite Coatings*

Fluorine ions are incorporated into magnesium-containing HA coatings to enhance the long-term stability. The as-deposited coatings are denoted by MgF_yHA ($Ca_9Mg_1(PO_4)_6F_y(OH)_{(2-y)}$, y varies from 0 to 2). The corresponding samples were labeled as MFA0, MFA1, MFA2, MFA3, and MFA4, or the corresponding coatings as MgF_0HA, $MgF_{0.5}HA$, MgF_1HA, $MgF_{1.5}HA$, and MgF_2HA for further discussion.

3.5.1 Co-Substitution of Mg and F Ions in the MgF_yHA Coatings

3.5.1.1 Phase Evolution

GIXRD patterns of as-deposited MgF_yHA coatings are shown in Figure 3.46. Without F incorporation (MFA0), β-TCMP is detected as the main phase, and HA phase exists as the secondary phase. Due to the presence of Mg ions, the transformation temperature from apatite to β-TCMP can be 100°C lower than pure β-TCP (700°C–800°C). Furthermore, Ca ion vacancies in the ideal β-TCP

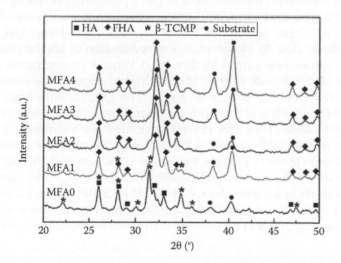

FIGURE 3.46
GIXRD patterns of as-deposited MgF_yHA coatings on Ti6Al4V substrates. (With kind permission from Springer Science+Business Media: *J. Mater. Sci. Mater. Med.*, Cai, Y., Zhang, S., Zeng, X., and Sun, D., Effect of fluorine incorporation on long-term stability of magnesium-containing hydroxyapatite coatings, 22, 2011, 1633.)

* With kind permission from Springer Science+Business Media: *J. Mater. Sci. Mater. Med.*, Cai, Y., Zhang, S., Zeng, X., and Sun, D., Effect of fluorine incorporation on long-term stability of magnesium-containing hydroxyapatite coatings, 22, 2011, 1633.

structure are too small to accommodate Ca ions, but allow for the inclusion of Mg ions, which thereby stabilizes the structure resulting from the difference in size of Ca^{2+} ions (0.99 Å) and Mg^{2+} ions (0.69 Å) [102,120]. Since the designed (Ca, Mg)/P ratio in MgF_0HA is 1.67, which exceeds the stoichiometric molar ratio of 1.5 for β-TCP, Mg ions preferentially incorporate into β-TCP lattice and the excess of Ca ions form HA phase.

When F is incorporated, FHA becomes the main phase in all the MgF_yHA coatings ($y > 0$), confirmed by the slight shifting of the main diffraction peaks toward higher degrees compared with that of HA [91]. Furthermore, it is worthy of noting that when F concentration increases from 0 to 0.5, i.e., $MgF_{0.5}HA$ (MFA1), β-TCMP is detected only with weak intensity; as F concentration further increases, no more β-TCMP forms. This is a clear indication of F ions' suppression ability of β-TCMP formation. In other words, F promotes the substitution of Ca^{2+} with Mg^{2+} in the HA lattice, in agreement of our previous findings (Section 3.4.1.1) as well as other researches [96].

3.5.1.2 Co-Substitution of Mg and F Ions in the MgF_yHA Coatings

The shift of (211) and (300) peaks and the shrinkage of *a*-axis take place when fluorine ions are incorporated into the HA crystal structure [69–71]. In the MgF_yHA coatings, the two-theta of (300) peak and calculated *a*-axis are listed in Table 3.3. As F concentration increases from 0 to 1.0, the (300) peak shifts toward higher degree due to the substitution of larger OH^- (1.68 Å) with smaller F^- (1.32 Å). However, at F concentration of 1.5, the two-theta of (300) peak remains the same with that of 1.0. When F concentration increases further to 2.0, (300) peak shifts a little to higher degree. The same trend is also observed from the change of *a*-axis. The gradual shrinkage of *a*-axis along with the increase of F concentration indicates the replacement of OH^- groups by F^- ions in the HA crystal structure. As mentioned in Section 3.4.1.2, the shift of (002) peak and the decrease in *c*-axis are indications of Mg substitution in the HA lattice. In pure HA, the two-theta of (002) is 25.88° and *c*-axis is 6.88 Å as shown in PDF#09-0432. In the MgF_yHA coatings, the (002) peak shifts to a higher degree of 26.02° or 26.04°, and calculated *c*-axis is ~6.84 Å. Since the designed Mg concentration is the same among the MgF_yHA coatings, there is no obvious difference between the *c*-axis values.

TABLE 3.3

Two-Theta of (300) Peak and Calculated *a*-Axis of MgF_yHA Coatings

Sample	MFA0	MFA1	MFA2	MFA3	MFA4
Two-theta (°)	33.00	33.16	33.22	33.22	33.24
a-axis (Å)	9.39	9.35	9.33	9.33	9.33

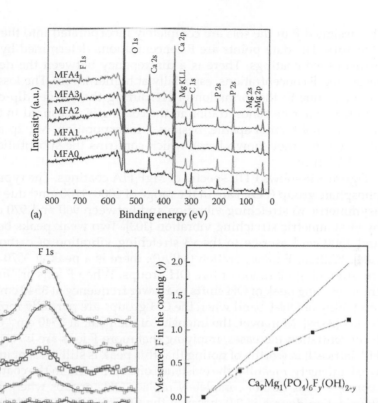

FIGURE 3.47
XPS analysis of MgF$_y$HA coatings: (a) XPS survey scans of MgF$_y$HA coatings, (b) XPS narrow scans of F 1s, (c) fluorine concentration in MgF$_y$HA coatings (straight line: designed; solid square: measured).

XPS analysis of MgF$_y$HA coatings (chemical states and F concentration) are shown in Figure 3.47. The survey scans (Figure 3.47a) present the existence of all the elements in MgF$_y$HA coatings including C 1s, which is the contaminant. Since F concentration varies in MgF$_y$HA coatings, F incorporation and concentration in the coatings are analyzed. Only one peak is detected in the F 1s narrow scans (Figure 3.47b) of the coatings. The binding energy is 684.3 ± 0.1 eV, which is the fingerprint of F in FHA or FA structure [8,75]. The result indicates the incorporation of F ions into the HA crystal structure. Moreover, the intensity of F 1s peak increases when F concentration increases, indicating that more F ions are incorporated. Figure 3.47c exhibits the comparison of the designed and the measured F concentration in the coatings. The straight line shows the ideal situation when all the designed F in the sols are completely incorporated into the HA crystal structure. The data points are F concentrations determined by XPS in the as-deposited coatings. There is a discrepancy between the designed and measured F concentrations, especially at higher levels. The loss of fluorine might be due to the formation of HF during the sol–gel dip-coating process [75]. Fluorine concentration in the coating is measured in the range of 0.39–1.16 mol F$^-$/mol apatite. Besides F incorporation, Mg 2p at 50.3 eV is also detected (not shown here), which confirms the substitution of Mg in the HA lattice.

Figure 3.48 shows FTIR spectra of MgF$_y$HA coatings. The typical peaks of phosphate group (PO$_4^{3-}$) are in the range of 1000–1200 cm^{-1} due to the triply asymmetric $v3$ stretching vibration and between 960 and 970 cm^{-1} caused by $v1$ symmetric stretching vibration [103]. Two weak peaks between 1400 and 1500 cm^{-1} ascribe to the $v3$ stretching vibration of carbonate (CO$_3^{2-}$) [104]. Without F incorporation (MFA0), there is a peak at 3570 cm^{-1} due to the stretching vibration of free OH groups. When F ions are incorporated, the stretching peak of OH shifts to a lower frequency at 3540 cm^{-1} due to the stretching of OH-F band when the OH groups are partially replaced by the F ions [38,69]. Moreover, the intensity of the peak at 3540 cm^{-1} decreases as F concentration increases, implying that more F ions are incorporated into HA lattice. It is worthy of noting that this peak is still present in MFA4 with weak intensity, meaning the existence of OH–F band in the coating. In other words, F$^-$ ions do not completely replace OH$^-$ groups when the designed fluoridation degree is 2.0, because the OH stretching band disappears in the case of pure FA [38]. This is in agreement with the measurement of F concentration in the coatings (Figure 3.47c).

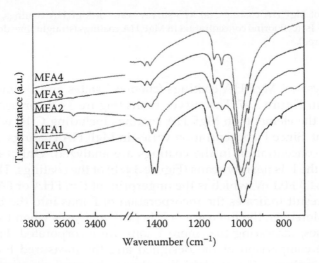

FIGURE 3.48
FTIR spectra of MgF$_y$HA coatings on Ti6Al4V substrates.

3.5.1.3 Surface and Interface

Surface and cross-sectional views of MgF_yHA coatings are similar to those of Mg_xFHA coatings, thus not shown here. The elemental distribution from the coating to the substrate is acquired by GDOES. A typical GDOES depth profile is plotted in Figure 3.49, showing the distribution of Ca, P, O, Mg, F, Ti, and Al along the depth direction. Three regions are identified at the cross section along the depth direction: the coating region (R_c), the transitional region (R_t), and the substrate region (R_s). Tangent method is used to determine the boundaries between two adjacent regions. In the coating region R_c, most of the elements in the coating are stable. At the boundary of R_c and R_t, the concentration of Ca, P, Mg, and F begins to decrease, whereas there is an abrupt increase in O concentration followed by a sharp decrease. The increase in O might be due to the presence of F ions because nc-CaF_2 formed in the dipping sol could be easily absorbed on the substrate surface through hydrogen bonds, which could attract more O near the interface [8]. In the region of R_t, all the elements of the coating as well as the substrate are present. Also the width of the R_t is about 500 nm. Thus, it is speculated that the mutual diffusion of the elements in the coating and the substrate takes place during the coating deposition process. At the boundary between the R_t and the R_s, Ti and Al stabilize, and Ca, P, O, Mg, and F disappear.

XPS narrow scans near the interface of the $MgF_{1.5}HA$ coating and the substrate are shown in Figure 3.50. The chemical states of Ti, Ca, O, and Mg

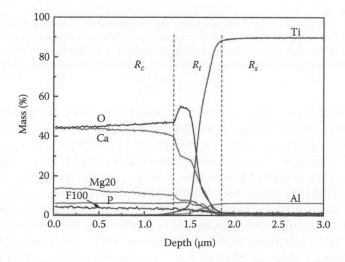

FIGURE 3.49
GDOES depth profile of MFA3 (Mg20: 20 times of Mg concentration; F100: 100 times of F concentration). (With kind permission from Springer Science+Business Media: *J. Mater. Sci. Mater. Med.*, Cai, Y., Zhang, S., Zeng, X., and Sun, D., Effect of fluorine incorporation on long-term stability of magnesium-containing hydroxyapatite coatings, 22, 2011, 1633.)

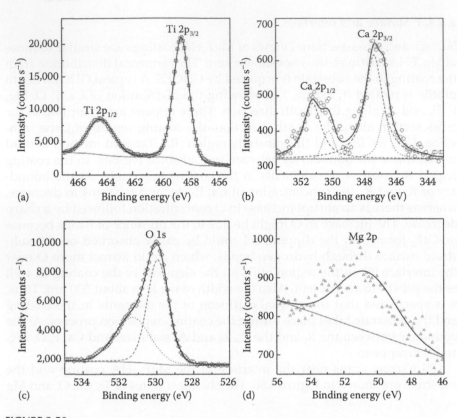

FIGURE 3.50
XPS spectra of Ti 2p (a), Ca 2p (b), O 1s (c), and Mg 2p (d) near the interface of the $MgF_{1.5}HA$ coating and the substrate. (With kind permission from Springer Science+Business Media: *J. Mater. Sci. Mater. Med.*, Cai, Y., Zhang, S., Zeng, X., and Sun, D., Effect of fluorine incorporation on long-term stability of magnesium-containing hydroxyapatite coatings, 22, 2011, 1633.)

are analyzed. Figure 3.50a shows Ti 2p spectrum. Only Ti^{4+} is detected, Ti $2p_{3/2}$ at 458.5 eV and Ti $2p_{1/2}$ at 464.2 eV, the typical binding energy in titanium dioxide (TiO_2) or titanate group (TiO_3^{2-}) [105,106]. Ca 2p spectrum in Figure 3.50b contains two different states: Ca $2p_{3/2}$ at 347.4 eV is the typical binding energy of Ca^{2+} in HA or FHA; Ca $2p_{3/2}$ at 346.7 eV can be attributed to $CaTiO_3$ [105,107]. Since the elemental interdiffusion takes place during the firing stage, it is speculated that $CaTiO_3$ forms near the interface of the coating and the substrate, i.e., Ca–O–Ti bonds form at the coating/substrate interface. O 1s spectrum is shown in Figure 3.50c. The main component at the binding energy of 529.9 eV can be assigned to O^{2-} species in TiO_2/ TiO_3^{2-} and that at 531.3 eV corresponds to O^{2-} in HA or FHA [105,107]. P 2p locating at 133.4 eV (not shown here) is attributable to PO_4^{3-} in HA or FHA [107], further proving the apatite formation. Figure 3.50d presents Mg 2p spectrum. Only one peak at 50.6 eV is detected, which is the typical binding energy of Mg^{2+}. Since Mg2p of pure Mg metal is around 49.8 eV, the peak shift indicates that Mg bonds with O and forms Mg–O bonds or Mg–O–Ti bonds.

3.5.2 Short-Term Osseointegration

3.5.2.1 *In Vitro Bioactivity*

After soaking in the SBF solutions for 28 days, the phases are indentified by GIXRD, shown in Figure 3.51. The GIXRD profiles are different from that of as-deposited MgF_yHA coatings (Figure 3.46). A broad peak is seen around 32° on all the samples, indicating a poorly crystallized apatite. Also main diffraction peaks of Ti6Al4V substrate are not observed, which should be at 40.5°, 38.5°, and 35.4°. The absence of the diffraction peaks of the substrate also suggests the formation of a new layer on the as-deposited coatings because the thickness of the detected layer is over the detection limit of GIXRD (usually several micrometers). Besides the apatite phase, β-TCMP is present in MFA0. Since β-TCMP is the main component of as-deposited MgF_0HA, it still exists after soaking in the SBF solution. Since the poorly crystallized apatite forms on all the MgF_yHA coatings, it suggests comparable bioactivity of MgF_yHA coatings regardless of the fluoridation degree.

Formation of new apatite phase is also supported by the surface morphology observation. Figure 3.52 shows typical surface morphologies of MgF_1HA coating after soaking in the SBF solution for 28 days.

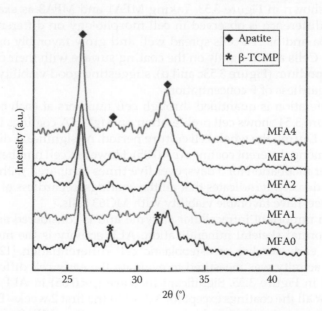

FIGURE 3.51
GIXRD profiles of MgF_yHA coatings on Ti6Al4V substrates after soaking in the SBF for 28 days.

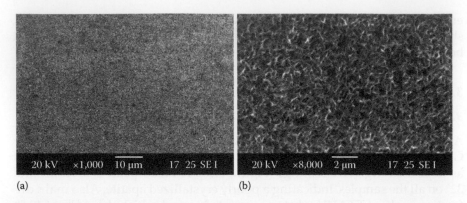

(a)   (b)

FIGURE 3.52
SEM images of MgF$_1$HA coating on Ti6Al4V substrates after soaking in the SBF for 28 days ([b] is the higher magnification of [a]).

After soaking, the surface is quite flat and homogeneous. From the high magnification (Figure 3.52b), this newly formed layer has submicroscale network structure, which is the typical microstructure of bone-like apatite grown in the SBF solution.

3.5.2.2 Cell Response

Typical SEM images of MG63 cells after cultured on MgF$_y$HA coatings for 3 days are shown in Figure 3.53. Taking MFA1 and MFA3 as examples, no significant difference is observed in cell morphology on different coatings (Figure 3.53a and b). All cells spread well and grow favorably on the coating surface. Cells attach tightly on the coating surface with their filopodium and lamellipodium (Figure 3.53c and d), suggesting good viability on all the coatings regardless of F concentration.

Cell proliferation is quantified through cell numbers at each cell culture period. Figure 3.54 shows cell proliferation on MgF$_y$HA coatings in terms of cell density. During the whole cell culture period, no significant difference is observed among different coatings ($p < 0.05$). However, cell numbers increase sharply after incubation for 5 days, over five times compared with the initial seeded cell density. It indicates that all the coatings, regardless of the fluoridation degree, have favorable viability with MG63 cells.

ALP is an important intracellular enzyme and is considered as a prerequisite for normal skeletal mineralization. ALP activity is the most widely recognized early marker of osteoblastic cell differentiation [121]. In this work, ALP activity was measured to evaluate the early cell differentiation, as revealed in Figure 3.55. Significant increase ($p < 0.05$) in ALP activity is observed for all the coatings except for MFA0 in the first 2 weeks. Both MFA2 and MFA3 have a significantly higher ALP activity than MFA0 after 14 days. Then, a downregulation is seen on most of the coatings. This trend in ALP

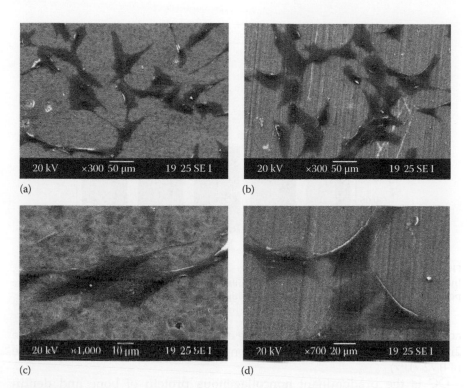

FIGURE 3.53
Cell morphology after cultured on MgF_yHA coatings for 3 days: MFA1 (a, c), MFA3 (b, d).

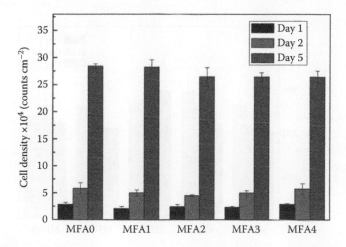

FIGURE 3.54
(See color insert.) Cell proliferation on MgF_yHA coatings in terms of cell density: no significant difference is observed on different coatings ($p < 0.05$).

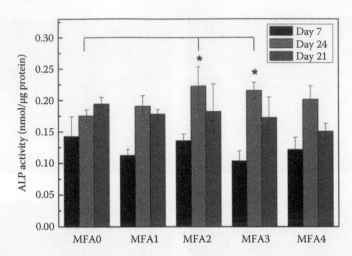

FIGURE 3.55
Intracellular ALP activities of MG63 cells cultured on MgF_yHA coatings over the culture period: *MFA2 and MFA3 have a significantly higher ALP activity than MFA0 ($p < 0.05$).

activity indicates that F ions stimulate early cell differentiation mostly at the 2nd week with a designed fluoridation degree of 1.0–1.5 (the measured one of 0.72–0.97).

OC is the predominant noncollagenous protein of bone and dentin and plays roles in bone formation and remodeling [122]. Late cell differentiation is assessed by OC expression, as shown in Figure 3.56. At day 7, MFA2 exhibits a significantly higher OC expression than all the other coatings ($p < 0.05$). However, no significant difference is observed

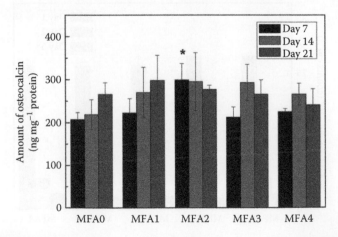

FIGURE 3.56
Intracellular OC expression of MG63 cells cultured on MgF_yHA coatings over the culture period: *MFA2 has a significantly higher OC expression than all the other coatings ($p < 0.05$).

among different coatings after 14 and 21 days. Although comparable levels of OC expression are found on different coatings after 14 and 21 days, the fastest entry of MFA2 to the late differentiation stage suggests that F incorporation in the coating can accelerate cells to late differentiation stage when the designed fluoridation degree is around 1.0 (the measured F concentration is 0.72).

It is well known that fluorine can directly stimulate cell proliferation and alkaline phosphate activity [123]. In our previous studies, when only F ions were incorporated into HA coatings, a significantly higher ($p < 0.05$) cell proliferation was observed on FHA coatings at the fluoridation degree of 0.8–1.1 [15]. However, this stimulation effect of F incorporation is not shown when Mg ions are co-substituted into HA coating. It might be due to the presence of Mg ions. Mg ions were proved to accelerate osteoblast cell adhesion on $FGMgCO_3Ap$–collagen composite [6]. Also in Section 3.4.2.2, Mg ions were shown to have more profound stimulation effect on cell proliferation than on early cell differentiation. Therefore, when F ions were co-substituted into HA with Mg ions, the effect of Mg ions on cell proliferation might overshadow that of F ions. Consequently, no significant difference is observed on cell proliferation with varied F concentrations.

The stimulating effect of fluorine is dose dependent as a higher dosage might cause toxic effect [123,124]. Generally, a moderate level of F content ($Ca_{10}(PO_4)_6F_{\sim 1.0}(OH)_{\sim 1.0}$) is recommended in clinical application. For example, in Cheng et al.'s work, when x is in the range of 0.67–1.68 ($Ca_{10}(PO_4)_6F_x(OH)_{2-x}$), sol–gel derived FHA coating is the most suitable for implantation [91]. Wang et al. further optimized the fluoridation degree, and they found that FHA coating exhibited the best dissolution resistance and cell activities at the molar level of fluorine ions from 0.8 to 1.1 [15]. Also in the case of electrochemical deposition, FHA coating with fluoridation degree from 1.0 to 1.25 had the most positive stimulating effects on cell proliferation and ALP activities [66]. In our study, the coatings with a measured fluoridation degree of 0.72–0.97 exhibit the most predominant stimulating effect on cell differentiation (ALP activity and OC expression), which is in agreement with the study mentioned earlier.

Moreover, it seems that the effect of F ions on cell behaviors varies at different stages. Inoue et al. investigated in vitro response of osteoblast-like and odontoblast-like cells to unsubstituted and substituted apatites, and they found that F ions contributed more to differentiation rather than to proliferation in the early period [125]. Similar results were also reported by Kim et al. They studied the effect of fluoridation of HA in hydroxyapatite–polycaprolactone (HA–PCL) composites on osteoblast activity and found that the proliferation of MG63 cells on FHA–PCL composite was at a similar level to those on HA–PCL composite. However, the ALP activity and OC production were increased significantly by the increase in fluoridation of HA [126]. Our study presents similar results: there is no significant difference on cell proliferation, but the stimulating effect is observed on both early and late stages of cell differentiation (significantly higher ALP activity and OC expression at the measured fluoridation degree of 0.72).

3.5.3 Long-Term Stability

Fluorine substitution in magnesium-containing HA coatings is expected to improve the long-term stability of the implant. In this part, the long-term stability of the MgF$_y$HA coatings is investigated in detail. The dissolution behavior is monitored by the change of Ca^{2+} concentration in the TBS solution, and the adhesion strength between the coating and the substrate is measured by the pull-off test.

3.5.3.1 Dissolution Rate

The substitution of foreign ions into the HA lattice disrupts the crystal structure and, in turn, alters the dissolution properties of the materials in solution. In other words, the dissolution rate can be traced by the appropriate substitution of foreign ions into the HA lattice. The dissolution behavior of MgF$_y$HA coatings with the fixed Mg content and the varied fluoridation degree is monitored by the change of Ca^{2+} concentration in the TBS solution, as shown in Figure 3.57. All the coatings release more Ca^{2+} ions with the increasing soaking time. However, MgF$_0$HA coating without F incorporation dissolves the fastest, exhibiting the highest Ca^{2+} concentration and the

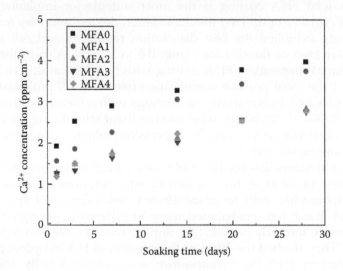

FIGURE 3.57
Dissolution behaviors of MgF$_y$HA coatings in the TBS solution: Ca^{2+} concentration as a function of both soaking time and F concentration. (With kind permission from Springer Science+Business Media: *J. Mater. Sci. Mater. Med.*, Cai, Y., Zhang, S., Zeng, X., and Sun, D., Effect of fluorine incorporation on long-term stability of magnesium-containing hydroxyapatite coatings, 22, 2011, 1633.)

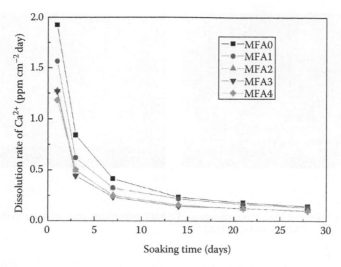

FIGURE 3.50
Dissolution rates of MgF$_y$HA coatings in the TBS solution as a function of both soaking time and F concentration.

fastest dissolution rate (Figure 3.58). When more F ions are incorporated, less Ca^{2+} ions are released from MgF$_{0.5}$HA, and the least Ca^{2+} ions from MgF$_1$HA, MgF$_{1.5}$HA, and MgF$_2$HA coatings. The dissolution behavior of MgF$_y$HA coatings reveals that MgF$_0$HA coating has the highest dissolution rate, and MgF$_1$HA, MgF$_{1.5}$HA, and MgF$_2$HA coatings have the lowest but comparable dissolution rate. The dissolution rate is dependent on the solubility of the phases in the coating. From Figure 3.46, the phases of MgF$_y$HA coatings are detected as follows: most β-TCMP and little HA in MgF$_0$HA, most FHA and little β-TCMP in MgF$_{0.5}$HA, single FHA in MgF$_1$HA, MgF$_{1.5}$HA, and MgF$_2$HA. Although β-TCMP is more stable and less soluble than β-TCP, and even that with 10.1 mol% Mg has a lower solubility than that of HA below pH 6.0, it is still more soluble than HA at pH 7.4 [127,128]. Thus, it is more soluble than FHA because the incorporation of F ions in the HA lattice lowers the solubility of HA [7,15]. Also it is reported that the resorption of calcium phosphates as implants takes place in this order: FHA < Mg-TCP (β-TCMP) < HA < β-TCP < α-TCP [129]. Therefore, MgF$_0$HA coating releases the most Ca^{2+} ions, less in MgF$_{0.5}$HA coating, and the least in MgF$_1$HA, MgF$_{1.5}$HA, and MgF$_2$HA coatings.

Surface morphology of MgF$_y$HA coatings after the dissolution test is observed (Figure 3.59). After soaking in the TBS solution for 28 days, all the coatings exhibit a rough surface caused by the rapid dissolution of more soluble phases leaving less soluble phases in the coating. However, different dissolution characteristics are found when F concentration varies. For example, large pores or grooves are seen in MgF$_0$HA (without F incorporation), but only small pores in MgF$_2$HA (with F incorporation).

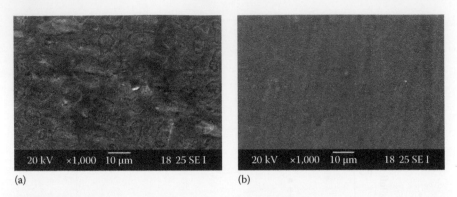

FIGURE 3.59
Surface morphology of MgF$_y$HA coatings on Ti6Al4V substrates after soaking in the TBS solutions for 28 days: (a) MFA0 and (b) MFA4.

The difference between the surface morphology is also reflected by the surface roughness (R_q). The R_q of MgF$_0$HA and MgF$_2$HA after the dissolution test is 0.845 ± 0.013 µm and 0.429 ± 0.032 µm, respectively. These kinds of dissolution characteristics correspond to the dissolution behaviors of MgF$_y$HA coatings mentioned earlier. The prior dissolution of β-TCMP in MgF$_0$HA leaves large grooves and pores on the coating, and slow dissolution of FHA in MgF$_2$HA only induces small pores.

The dissolution kinetics and mechanism of FHA have been widely studied at different temperature and pH values [130–134]. It was found that the dissolution rates of FHA are pH independent at neutral and basic conditions ($7 \leq pH \leq 10$). The dissolution seems to be initiated by the relatively rapid removal of F and/or OH and Ca near the surface. After the initial preferential release of F, OH, and Ca, stoichiometric release of Ca, P, and F takes place [132]. For pure HA, the presence of water induces the onset of dissolution through the escape of hydroxyl groups from the surface of apatite. The surface calcium ions also show the onset of dissolution at a lower extent: the bonds between calcium and lattice oxygen ions are lengthened by the surface water molecules. However, when fluorine ions replace hydroxyl groups partially in HA lattice, hydroxyl groups may dissolve, but fluorine ions at the surface remain in the crystal. Moreover, the surface calcium ions are anchored on the surface by the strong interaction with fluorine ions [67]. Therefore, the incorporation of fluorine ions into HA lattice makes the material more dissolution resistant. It can be explained from the change of HA crystal structure. In pure HA lattice, there is a certain degree of disorder with randomly oriented OH groups because OH groups are diatomic and located off-center from the symmetry position in the c-direction. When OH groups are replaced by F ions, there will be no elastic strain caused by the substitution of OH groups by F ions because F$^-$ (1.32 Å) ion is smaller than OH$^-$ group (1.68 Å). Also F ions can locate on the symmetry positions in HA lattice [67]. The substitution of OH group by F ions stabilizes randomization of OH groups in the columns by the reversion of the OH alignment within the column [135]. Moreover, fluorine is very electronegative and has an even greater capacity than oxygen in forming hydrogen bonds. Hydrogen bonds formed between F ions and adjacent OH groups further stabilize the crystal structure. Therefore, the crystal structure of FHA is more stable than HA, resulting in lower dissolution rate than that of HA. This is also supported by the calculated excess heats of solid solution that the replacement of OH groups in HA by F ions is an exothermic process [135].

3.5.3.2 Adhesion Strength

After the pull-off test, the surface morphology is observed by SEM as shown in Figure 3.60. Typical adhesive failure is seen on MFA0 (without F incorporation) (Figure 3.60a): big pieces of the coating are torn from the substrate. While on MFA3 (with F incorporation) (Figure 3.60b), the coating still adheres well on the substrate without or with only a little detached coating. The fraction of the adhesive failure on the whole surface of the coating is estimated by the Image-Pro Plus 6.0 after capturing the images of the whole coating, listed in Table 3.4. The percentage of the adhesive failure in MFA0, MFA1, and MFA2 is more than 20% but that in MFA3 and MFA4 is less than 10%. Thus, it can be speculated that the adherence of the coating on the substrate of MFA3 and MFA4 is stronger than that on others.

Correspondingly, the pull-off adhesion strength is compared in Figure 3.61. Without F incorporation (MFA0), the pull-off adhesion strength is about 16 MPa. When F is incorporated at the fluoridation degree of 0.5–1.0, no significant difference is observed. However, in the range of 1.5–2.0, the pull-off adhesion strength increases significantly ($p < 0.05$) to 29 and 24 MPa. It suggests that F incorporation into Mg-containing HA coating enhances the adhesion strength at higher F concentration. The trend of the

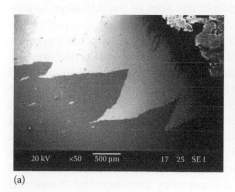
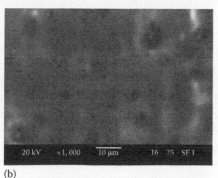

(a) (b)

FIGURE 3.60
Surface morphology of MgF_yHA coatings after the pull-off test: (a) MFA0 and (b) MFA3.

TABLE 3.4

Fraction of Adhesive Failure on MgF$_y$HA Coatings after the
Pull-Off Test

Sample	MFA0	MFA1	MFA2	MFA3	MFA4
Adhesive failure (%)	23.4±4.9	24.1±8.6	30.1±7.8	2.7±4.3	9.0±7.5

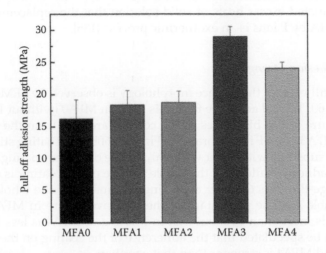

FIGURE 3.61

Pull-off adhesion strength between MgF$_y$HA coatings and Ti6Al4V substrates: significantly higher pull-off adhesion strength on MFA3 and MFA4 ($p < 0.05$). (With kind permission from Springer Science + Business Media: *J. Mater. Sci. Mater. Med.*, Cai, Y., Zhang, S., Zeng, X., and Sun, D., Effect of fluorine incorporation on long-term stability of magnesium-containing hydroxyapatite coatings, 22, 2011, 1633.)

calculated pull-off adhesion strength quite matches that of the fraction of the adhesive failure. Moreover, the adhesion strength between the coating and the substrate should be much higher if the area of the adhesive failure is considered.

The influence of F incorporation on the adhesion strength between the coating and the substrate owes to the changes of HA crystal structure, the incorporated F concentration, the reduction of the coefficient of thermal expansion, and the chemical bonding. The incorporation of F ions into HA lattice stabilizes the crystal structure by reversing the OH alignment within the column and forming hydrogen bonds between OH groups and nearby F ions. Theoretically, the structure is the most stable at around 50% substitution due to the alternating arrangement of the F ions between each pair of OH groups [68]. The fluorine concentration measured by XPS is shown in Figure 3.47c. There is a discrepancy between the measured and designed F concentrations. In MgF$_{1.5}$HA and MgF$_2$HA, the measured

F concentration is around 1.0 or nearly 50% substitution, thus they have the most stable crystal structure. Correspondingly, significantly higher pull-off adhesion strength is observed in MFA3 and MFA4. Another effect of F ions is the reduction of the coefficient of thermal expansion. The coefficient of thermal expansion of pure HA, pure FA, and Ti6Al4V is $15 \times 10^{-6}/K$, $9.1 \times 10^{-6}/K$, and $8.9 \times 10^{-6}/K$, respectively. The substitution of F ions in HA lattice reduces the coefficient of thermal expansion, so does the thermal mismatch between the coating and the substrate, thus the residual stress in the coating is reduced [8]. Furthermore, the chemical bonds among Ti, O, Ca, Mg, P, and F at the interface formed during the drying and firing process, such as Ca-O-Ti and Mg-O-Ti bonds, also contribute to the enhancement of the adhesion strength between the coating and the substrate.

3.6 Mg1.5FHA/MgF1.5HA Bilayer Structured HA Coating

As discussed in Sections 3.4 and 3.5, Mg_xFHA and MgF_yHA coatings exhibited distinct properties in the short-term osseointegration and the long-term stability. Mg incorporation in Mg_xFHA coatings took a great effect on the short-term osseointegration, but not significant on the long-term stability. $Ca_{8.5}Mg_{1.5}(PO_4)_6F(OH)$ ($Mg_{1.5}FHA$) showed the best in vitro bioactivity and cellular response, i.e., the best short-term osseointegration. However, F incorporation in MgF_yHA coatings enhanced the long-term stability much, but the effects on the short-term osseointegration were comparable. $Ca_9Mg_1(PO_4)_6F_{1.5}(OH)_{0.5}$ ($MgF_{1.5}HA$) presented the lowest dissolution rate and the highest adhesion strength, i.e., the long-term stability. Therefore, a bilayer structured HA coating was developed on the basis of systematical investigation of both Mg_xFHA and MgF_yHA coatings, i.e., $Mg_{1.5}FHA$ was selected as the upper layer and $MgF_{1.5}HA$ as the bottom layer. This bilayer structured HA coating is denoted by BHA coating. The physical and chemical properties as well as the short-term osseointegration and the long-term stability were evaluated compared with pure HA coating.

3.6.1 Physical and Chemical Properties

Phase comparison between pure HA and BHA is done by GIXRD and shown in Figure 3.62. Typical diffraction peaks of HA are detected on pure HA deposited sample: main peaks of (211), (112), (300), and (002) locate at 31.7°, 32.2°, 32.8°, and 25.9°. On the other hand, for BHA, only FHA phase is detected. Compared with HA, all the main diffraction peaks of BHA shift a little to higher degree: (211) at 32.0°, (300) at 33.2°, and (002) at 25.94°.

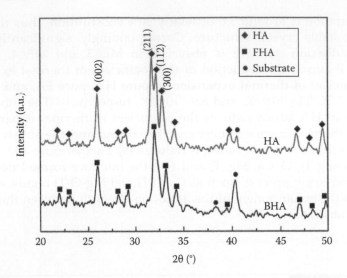

FIGURE 3.62
Phase comparison between HA and BHA by GIXRD.

TABLE 3.5

Calculated *a*-Axis and *c*-Axis
of HA and BHA Coatings

	a-Axis (Å)	*c*-Axis (Å)
HA	9.44	6.87
BHA	9.35	6.85

The shifting of diffraction peaks to higher degree corresponds to the shrinkage of *a*- and *c*-axes as calculated in Table 3.5, indicating the incorporation of Mg and F ions in the HA lattice as discussed in Sections 3.4.1 and 3.5.1.

The phase formation of BHA coating is illustrated by comparing the phases of bottom layer with MFA3 ($MgF_{1.5}HA$) (Figure 3.63a), and BHA with FMA3 ($Mg_{1.5}FHA$) (Figure 3.63b). Since the designed composition of bottom layer is the same as that of MFA3 ($MgF_{1.5}HA$) and the only difference is the deposition times, the phases of the bottom layer and MFA3 are identical as shown in Figure 3.63a. In other words, the phase formed on the substrate is independent of the thickness of the coating. When the upper layer is deposited on the bottom layer to form BHA coating, a little difference in phases is observed in Figure 3.63b: a weak peak of β-TCMP detected in FMA3 is not present in BHA. This is because the gradient of Mg concentration in the upper and bottom layers from Mg1.5 to Mg1.0. Mg diffusion during the firing process results in a lack of Mg to form β-TCMP, thus only FHA phase forms after the upper layer is deposited.

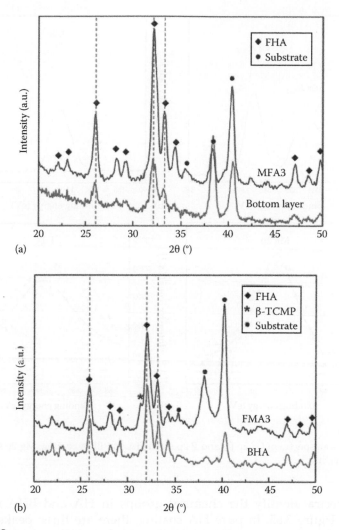

FIGURE 3.63
Illustration of phase formation of BHA coating. (a) Comparison of MFA3 and bottom layer of BHA. (b) Comparison of FMA3 and BHA.

The chemical states of the main elements in the coatings are measured by XPS, as shown in Figure 3.64. HA and BHA coatings have the same peaks of Ca 2p and P 2p, respectively. The binding energy of Ca $2p_{1/2}$, Ca $2p_{3/2}$, and P 2p locates at 350.7 ± 0.1, 347.2 ± 0.1, and 133.3 eV, which are the typical binding energy of Ca 2p and P 2p in HA and FHA [105,107]. In BHA coating, Mg 2p and F 1s are detected at the binding energy of 50.6 and 684.3 eV, indicating the substitution of Mg and F ions in the HA lattice as discussed in Mg_xFHA and MgF_yHA coatings.

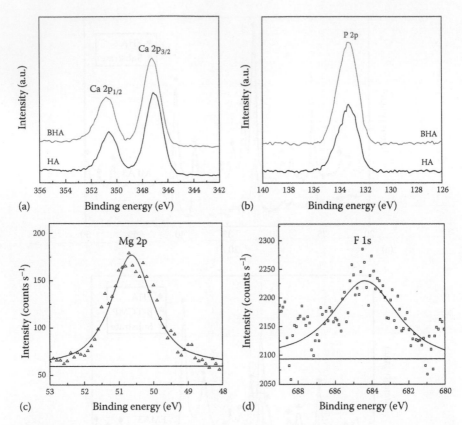

FIGURE 3.64
XPS narrow scan spectra of Ca 2p (a) and P 2p (b) in HA and BHA coating, Mg 2p (c) and F 1s (d) in BHA coating.

FTIR spectra identify the chemical groups in HA and BHA coatings, shown in Figure 3.65. In pure HA coating, there are three peaks belonging to the phosphate group (PO_4^{3-}) at 959, 1005, and 1086 cm^{-1}. The peaks of PO_4^{3-} in BHA coating are very close to those of pure HA coating, located at 966, 1012, and 1093 cm^{-1}, indicating the formation of apatite structure [103]. The weak peaks in the range of 1400–1500 cm^{-1} ascribe to carbonate (CO_3^{2-}) [104]. In pure HA coating, the stretching vibration of OH group is at 3570 cm^{-1}; however, this peak shifts to 3540 cm^{-1} in BHA coating due to the vibration of OH–F, suggesting the partial substitution of OH groups with F ions in the HA crystal structure [38,69].

Figure 3.66 shows surface morphology and cross-sectional view of BHA coating. The surface of the coating is very homogeneous without any cracks or pores (Figure 3.66a). The cross-sectional view (Figure 3.66b) exhibits a well bonding between the coating and the substrate without

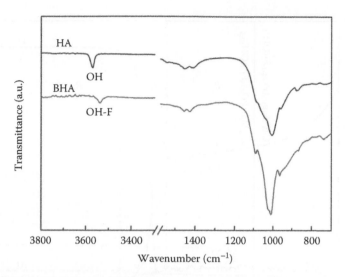

FIGURE 3.65
FTIR spectra of HA and BHA coatings.

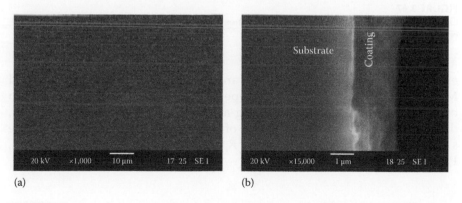

(a)   (b)

FIGURE 3.66
(a) Surface and (b) cross-sectional view of BHA coating on Ti6Al4V substrate.

any delamination. Also it is observed that the coating thickness is about 1.5 μm. The depth profiling of BHA is carried out by GDOES, as shown in Figure 3.67. As analyzed in Mg_xFHA and MgF_yHA, the coating region (R_c), the transitional region (R_t), and the substrate region (R_s) are divided. In the coating region R_c, it is roughly set to the upper layer ($Ca_{8.5}Mg_{1.5}(PO_4)_6F(OH)$) and the bottom layer ($Ca_9Mg_1(PO_4)_6F_{1.5}(OH)_{0.5}$) according to the deposition times though there is no obvious boundary between them. Mg concentration slightly decreases along the depth direction, but F concentration

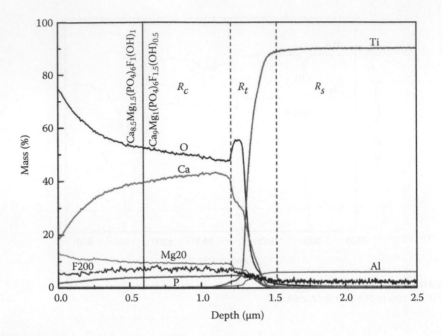

FIGURE 3.67
GDOES depth profile of BHA (Mg20: 20 times of Mg concentration; F200: 200 times of F concentration).

exhibits the contrary trend, which are in accordance with the designed Mg and F concentrations in the upper and bottom layers. In the region of R_t, all the elements of the coating as well as the substrate are present, demonstrating the mutual diffusion of the elements in the coating and the substrate. At the boundary between R_t and R_s, Ti and Al stabilize, and Ca, P, O, Mg, and F disappear.

3.6.2 Short-Term Osseointegration

3.6.2.1 In Vitro Bioactivity

In vitro bioactivity of HA and BHA coating is evaluated by immersing the samples in the SBF solution for different days. After 7 days, the phase evolution is examined by GIXRD (Figure 3.68). A broad peak is observed around 32° on both HA and BHA due to the formation of poorly crystallized apatite. Moreover, the absence of the diffraction peaks of Ti6Al4V substrate suggests that the total thickness of the newly formed layer and the primary coating is more than the detection depth of GIXRD (usually several micrometers). Since similar phase evolution happens on both HA and BHA, it suggests comparable in vitro bioactivity.

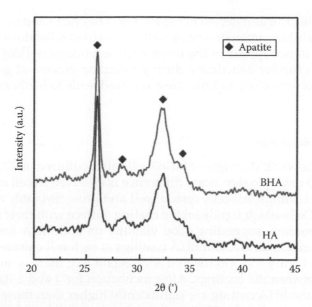

FIGURE 3.68
GIXRD profiles of HA and BHA coatings on Ti6Al4V substrates after soaking in the SBF for 7 days.

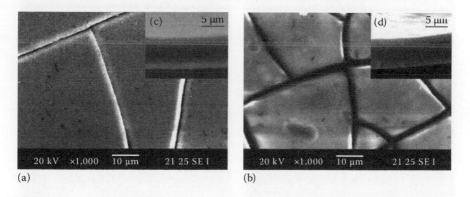

FIGURE 3.69
SEM images of BHA (a) and HA (b) coatings on Ti6Al4V substrates after soaking in the SBF for 7 days ([c, d] the cross-sectional views of [a, b]).

Surface morphology observation is another proof to demonstrate the formation of new apatite layer. Figure 3.69 shows surface morphologies of BHA and HA coatings after soaking in the SBF solution for 7 days. Many cracks are seen in Figure 3.69a and b. Some pieces of the layer are peeled off from the primary coating as shown in Figure 3.69c and d, revealing that the thickness of the peeled layer is around 10 μm, much thicker than the primary coating (~1.5 μm). Formation of a much thicker layer is an evidence of newly formed apatite layer as detected in Figure 3.68. The cracks and peeling off are due to the shrinkage upon drying as well as the stress relief during dehydration. They will not appear in the moist body environment [136]. Moreover, the coating is further dehydrated during pumping process of gold sputtering for SEM observation, and the stress released leads to further cracking of the coating.

3.6.2.2 Cell Response

Figure 3.70 shows SEM images of MG63 cells after cultured on BHA and HA coatings for 3 days. No significant difference is observed in cell morphology between the coatings. All cells spread well and grow favorably on the coating surface. Cells attach tightly on the coating surface with their filopodium and lamellipodium, suggesting good viability on both BHA and HA coatings. Cell numbers on HA and BHA coatings at each cell culture period are compared in Figure 3.71. On the first and second day, there is no significant difference between the coatings. After incubation for 3 and 5 days, the cell numbers on the BHA coating are significantly higher than those on the HA coating. It suggests that the BHA coating stimulates cell proliferation more.

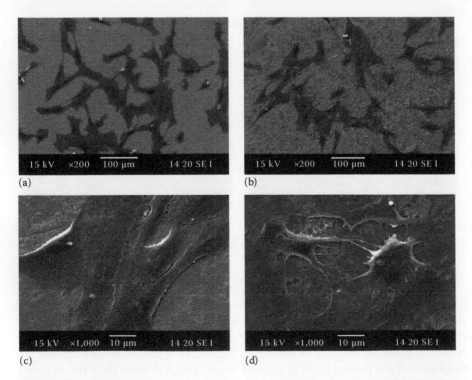

(a)   (b)

(c)   (d)

FIGURE 3.70
Cell morphology after cultured on BHA (a, c) and HA (b, d) coatings for 3 days.

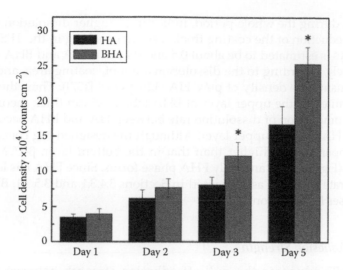

FIGURE 3.71
Cell proliferation on HA and BHA coatings in terms of cell density: significantly higher cell density on BHA coating at days 3 and 5 ($p < 0.05$).

3.6.3 Long-Term Stability

3.6.3.1 Dissolution Rate

Dissolution behaviors of HA and BHA coatings are compared by the change of Ca^{2+} concentrations in the TBS solution, as shown in Figure 3.72. Ca^{2+} concentrations increase with the soaking time, but HA coating releases more

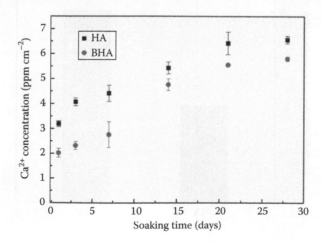

FIGURE 3.72
Comparison of dissolution behaviors of HA and BHA coatings in the TBS solution: Ca^{2+} concentration as a function of soaking time.

Ca^{2+} ions during the whole period, indicating a higher dissolution rate. The average reduction of the coating thickness after soaking in the TBS solution for 28 days is estimated to be about 0.5 and 0.4 µm in HA and BHA coatings, respectively, according to the dissolution amount, coating area, and coating density (using the density of pure HA, 3.16 g cm^{-3} [137]). Thus, the dissolution is limited in the upper layer of BHA (about 0.6 µm thick theoretically), and the comparison of dissolution rate between HA and BHA becomes that between HA and the upper layer. Although the designed Mg concentration in the upper layer is higher than that in the bottom layer, β-TCMP is not detected (Figure 3.68), and only FHA phase forms. Since FHA has lower dissolution rate than HA as discussed in Sections 3.4.3.1 and 3.5.3.1, BHA coating releases less Ca^{2+} ions.

3.6.3.2 Adhesion Strength

Figure 3.73 compares the pull-off adhesion strength between HA and BHA coating on Ti6Al4V substrates. Without distinguishing the adhesive and cohesive failures, the pull-off adhesion strength between HA coating and Ti6Al4V substrate is about 10 MPa, and about 20 MPa for BHA coating, which is nearly two times higher. Compared with MFA3 (the designed composition of the bottom layer, of which the pull-off adhesion strength is 29 MPa), the pull-off adhesion strength of BHA is much lower.

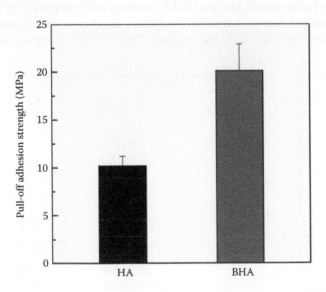

FIGURE 3.73
Pull-off adhesion strength between HA and BHA coating on Ti6Al4V substrates: the pull-off adhesion strength of BHA is two times higher than that of HA.

It is caused by the elemental interdiffusion during the drying and firing process, thus the elemental concentrations in BHA coating are not identical to the designed ones. Although it is much lower than that of MFA3, it is still significantly higher than that of pure HA, a lot more than the minimum 15 MPa stipulated by ISO standards (ISO 13779) for biomedical applications [88].

3.7 Summary

In this work, a bilayer structured magnesium (Mg) and fluorine (F) ions co-substituted HA coating on Ti6Al4V substrate is developed to achieve an integration of short-term osseointegration and long-term stability. GIXRD, XPS, and FTIR spectroscopy results indicate that Ca ions and OH groups in HA crystal structure are substituted with Mg and F ions, respectively. Table 3.6 summarizes the compositional dependence of co-substitution of Mg and F on in vitro bioactivity, cell response, dissolution rate, and adhesion strength. Details are as follows.

High Mg concentrations ($1.0 \leq x \leq 2.0$) in the Mg_xFHA coatings improve the in vitro bioactivity a lot. However, F concentration does not influence the bioactivity of the MgF_yHA coatings. MG63 cells attach and spread well on all the Mg_xFHA and MgF_yHA coatings. In the Mg_xFHA coatings, Mg ions show a significant stimulating effect on cell proliferation and late cell differentiation when x is 1.5. F concentration studied does not affect cell numbers grown on the MgF_yHA coatings. However, it maintains a positive stimulating effect on cell differentiation.

The incorporation of Mg increases the dissolution rate of Mg_xFHA coating in the TBS solution, and the maximum is achieved at $x = 1.5$. F incorporation decreases the dissolution rate of MgF_yHA coating in the TBS solution, especially at the measured fluoridation degree of 0.72–1.16. Mg substitution enhances the adhesion strength, but comparable among different Mg concentrations. Greatly enhanced adhesion strength is achieved by F incorporation at the measured F concentration around 1.0.

Finally, $Mg_{1.5}FHA/MgF_{1.5}HA$ bilayer structured HA coating is developed. Single FHA phase forms in the coating with the substitution of Mg and F ions in HA lattice. It shows comparable in vitro bioactivity with pure HA coating, but more significant cell proliferation. The long-term stability of the bilayer structured HA coating is much better than pure HA coating, exhibiting lower dissolution rate and higher adhesion strength. Thus, a combination of both the short-term osseointegration and the long-term stability is achieved in this one coating/substrate system, which will have potential applications in the biomedical fields.

TABLE 3.6

Co-substitution Effect of Mg and F Ions on Osseointegration and Stability

	In Vitro Bioactivity	Cell Responses			Dissolution Rate	Adhesion Strength
		Cell Morphology	Cell Proliferation	Cell Differentiation		
Mg incorporation in FHA coating	↑	↔	↑	↑	↑	↔
F substitution in MgHA coating	↔	↔	↔	↑	↓	↑

"↑" means increasing or better, "↓" means decreasing or worse, "↔" means unchanged or insignificant effect.

References

1. Albrektsson T, Johansson C. Osteoinduction, osteoconduction and osseointegration. *European Spine Journal* 2001;10:S96.
2. Kasemo B. Biological surface science. *Surface Science* 2002;500:656.
3. Bertoni E, Bigi A, Cojazzi G, Gandolfi M, Panzavolta S, Roveri N. Nanocrystals of magnesium and fluoride substituted hydroxyapatite. *Journal of Inorganic Biochemistry* 1998;72:29.
4. Kannan S, Rocha JHG, Ferreira JMF. Synthesis and thermal stability of sodium, magnesium co-substituted hydroxyapatites. *Journal of Materials Chemistry* 2006;16:286.
5. Kalita SJ, Bhardwaj A, Bhatt HA. Nanocrystalline calcium phosphate ceramics in biomedical engineering. *Materials Science and Engineering C* 2007;27:441.
6. Yamasaki Y, Yoshida Y, Okazaki M, Shimazu A, Kubo T, Akagawa Y, Uchida T. Action of FGMgCO$_3$Ap-collagen composite in promoting bone formation. *Biomaterials* 2003;24:4913.
7. Kim HW, Kim HE, Knowles JC. Fluor-hydroxyapatite sol-gel coating on titanium substrate for hard tissue implants. *Biomaterials* 2004;25:3351.
8. Zhang S, Zeng X, Wang Y, Cheng K, Weng W. Adhesion strength of sol-gel derived fluoridated hydroxyapatite coatings. *Surface and Coatings Technology* 2006;200:6350.
9. Ginty F, Flynn A, Cashman KD. The effect of dietary sodium intake on biochemical markers of bone metabolism in young women. *British Journal of Nutrition* 1998;79:343.
10. Itoh R, Suyama Y. Sodium excretion in relation to calcium and hydroxyproline excretion in a healthy Japanese population. *American Journal of Clinical Nutrition* 1996;63:735.
11. Bigi A, Foresti E, Gregorini R, Ripamonti A, Roveri N, Shah J. The role of magnesium on the structure of biological apatites. *Calcified Tissue International* 1992;50:439.
12. Rude RK, Gruber HE. Magnesium deficiency and osteoporosis: Animal and human observations. *Journal of Nutritional Biochemistry* 2004;15:710.
13. Wiesmann HP, Plate U, Zierold K, Höhling HJ. Potassium is involved in apatite biomineralization. *Journal of Dental Research* 1998;77:1654.
14. Kannan S, Ferreira JMF. Synthesis and thermal stability of hydroxyapatite-β-tricalcium phosphate composites with cosubstituted sodium, magnesium, and fluorine. *Chemistry of Materials* 2006;18:198.
15. Wang YS, Zhang S, Zeng XT, Ma LL, Weng WJ, Yan WQ, Qian M. Osteoblastic cell response on fluoridated hydroxyapatite coatings. *Acta Biomaterialia* 2007;3:191.
16. Schlesinger PH, Blair HC, Teitelbaum SL, Edwards JC. Characterization of the osteoclast ruffled border chloride channel and its role in bone resorption. *Journal of Biological Chemistry* 1997;272:18636.
17. Kannan S, Rebelo A, Ferreira JMF. Novel synthesis and structural characterization of fluorine and chlorine co-substituted hydroxyapatites. *Journal of Inorganic Biochemistry* 2006;100:1692.
18. Cazalbou S, Combes C, Eichert D, Rey C. Adaptative physico-chemistry of bio-related calcium phosphates. *Journal of Materials Chemistry* 2004;14:2148.

19. LeGeros RZ. Calcium phosphates in oral biology and medicine. *Monographs in Oral Science* 1991;15:1.
20. Suchanek WL, Byrappa K, Shuk P, Riman RE, Janas VF, Tenhuisen KS. Preparation of magnesium-substituted hydroxyapatite powders by the mechanochemical-hydrothermal method. *Biomaterials* 2004;25:4647.
21. Lilley KJ, Gbureck U, Knowles JC, Farrar DF, Barralet JE. Cement from magnesium substituted hydroxyapatite. *Journal of Materials Science: Materials in Medicine* 2005;16:455.
22. LeGeros RZ, Bleiwas CB, Retino M, Rohanizadeh R, LeGeros JP. Zinc effect on the in vitro formation of calcium phosphates: Relevance to clinical inhibition of calculus formation. *American Journal of Dentistry* 1999;12:65.
23. Miao S, Weng W, Cheng K, Du P, Shen G, Han G, Zhang S. Sol-gel preparation of Zn-doped fluoridated hydroxyapatite films. *Surface and Coatings Technology* 2005;198:223.
24. Miao S, Weng W, Cheng K, Du P, Shen G, Han G, Huang X, Yan W, Zhang S. In vitro bioactivity and osteoblast-like cell test of zinc containing fluoridated hydroxy-apatite films. *Journal of Materials Science: Materials in Medicine* 2007;18:2101.
25. Oliveira AL, Reis RL, Li P. Strontium-substituted apatite coating grown on Ti6Al4V substrate through biomimetic synthesis. *Journal of Biomedical Materials Research—Part B Applied Biomaterials* 2007;83:258.
26. Agathopoulos S, Tulyaganov DU, Ventura JMG, Kannan S, Karakassides MA, Ferreira JMF. Formation of hydroxyapatite onto glasses of the CaO-MgO-SiO$_2$ system with B$_2$O$_3$, Na$_2$O, CaF$_2$ and P$_2$O$_5$ additives. *Biomaterials* 2006;27:1832.
27. Ellis DE, Terra J, Warschkow O, Jiang M, González GB, Okasinski JS, Bedzyk MJ, Rossi AM, Eon JG. A theoretical and experimental study of lead substitution in calcium hydroxyapatite. *Physical Chemistry Chemical Physics* 2006;8:967.
28. Sugiyama S, Abe K, Hayashi H, Moffat JB. Nitrous oxide as oxidant for methane conversion on calcium hydroxyapatites. *Applied Catalysis A: General* 1999;183:135.
29. Boanini E, Bigi A. Biomimetic synthesis of carbonated hydroxyapatite thin films. *Thin Solid Films* 2006;497:53.
30. Jokanović V, Izvonar D, Dramićanin MD, Jokanović B, Živojinović V, Marković D, Dačić B. Hydrothermal synthesis and nanostructure of carbonated calcium hydroxyapatite. *Journal of Materials Science: Materials in Medicine* 2006;17:539.
31. He QJ, Huang ZL, Cheng XK, Yu J. Thermal stability of porous A-type carbonated hydroxyapatite spheres. *Materials Letters* 2008;62:539.
32. LeGeros RZ. Apatites in biological systems. *Progress in Crystal Growth and Characterization* 1981;4:1.
33. Elliott JC. The problems of the composition and structure of the mineral components of the hard tissues. *Clinical Orthopaedics and Related Research* 1973;93:313.
34. Kim SR, Lee JH, Kim YT, Riu DH, Jung SJ, Lee YJ, Chung SC, Kim YH. Synthesis of Si, Mg substituted hydroxyapatites and their sintering behaviors. *Biomaterials* 2003;24:1389.
35. Gibson IR, Best SM, Bonfield W. Chemical characterization of silicon-substituted hydroxyapatite. *Journal of Biomedical Materials Research* 1999;44:422.
36. Tang XL, Xiao XF, Liu RF. Structural characterization of silicon-substituted hydroxyapatite synthesized by a hydrothermal method. *Materials Letters* 2005;59:3841.

37. Shirkhanzadeh M, Azadegan M, Stack V, Schreyer S. Fabrication of pure hydroxyapatite and fluoridated-hydroxyapatite coatings by electrocrystallisation. *Materials Letters* 1994;18:211.

38. Lee E-J, Lee S-H, Kim H-W, Kong Y-M, Kim H-E. Fluoridated apatite coatings on titanium obtained by electron-beam deposition. *Biomaterials* 2005;26:3843.

39. Cavalli M, Gnappi G, Montenero A, Bersani D, Lottici PP, Kaciulis S, Mattogno G, Fini M. Hydroxy- and fluorapatite films on Ti alloy substrates: Sol-gel preparation and characterization. *Journal of Materials Science* 2001;36:3253.

40. Legeros RZ. Variations in crystalline components of human dental calculus.1. Crystallographic and spectroscopic methods of analysis. *Journal of Dental Research* 1974;53:45.

41. Kannan S, Rebelo A, Lemos AF, Barba A, Ferreira JMF. Synthesis and mechanical behaviour of chlorapatite and chlorapatite/β-TCP composites. *Journal of the European Ceramic Society* 2007;27:2287.

42. Matsunaga K et al. Theoretical calculations of the thermodynamic stability of ionic substitutions in hydroxyapatite under an aqueous solution environment. *Journal of Physics: Condensed Matter* 2010;22:384210.

43. Matsunaga K, Inamori H, Murata H. Theoretical trend of ion exchange ability with divalent cations in hydroxyapatite. *Physical Review B* 2008;78.094101.

44. Yasukawa A, Ouchi S, Kandori K, Ishikawa T. Preparation and characterization of magnesium-calcium hydroxyapatites. *Journal of Materials Chemistry* 1996;6:1401.

45. Ma QY, Traina SJ, Logan TJ, Ryan JA. In situ lead immobilization by apatite. *Environmental Science & Technology* 1993;27:1803.

46. O'Donnell MD, Fredholm Y, de Rouffignac A, Hill RG. Structural analysis of a series of strontium-substituted apatites. *Acta Biomaterialia* 2008;4:1455.

47. Bigi A, Falini G, Foresti E, Ripamonti A, Gazzano M, Roveri N. Magnesium influence on hydroxyapatite crystallization. *Journal of Inorganic Biochemistry* 1993;49:69.

48. Ren F, Leng Y, Xin R, Ge X. Synthesis, characterization and ab initio simulation of magnesium-substituted hydroxyapatite. *Acta Biomaterialia* 2010;6:2787.

49. Qi G, Zhang S, Khor KA, Lye SW, Zeng X, Weng W, Liu C, Venkatraman SS, Ma LL. Osteoblastic cell response on magnesium-incorporated apatite coatings. *Applied Surface Science* 2008;255:304.

50. Patel PN. Mangnesium calcium hydroxylapatite solid solutions: Preparation, IR and lattice constant measurements. *Journal of Inorganic and Nuclear Chemistry* 1980;42:1129.

51. Li ZY, Lam WM, Yang C, Xu B, Ni GX, Abbah SA, Cheung KMC, Luk KDK, Lu WW. Chemical composition, crystal size and lattice structural changes after incorporation of strontium into biomimetic apatite. *Biomaterials* 2007;28:1452.

52. Kalita SJ, Bhatt HA. Nanocrystalline hydroxyapatite doped with magnesium and zinc: Synthesis and characterization. *Materials Science and Engineering C* 2007;27:837.

53. Benayoun S, Fouilland-Paill L, Hantzpergue JJ. Microscratch test studies of thin silica films on stainless steel substrates. *Thin Solid Films* 1999;352:156.

54. Forsgren J, Svahn F, Jarmar T, Engqvist H. Formation and adhesion of biomimetic hydroxyapatite deposited on titanium substrates. *Acta Biomaterialia* 2007;3:980.

55. Julia-Schmutz C, Hintermann HE. Microscratch testing to characterize the adhesion of thin layers. *Surface and Coatings Technology* 1991;48:1.

56. Clemens JAM, Wolke JGC, Klein CPAT, de Groot K. Fatigue behavior of calcium phosphate coatings with different stability under dry and wet conditions. *Journal of Biomedical Materials Research* 1999;48:741.

57. ASTM. Standard test method for adhesion or cohesion strength of thermal spray coatings. *ASTM* R2008;C 633-01:2001.03.10.

58. Qi G, Zhang S, Khor KA, Weng W, Zeng X, Liu C. An interfacial study of sol-gel-derived magnesium apatite coatings on Ti6Al4V substrates. *Thin Solid Films* 2008;516:5172.

59. Yan X, Huang X, Yu C, Deng H, Wang Y, Zhang Z, Qiao S, Lu G, Zhao D. The in-vitro bioactivity of mesoporous bioactive glasses. *Biomaterials* 2006;27:3396.

60. Jallot E, Nedelec JM, Grimault AS, Chassot E, Grandjean-Laquerriere A, Laquerriere P, Laurent-Maquin D. STEM and EDXS characterisation of physico-chemical reactions at the periphery of sol-gel derived Zn-substituted hydroxyapatites during interactions with biological fluids. *Colloids and Surfaces B: Biointerfaces* 2005;42:205.

61. Landi E, Tampieri A, Mattioli-Belmonte M, Celotti G, Sandri M, Gigante A, Fava P, Biagini G. Biomimetic Mg- and Mg,CO_3-substituted hydroxyapatites: Synthesis characterization and in vitro behaviour. *Journal of the European Ceramic Society* 2006;26:2593.

62. Ni GX, Lu WW, Xu B, Chiu KY, Yang C, Li ZY, Lam WM, Luk KDK. Interfacial behaviour of strontium-containing hydroxyapatite cement with cancellous and cortical bone. *Biomaterials* 2006;27:5127.

63. Ni GX, Lu WW, Chiu KY, Li ZY, Fong DYT, Luk KDK. Strontium-containing hydroxyapatite (Sr-HA) bioactive cement for primary hip replacement: An in vivo study. *Journal of Biomedical Materials Research Part B: Applied Biomaterials* 2006;77B:409.

64. Xue W, Hosick HL, Bandyopadhyay A, Bose S, Ding C, Luk KDK, Cheung KMC, Lu WW. Preparation and cell-materials interactions of plasma sprayed strontium-containing hydroxyapatite coating. *Surface and Coatings Technology* 2007;201:4685.

65. Qi G, Zhang S, Khor KA, Liu C, Zeng X, Weng W, Qian M. In vitro effect of magnesium inclusion in sol-gel derived apatite. *Thin Solid Films* 2008;516:5176.

66. Wang J, Chao Y, Wan Q, Zhu Z, Yu H. Fluoridated hydroxyapatite coatings on titanium obtained by electrochemical deposition. *Acta Biomaterialia* 2009;5:1798.

67. de Leeuw NH. A computer modelling study of the uptake and segregation of fluoride ions at the hydrated hydroxyapatite (0001) surface: Introducing a $Ca_{10}(PO_4)_6(OH)_2$ potential model. *Physical Chemistry Chemical Physics* 2004;6:1860.

68. Chen Y, Miao X. Thermal and chemical stability of fluorohydroxyapatite ceramics with different fluorine contents. *Biomaterials* 2005;26:1205.

69. Kim HW, Knowles JC, Salih V, Kim HE. Hydroxyapatite and fluor-hydroxyapatite layered film on titanium processed by a sol-gel route for hard-tissue implants. *Journal of Biomedical Materials Research—Part B Applied Biomaterials* 2004;71:66.

70. Wei M, Evans JH, Bostrom T, Grøndahl L. Synthesis and characterization of hydroxyapatite, fluoride-substituted hydroxyapatite and fluorapatite. *Journal of Materials Science: Materials in Medicine* 2003;14:311.

71. Rodríuez-Lorenzo LM, Hart JN, Gross KA. Influence of fluorine in the synthesis of apatites. Synthesis of solid solutions of hydroxy-fluorapatite. *Biomaterials* 2003;24:3777.

72. Okazaki M, Miake Y, Tohda H, Yanagisawa T, Matsumoto T, Takahashi J. Functionally graded fluoridated apatites. *Biomaterials* 1999;20:1421.
73. Tanaka H, Yasukawa A, Kandori K, Ishikawa T. Surface structure and properties of fluoridated calcium hydroxyapatite. *Colloids and Surfaces A: Physicochemical and Engineering Aspects* 2002;204:251.
74. Tanizawa Y, Tsuchikane H, Sawamura K, Suzuki T. Reaction characteristics of hydroxyapatite with F^- and PO_3F^{2-} ions. Chemical states of fluorine in hydroxyapatite. *Journal of the Chemical Society, Faraday Transactions* 1991;87:2235.
75. Cheng K, Zhang S, Weng WJ. The F content in sol-gel derived FHA coatings: An XPS study. *Surface and Coatings Technology* 2005;198:237.
76. Fulmer MT, Ison IC, Hankermayer CR, Constantz BR, Ross J. Measurements of the solubilities and dissolution rates of several hydroxyapatites. *Biomaterials* 2002;23:751.
77. Mir NA, Higuchi WI, Hefferren JJ. The mechanism of action of solution fluoride upon the demineralization rate of human enamel. *Archives of Oral Biology* 1969;14:901.
78. Driessens FCM. Relation between apatite solubility and anti-cariogenic effect of fluoride. *Nature* 1973;243:420.
79. Higuchi WI, Valvani SC, Hefferren JJ. The kinetics and mechanisms of reactions of human tooth enamel in buffered solutions of high fluoride concentrations. *Archives of Oral Biology* 1974;19:737.
80. Moreno EC, Kresak M, Zahradnik RT. Fluoridated hydroxyapatite solubility and caries formation. *Nature* 1974;247:64.
81. Okazaki M, Miake Y, Tohda H, Yanagisawa T, Takahashi J. Fluoridated apatite synthesized using a multi-step fluoride supply system. *Biomaterials* 1999;20:1303.
82. Okazaki M, Tohda H, Yanagisawa T, Taira M, Takahashi J. Differences in solubility of two types of heterogeneous fluoridated hydroxyapatites. *Biomaterials* 1998;19:611.
83. Bhadang KA, Gross KA. Influence of fluorapatite on the properties of thermally sprayed hydroxyapatite coatings. *Biomaterials* 2004;25:4935.
84. de Leeuw NH. Resisting the onset of hydroxyapatite dissolution through the incorporation of fluoride. *Journal of Physical Chemistry B* 2004;108:1809.
85. Gross KA, Rodríuez-Lorenzo LM. Sintered hydroxyfluorapatites. Part II: Mechanical properties of solid solutions determined by microindentation. *Biomaterials* 2004;25:1385.
86. Eslami H, Solati-Hashjin M, Tahriri M. The comparison of powder characteristics and physicochemical, mechanical and biological properties between nano-structure ceramics of hydroxyapatite and fluoridated hydroxyapatite. *Materials Science and Engineering: C* 2009;29:1387.
87. Zhang S, Wang YS, Zeng XT, Cheng K, Qian M, Sun DE, Weng WJ, Chia WY. Evaluation of interfacial shear strength and residual stress of sol-gel derived fluoridated hydroxyapatite coatings on Ti6Al4V substrates. *Engineering Fracture Mechanics* 2007;74:1884.
88. Zhang S, Wang YS, Zeng XT, Khor KA, Weng W, Sun DE. Evaluation of adhesion strength and toughness of fluoridated hydroxyapatite coatings. *Thin Solid Films* 2008;516:5162.
89. Ellingsen JE, Thomsen P, Lyngstadaas SP. Advances in dental implant materials and tissue regeneration. *Periodontology 2000*, 2006;41:136.

90. Harrison J, Melville AJ, Forsythe JS, Muddle BC, Trounson AO, Gross KA, Mollard R. Sintered hydroxyfluorapatites—IV: The effect of fluoride substitutions upon colonisation of hydroxyapatites by mouse embryonic stem cells. *Biomaterials* 2004;25:4977.

91. Cheng K, Weng W, Wang H, Zhang S. In vitro behavior of osteoblast-like cells on fluoridated hydroxyapatite coatings. *Biomaterials* 2005;26:6288.

92. Qu H, Wei M. The effect of fluoride contents in fluoridated hydroxyapatite on osteoblast behavior. *Acta Biomaterialia* 2006;2:113.

93. Gineste L, Gineste M, Ranz X, Ellefterion A, Guilhem A, Rouquet N, Frayssinet P. Degradation of hydroxylapatite, fluorapatite, and fluorhydroxyapatite coatings of dental implants in dogs. *Journal of Biomedical Materials Research* 1999;48:224.

94. Savarino L, Stea S, Ciapetti G, Granchi D, Donati ME, Cervellati M, Visentin M, Moroni A, Pizzoferrato A. The interface of bone microstructure and an innovative coating: An x-ray diffraction study. *Journal of Biomedical Materials Research* 1998;40:86.

95. Caulier H, Vercaigne S, Naert I, van der Waerden JPCM, Wolke JGC, Kalk W, Jansen JA. The effect of Ca-P plasma-sprayed coatings on the initial bone healing of oral implants: An experimental study in the goat. *Journal of Biomedical Materials Research* 1997;34:121.

96. LeGeros RZ, Kijowska R, Jia W, LeGeros JP. Fluoride-cation interactions in the formation and stability of apatites. *Journal of Fluorine Chemistry* 1988;41:53.

97. Gibson IR, Bonfield W. Preparation and characterization of magnesium/carbonate co-substituted hydroxyapatites. *Journal of Materials Science: Materials in Medicine* 2002;13:685.

98. Okazaki M. Magnesium-containing fluoridated apatites. *Journal of Fluorine Chemistry* 1988;41:45.

99. Okazaki M. Crystallographic properties of heterogeneous Mg-containing fluoridated apatites synthesized with a two-step supply system. *Biomaterials* 1995;16:703.

100. Hidouri M, Bouzouita K, Kooli F, Khattech I. Thermal behaviour of magnesium-containing fluorapatite. *Materials Chemistry and Physics* 2003;80:496.

101. Sun ZP, Ercan B, Evis Z, Webster TJ. Microstructural, mechanical, and osteocompatibility properties of Mg^{2+}/F^--doped nanophase hydroxyapatite. *Journal of Biomedical Materials Research Part A* 2010;94A:806.

102. Kannan S, Ventura JM, Ferreira JMF. Aqueous precipitation method for the formation of Mg-stabilized β-tricalcium phosphate: An x-ray diffraction study. *Ceramics International* 2007;33:637.

103. Hou X, Choy K-L, Leach SE. Processing and in vitro behavior of hydroxyapatite coatings prepared by electrostatic spray assisted vapor deposition method. *Journal of Biomedical Materials Research Part A* 2007;83A:683.

104. Chang MC, Tanaka J. FT-IR study for hydroxyapatite/collagen nanocomposite cross-linked by glutaraldehyde. *Biomaterials* 2002;23:4811.

105. Milella E, Cosentino F, Licciulli A, Massaro C. Preparation and characterisation of titania/hydroxyapatite composite coatings obtained by sol-gel process. *Biomaterials* 2001;22:1425.

106. Atuchin VV, Kesler VG, Pervukhina NV, Zhang Z. Ti 2p and O 1s core levels and chemical bonding in titanium-bearing oxides. *Journal of Electron Spectroscopy and Related Phenomena* 2006;152:18.

107. Kačiulis S, Mattogno G, Pandolfi L, Cavalli M, Gnappi G, Montenero A. XPS study of apatite-based coatings prepared by sol-gel technique. *Applied Surface Science* 1999;151:1.
108. Surendran KP, Wu A, Vilarinho PM, Ferreira VM. Ni and Zn doped $MgTiO_3$ thin films: Structure, microstructure, and dielectric characteristics. *Journal of Applied Physics* 2010;107:114112.
109. Zhang W, Huang ZL, Liao SS, Cui FZ. Nucleation sites of calcium phosphate crystals during collagen mineralization. *Journal of the American Ceramic Society* 2003;86:1052.
110. Danil'chenko S, Kulik A, Bugai A, Pavlenko P, Kalinichenko T, Ul'yanchich N, Sukhodub L. Determination of the content and localization of magnesium in bioapatite of bone. *Journal of Applied Spectroscopy* 2005;72:899.
111. Barrère F, Layrolle P, Van Blitterswijk CA, De Groot K. Biomimetic calcium phosphate coatings on Ti6Al4V: A crystal growth study of octacalcium phosphate and inhibition by Mg^{2+} and HCO_3^-. *Bone* 1999;25:107S.
112. Gronowicz G, McCarthy MB. Response of human osteoblasts to implant materials: Integrin-mediated adhesion. *Journal of Orthopaedic Research* 1996;14:878.
113. Lange TS, Bielinsky AK, Kirchberg K, Bank I, Herrmann K, Krieg T, Scharffetter-Kochanek K. Mg^{2+} and Ca^{2+} differentially regulate β_1 integrin-mediated adhesion of dermal fibroblasts and keratinocytes to various extracellular matrix proteins. *Experimental Cell Research* 1994;214:381.
114. Zreiqat H, Howlett CR, Zannettino A, Evans P, Schulze-Tanzil G, Knabe C, Shakibaei M. Mechanisms of magnesium-stimulated adhesion of osteoblastic cells to commonly used orthopaedic implants. *Journal of Biomedical Materials Research* 2002;62:175.
115. Dorozhkin SV. Calcium orthophosphates. *Journal of Materials Science* 2007;42:1061.
116. Gutowska I, Machoy Z, Machaliński B. The role of bivalent metals in hydroxyapatite structures as revealed by molecular modeling with the HyperChem software. *Journal of Biomedical Materials Research Part A* 2005;75A:788.
117. Martini D, Fini M, Franchi M, Pasquale VD, Bacchelli B, Gamberini M, Tinti A, Taddei P, Giavaresi G, Ottani V, Raspanti M, Guizzardi S, Ruggeri A. Detachment of titanium and fluorohydroxyapatite particles in unloaded endosseous implants. *Biomaterials* 2003;24:1309.
118. Orly I, Gregoire M, Menanteau J, Heughebaert M, Kerebel B. Chemical changes in hydroxyapatite biomaterial under in vivo and in vitro biological conditions. *Calcified Tissue International* 1989;45:20.
119. Blackwood DJ, Seah KHW. Influence of anodization on the adhesion of calcium phosphate coatings on titanium substrates. *Journal of Biomedical Materials Research Part A* 2010;93A:1551.
120. Dorozhkin S. Calcium orthophosphates in nature, biology and medicine. *Materials* 2009;2:399.
121. Stucki U, Schmid J, Hämmerle CF, Lang NP. Temporal and local appearance of alkaline phosphatase activity in early stages of guided bone regeneration. *Clinical Oral Implants Research* 2001;12:121.
122. Boskey AL, Gadaleta S, Gundberg C, Doty SB, Ducy P, Karsenty G. Fourier transform infrared microspectroscopic analysis of bones of osteocalcin-deficient mice provides insight into the function of osteocalcin. *Bone* 1998;23:187.

123. Farley JR, Jon EW, Baylink DJ. Fluoride directly stimulates proliferation and alkaline phosphatase activity of bone-forming cells. *Science* 1983;222:330.
124. Mehta S, Reed B, Antich P. Effects of high levels of fluoride on bone formation: An in vitro model system. *Biomaterials* 1995;16:97.
125. Inoue M, LeGeros RZ, Inoue M, Tsujigiwa H, Nagatsuka H, Yamamoto T, Nagai N. In vitro response of osteoblast-like and odontoblast-like cells to unsubstituted and substituted apatites. *Journal of Biomedical Materials Research Part A* 2004;70A:585.
126. Kim H-W, Lee E-J, Kim H-E, Salih V, Knowles JC. Effect of fluoridation of hydroxyapatite in hydroxyapatite-polycaprolactone composites on osteoblast activity. *Biomaterials* 2005;26:4395.
127. Li X, Ito A, Sogo Y, Wang X, LeGeros RZ. Solubility of Mg-containing β-tricalcium phosphate at 25°C. *Acta Biomaterialia* 2009;5:508.
128. Xue W, Dahlquist K, Banerjee A, Bandyopadhyay A, Bose S. Synthesis and characterization of tricalcium phosphate with Zn and Mg based dopants. *Journal of Materials Science: Materials in Medicine* 2008;19:2669.
129. Tardei C, Grigore F, Pasuk I, Stoleriu S. The study of Mg^{2+}/Ca^{2+} substitution of β-tricalcium phosphate. *Journal of Optoelectronics and Advanced Materials* 2006;8:568.
130. Dorozhkin SV. Acidic dissolution mechanism of natural fluorapatite. I. Milli- and microlevels of investigations. *Journal of Crystal Growth* 1997;182:125.
131. Dorozhkin SV. Acidic dissolution mechanism of natural fluorapatite. II. Nanolevel of investigations. *Journal of Crystal Growth* 1997;182:133.
132. Chaïrat C, Schott J, Oelkers EH, Lartigue J-E, Harouiya N. Kinetics and mechanism of natural fluorapatite dissolution at 25°C and pH from 3 to 12. *Geochimica et Cosmochimica Acta* 2007;71:5901.
133. Dorozhkin SV. A review on the dissolution models of calcium apatites. *Progress in Crystal Growth and Characterization of Materials* 2002;44:45.
134. Dorozhkin SV. Surface reactions of apatite dissolution. *Journal of Colloid and Interface Science* 1997;191:489.
135. de Leeuw NH. Density functional theory calculations of local ordering of hydroxy groups and fluoride ions in hydroxyapatite. *Physical Chemistry Chemical Physics* 2002;4:3865.
136. Gu YW, Khor KA, Pan D, Cheang P. Activity of plasma sprayed yttria stabilized zirconia reinforced hydroxyapatite/Ti-6Al-4V composite coatings in simulated body fluid. *Biomaterials* 2004;25:3177.
137. Dorozhkin SV, Epple M. Biological and medical significance of calcium phosphates. *Angewandte Chemie—International Edition* 2002;41:3130.
138. Cai Y, Zhang S, Zeng X, Wang Y, Qian M, Weng W. Improvement of bioactivity with magnesium and fluorine ions incorporated hydroxyapatite coatings via sol-gel deposition on Ti6Al4V alloys. *Thin Solid Films* 2009;517:5347.
139. Cai Y, Zhang S, Zeng X, Qian M, Sun D, Weng W. Interfacial study of magnesium-containing fluoridated hydroxyapatite coatings. *Thin Solid Films* 2011;519:4629.
140. Cai Y, Zhang J, Zhang S, Mondal D, Venkatraman SS, Zeng X. Osteoblastic cell response on fluoridated hydroxyapatite coatings: The effect of magnesium incorporation. *Biomedical Materials* 2010;5:054114.
141. Cai Y, Zhang S, Zeng X, Sun D. Effect of fluorine incorporation on long-term stability of magnesium-containing hydroxyapatite coatings. *Journal of Materials Science: Materials in Medicine* 2011;22:1633.

142. El Feki H, Amami M, Ben Salah A, Jemal M. Synthesis of potassium chloroapatites, IR, X-ray and Raman studies. *Physica Status Solidi (C)* 2004;1:1985.
143. Mayer I, Schlam R, Featherstone JDB. Magnesium-containing carbonate apatites. *Journal of Inorganic Biochemistry* 1997;66:1.
144. Yamasaki Y, Yoshida Y, Okazaki M, Shimazu A, Uchida T, Kubo Y, Akagawa Y, Hamada Y, Takahashi J, Matsuura N. Synthesis of functionally graded $MgCO_3$ apatite accelerating osteoblast adhesion. *Journal of Biomedical Materials Research* 2002;62:99.
145. Okazaki M. Crystallographic behaviour of fluoridated hydroxyapatites containing Mg^{2+} and CO_3^{2-} ions. *Biomaterials* 1991;12:831.
146. Kannan S, Lemos AF, Ferreira JMF. Synthesis and mechanical performance of biological-like hydroxyapatites. *Chemistry of Materials* 2006;18:2181.
147. Miao S, Cheng K, Weng W, Du P, Shen G, Han G, Yan W, Zhang S. Fabrication and evaluation of Zn containing fluoridated hydroxyapatite layer with Zn release ability. *Acta Biomaterialia* 2008;4:441.
148. Jeyachandran YL, Narayandass SK, Mangalaraj D, Bao CY, Li W, Liao YM, Zhang CL, Xiao LY, Chen WC. A study on bacterial attachment on titanium and hydroxyapatite based films. *Surface and Coatings Technology* 2006;201:3462.

4

Zinc- and Fluorine-Doped HA Coatings via Sol–Gel Method

Kui Cheng and Wenjian Weng

CONTENTS

4.1 Introduction

Hydroxyapatite (HA)-coated metallic implants have been widely investigated due to their fast integration with bone compared with pure metal [1,2]. In order to improve their holistic performance in vivo, researches have focused on the following respects: (1) inducing active interaction between the film and tissue through promotion of osteoblast cell attachment, proliferation, and differentiation, resulting in fast new bone formation and osteointegration [3,4] and (2) making films more dissolution resistant through reduction of intrinsic solubility of the film materials, so that better long-term effectiveness of the implants could be obtained [5,6].

Compared with HA, fluoridated hydroxyapatite (FHA) has proven to have a stronger dissolution resistance and comparable biocompatibility [7–12]. FHA can therefore be applied as a thinner coating, reducing the risk of coating delamination frequently observed with thicker coatings [13], thus providing a longer-term stability in osseointegration. However, significant improvements in HA coatings are still necessary in early bone–implant osseointegration processes to promote rapid physiologic loading and to further improve the repair quality of an implant. Osseointegration of an implant into a bone site depends primarily on the activity of the bone cells or their precursors. The ability of bone cells to migrate to the implant and to adhere, proliferate, and differentiate on the implant surface is a major factor in determining the success of osseointegration. Attempts have been made to combine bone-promoting pharmaceuticals with the implants, and the release of these pharmaceuticals directly at the bone–implant interface has high therapeutic efficiency. A possible approach is to apply osteoinductive coatings that are supplemented with beneficial trace elements that promote early osseointegration [14–16]. The pharmaceutical effects of many trace elements on bone cells, such as Sr, Si, and Zn, have been widely studied [17–19]. Zinc is well established as an essential trace element that has stimulatory effects on osteoblast activity and bone formation at low concentration [20–22]. Several studies have revealed that the pharmacological performance of Zn-containing implants can be significantly influenced by the mode of Zn release. In general, slow and sustained release of the Zn-incorporated implant material promoted the best bone formation around the implant [23].

To satisfy both the requirements of dissolution resistancy and bone formation stimulatory, fluorine- and zinc-substituted HAs could be a good alternative, if they could be incorporated in an appropriate manner. In this chapter, the incorporation of fluorine and zinc as well as the effects of such incorporation are introduced and discussed.

4.2 Sol–Gel-Derived Fluorine-Doped Hydroxyapatite Coatings

Although many methods have been developed to prepare HA coatings, sol–gel method remains one of the most prevalent. Compared with other methods, sol–gel preparation possesses many advantages, such as low cost, simple equipment, and easy control of chemical composition. Generally, the coating material or its precursors are first transformed into sol and then coated on the substrate surface uniformly; the rapid solvent evaporation and subsequent condensation reactions will lead to the formation of the gel film. After drying and calcining, the required film is finally prepared.

The pioneer work of sol–gel preparation of bioactive materials was actually carried out by Hench on bioglass [24]. In the past decades, owing to the excellent performance of apatite and the development of bioactive films, the sol–gel technique has begun to be widely used in the preparation of HA films.

According to the type of sol, the sol–gel technique can be divided into two categories. First, HA particles are directly prepared and mixed with proper solvent to form HA sol after dispersion. Second, calcium- and phosphorus-containing reagents are dissolved to be Ca and P precursors, respectively, and then gelation happens as a result of solvent evaporation and probably some polycondensation reactions. HA films will be obtained after calcining of the dried gel.

Mavis et al. prepared the HA films using HA particles [25]. The process was as follows: first, HA fine powder was obtained through coprecipitation method and calcined to improve crystallinity; then the HA powder was ultrasonically dispersed into water to form HA sol. After dip coating and 840°C heat treatment, HA films with thickness of about 25 μm and bonding strength of about 30 MPa were finally obtained. By this method, although the size of HA particles is much smaller than that prepared through plasma spray, the corresponding films are still thick, which may affect the bonding strength and long-term stability. In fact, many researchers prefer the precursor ways to prepare HA films instead of directly using HA particles. Given the characteristics of the technique itself, the differences between various methods mainly lie in the choice of precursors. It is rather important to find the most suitable sol system for the preparation of bioactive apatite coatings although a variety of sol–gel systems have been developed.

4.2.1 Basics of Sol–Gel-Derived Films

4.2.1.1 Dip Coating

Dip coating is the most commonly used technique to prepare films or coatings utilizing a sol–gel process. The substrates were first immersed in the precursor solution and then vertically pulled upward with a constant pulling speed [26]. During this process, the sol attaches to the substrate to be a liquid sol film. There are two liquid layers in the sol film: the outer layer of the liquid flows back into the sol pool, and the inner layer remains on the substrate after the solvent evaporates. The liquid film is similar to a wedge, the starting point of which begins with a clear-cut line called the drying line. If the downward speed of the drying line is equal to the substrate pulling speed, then the film deposition process is relatively stable.

In the presence of inorganic materials, the inorganic is the first to move from the sol to the drying line and then precipitates owing to the evaporation of the solvent. The deposition time depends on the thickness of the film and evaporation rate. For a precursor solution that usually contains ethanol, the steady state stays only a few seconds.

4.2.1.2 Drying of Films

The dip-coating process accompanies the drying of films. That actually determines the shape of the liquid film. The drying process increases the concentration of the liquid and leads to the formation of an elastic or viscoelastic gel. Further evaporation results in the capillary pressure, which balances with the surrounding solid phase pressure and further contracts.

Scherer divided the drying process into two phases [27]: constant phase (CRP) and deceleration. In the constant phase, mass transferring rate depends on the surface convection and the speed of gas taken away; in deceleration phase, mass transferring rate depends on the internal permeation of the gel. According to this theory, most of the drying process belongs to constant one that proves the wedge shape of the film.

4.2.1.3 Calcination of Films

After drying, the films may need calcinations to discharge the residual organic and then crystallize. In this process, as films have lost mobility, it is prone to stress and cracks. In fact, this process has already begun in the drying stage. The film shrinks with drying. However, when attached to the substrate, the shrinkage is restrained. That means the volume contraction can only be achieved by the reduction of film thickness. As a result, tensile stress in the substrate surface will occur. Theoretically, if the film adheres to the substrate well, the critical thickness of the cracks that begin to proliferate or holes that begin to grow is determined as follows:

$$h_C = \left(\frac{K_{IC}}{\sigma}\right)^2$$

where
  K_{IC} is the critical stress intensity factor
  σ represents the ratio of elastic modulus of the film and the substrate

For films less thick than h_C, the energy required for crack extension is smaller than that required for the crack released from the vicinal stress, and the film remains crack-free.

4.2.2 Systems for Fluorine-Doped Hydroxyapatite Films

HA films made by sol–gel technique have been much improved compared with traditional methods. However, due to the fact that the composition is still mainly HA, the bioactive film dissolves rather quickly when implanted, which affects the long-term stability of the implant or even leads to complete failure of implantation. Thus, it is extremely necessary to adjust components of the film. As fluorine can improve the solubility of HA and play an important role in biological mineralization as well, FHA (fluorine-doped HA) has attracted widespread concern. Fluorine can be introduced into the film via fluorine-containing reagents, such as trifluoroacetic acid (CF_3COOH), hexafluorophosphoric acid (HPF_6), and ammonium hexafluorophosphate (NH_4PF_6). Depending on the chosen fluorine-containing compound, they could be divided into different systems.

4.2.2.1 *Ca(NO₃)₂–PO(OH)ₓ(OEt)₃₋ₓ–CF₃COOH System*

In this system, a designed amount of triethanolamine ($N(CH_2CH_2OH)_3$ or $N(EtOH)_3$ is always added to reduce the acidity of precursor solution caused by CF_3COOH. The effect of $N(EtOH)_3$ on the solutions containing CF_3COOH can be described as follows.

When $N(EtOH)_3$ is added into CF_3COOH solution, being an organic base, it will react with the organic acid, CF_3COOH. The reaction tendency leads to the complete transformation of dimeric CF_3COOH species to monomeric CF_3COOH species and enhances the dissociation of monomeric CF_3COOH species to form trifluoroacetate or an organic salt (Equation 4.1):

$$CF_3COOH + N(EtOH)_3 \Leftrightarrow CF3COO^- + HN(EtOH)_3^+$$

$$\Leftrightarrow \{CF_3COO^-\} \cdot \left\{HN(EtOH)_3^+\right\} \tag{4.1}$$

When $Ca(NO_3)_2 \cdot 4H_2O$ is added into CF_3COOH solution, the H_2O coming along with $Ca(NO_3)_2$ will increase the polarity of the solvent. The polarity increase leads to the formation of monomeric CF_3COOH species and further causes the monomeric CF_3COOH species to be dissociated into CF_3COO^- and H^+ species. The CF_3COO^- reacts with Ca species to form a complex species as follows (Equation 4.2):

$$xCF_3COOH + H_2O + Ca(NO_3)_2 \Leftrightarrow Ca(OOCF_3C)_x(NO_3)_{2-x}$$

$$+ xH^+ + xNO_3^- + H_2O \tag{4.2}$$

Under an acidic condition, the CF_3COO^- does not have a strong ability to coordinate with Ca ion by replacing nitrate group. Hence, free CF_3COO^- could be dominant in the solution. When the basicity of the solution increases, Equation 4.2 is greatly enhanced, leading to a high concentration of the complex species, $Ca(OOCF_3C)_x(NO_3)_{2-x}$, as the dominant species in the solution.

To form FHA finally, the mixed solution needs to be dried and fired. If a mixed solution containing CF_3COOH but without $N(EtOH)_3$ is heated, the species in the solution are concentrated during the evaporation of the solvent, and the decrease of the solvent will lower the dissociation degree of CF_3COOH. Since the boiling point of CF_3COOH is 72.4°C, it is easy to be evaporated. As a result, very little amount of fluorine-containing species could remain to take part in the formation reactions of apatite. Hence, FHA can be hardly formed. If $N(EtOH)_3$ is added into the mixed solution, upon drying and firing, the $Ca(OOCF_3C)_x(NO3)_{2-x}$ species could remain during drying because $N(EtOH)_3$ has a higher boiling point of 277°C and keeps $Ca(OOCF_3C)_x(NO_3)_{2-x}$ stable. Hence, fluorine-containing species will effectively take part in the formation reactions of FHA.

Besides promoting FHA formation, the existence of $N(EtOH)_3$ with high boiling point will react with decomposed products of nitrates to produce a flame. During the flaming, FHA could be directly formed. Moreover, according to Equations 4.1 and 4.2, ammonium nitrate may be produced as follows:

$$HN(CH_2CH_2OH)_3^+ + NO_3^- \rightarrow NH(CH_2CH_2OH)_3NO_3$$

In the fast drying process, $NH(CH_2CH_2OH)_3NO_3$ will oxidize and decompose, leading to the release of large amounts of gas. That may result in loose and cracked film.

4.2.2.2 $Ca(NO_3)_2–PO(OH)_x(OEt)_{3-x}–PF_6^-$ System

When HPF_6 is added into the initial mixed precursor solution, the resulting reactions could be mainly alcoholysis and hydrolysis, and the reactions are proposed as follows:

$$HPF_6 + EtOH \Leftrightarrow (OEt)POF_2 + HF \qquad (4.3)$$

$$(OEt)POF_2 + EtOH \Leftrightarrow (OEt)POF + HF \qquad (4.4)$$

$$(OEt)POF_2 + H_2O \Leftrightarrow HPO_2(OEt)F + HF \qquad (4.5)$$

$$PO(OEt)_{2-x}(OH)_xF + EtOH \Leftrightarrow PO(OEt)_{3-x}(OH)_x + HF \qquad (4.6)$$

In addition, Ca species have a strong tendency to react with HF to form CaF_2. The HF produced in reactions (4.3) through (4.6) is rapidly consumed by Ca species in the previous refluxing mixture to form fine CaF_2 nanoparticles. The formation of CaF_2 nanoparticles causes the mixture to be cloudy or even sedimentated.

During refluxing, the increase in temperature and the formation of solid CaF_2 as a reaction product will promote and lead to the completion of alcoholysis and hydrolysis reactions (Equations 4.3 through 4.6), resulting in the refluxed mixture (dipping sol) containing only the solid CaF_2 particles and the soluble $PO(OEt)_{3-x}(OH)_x$ species as the detected reaction products. Since the solid CaF_2 particles are very fine, the sol could be kept stable for a long time period, if sedimentation did not take place within 4 months.

Adding NH_4PF_6 to the initial mixed precursor solution would lead to similar alcoholysis and hydrolysis reactions to those suggested in the case of HPF_6 with the formation of fine CaF_2 particles. However, compared with the mixtures with HPF_6, the substitution of H^+ for NH_4^+ decreases the acidity of the mixture, which could favor reactions (4.3) through (4.6), and further accelerates the reaction between Ca and F. Hence, a faster precipitation process occurs and CaF_2 particles grow more easily in the mixture with NH_4PF_6 than in that with HPF_6. Besides, a decrease in acidity will also favor the formation of apatite precipitates. As a consequence, the dipping sol is less stable against sedimentation.

Since dipping sols with HPF_6 or NH_4PF_6 contain fine and non-evaporable CaF_2 particles, the incorporation of fluorine into HA film becomes easy. As the amounts of HPF_6 or NH_4PF_6 in the initial mixed precursor solution increase, the resulting sols become rich in F-containing products (CaF_2), and the F content in the films tends to increase. In addition, the incorporation of F into HA enhances the development of the apatite phase. Compared with $Ca(NO_3)_2–PO(OH)_x(OEt)_{3-x}–CF_3COOH$ system, gas released during the heat treatment is much less, which is also favorable for high-quality films. For the morphology of films, the film prepared by HPF_6 is dense and smooth, while that prepared by NH_4PF_6 is dense but contains many embedded particles with a few hundreds of nanometers.

4.2.3 Microstructures of FHA Coatings

The formation of FHA films can be described as follows: first, a liquid layer containing $Ca(NO_3)_2$ or CaF_2, $PO(OH)_x(OEt)_{3-x}$, and solvent (C_2H_5OH) attach to the substrate, and second, a series of reactions occur in the subsequent drying and calcining process: the film loses mobility after solvent evaporation, and then its composition transforms into $Ca(NO_3)_2$ or CaF_2 crystal phase and a large amount of amorphous calcium phosphate as the oxidation of the organic matter in $PO(OH)(OEt)_{3-x}$ occurs. As the calcining temperature increases, the nucleation and growth of FHA begins. Eventually a crystalline FHA film forms.

Figure 4.1 shows the x-ray diffraction (XRD) patterns of HA films prepared by sol–gel method at different calcining temperature without any F-containing compound added. As is shown in the figure, all samples contained apatite phase. Increased calcining temperature (Figure 4.1) led to a better developed apatite crystalline phase. However, films treated at 800°C had a small amount of β-TCP phase, which was a result of the decomposition of HA.

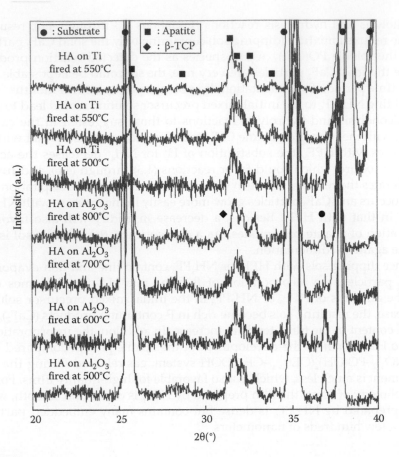

FIGURE 4.1
XRD patterns of films without fluorine (HAA series). (Reprinted from Cheng, K., Solgel preparation of fluoridated hydroxyapatite films and their in vitro properties, PhD dissertation.)

When HPF_6 was added in the preparation, as discussed previously, FHA films were obtained. Figure 4.2 shows the XRD results of films prepared at different calcining temperatures but with the same amount HPF_6. As demonstrated earlier, the intensity of the XRD peaks belonging to apatite enhances with the increase of calcining temperature. Nevertheless, no trace of CaO, CaF_2, or other calcium phosphate phase presented (Figure 4.2). With the increase of HPF_6 (Figure 4.2), a small amount of rutile titanium dioxide phase appeared.

FHA films could also be obtained if NH_4PF_6 was added instead of HPF_6 (Figure 4.3). All films showed to be apatite, just as those with HPF_6 addition. Among them, the intensity of the XRD peaks belonging to apatite was enhanced with the increase of calcining temperature. With the increase in the amount of NH_4PF_6, the rutile phase was also observed.

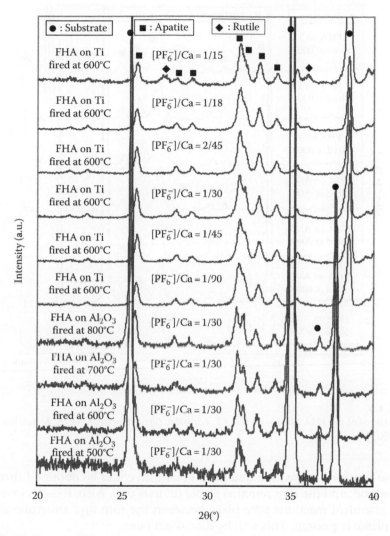

FIGURE 4.2
XRD patterns of the FHA films with HPF_6 addition. (From Cheng, K. et al., *J. Biomed. Mater. Res. B Appl. Biomater,* 69B, 33, 2004; Cheng, K. et al., *Mater. Res. Bull.,* 38, 89, 2003.)

Figure 4.4 shows the FTIR spectra of typical HA and FHA films. All the spectra show the existence of apatite, whether HPF_6 or NH_4PF_6 was added: PO_4^{3-} group of v1 (965 cm⁻¹), v3 (1019 cm⁻¹, 1087 cm⁻¹), and v4 (597 cm⁻¹, 560 cm⁻¹) stretching bands and CO_3^{3-} group of v2 (879 cm⁻¹) stretching band. HA film shows the OH librational band ρOH at 632 cm⁻¹ and stretching band vOH at 3571 cm⁻¹, while the ρOH and vOH of other films (FHA) shift to 748 and 3540 cm⁻¹, respectively. According to existing FTIR results on HA, such shift was ascribed to the variation of OH chains in the HA structure.

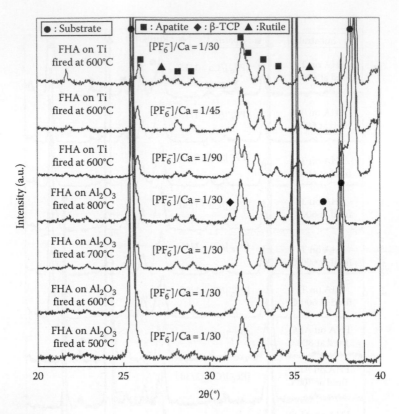

FIGURE 4.3
XRD patterns of the FHA films with NH$_4$PF$_6$ addition. (From Cheng, K. et al., *Mater. Res. Bull.*, 38, 89, 2003.)

These results demonstrate that pure FHA film could be obtained through this method, and the film remains stable up to 800°C. Also, it is obvious that active chemical reactions take place between the film and substrate as the rutile phase is present. This will be discussed later.

Many factors may have an effect on the microstructure of FHA films. Among them, calcining temperature and F content play important roles. Generally speaking, the calcining temperature affects the crystallinity, phase composition, and morphology of the films.

With the increase of calcining temperature, it is more likely for amorphous gels to crystallize into HA phase, resulting in the increase of the relative content of the crystalline phase. However, a small amount of β-TCP, which is a product of apatite decomposition, also appears with increasing calcining temperature, just as indicated in Figures 4.1 through 4.3.

The surface morphology of the films is shown in Figure 4.5. For HA films, the surfaces were very smooth, virtually no pores and cracks. When the calcining temperature increased to 600°C, there was basically no change

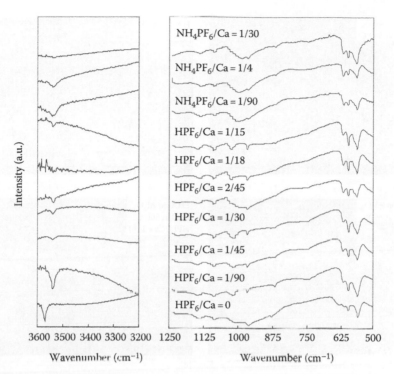

FIGURE 4.4
FTIR reflectance spectra of the FHA. (From Cheng, K., Sol–gel preparation of fluoridated hydroxyapatite films and their in vitro properties, PhD dissertation, 2003.)

in morphology; when the calcining temperature continued to increase, a small amount of pores and cracks appeared and grew in number. While for FHA prepared with the addition of HPF_6, the films were basically smooth with a small amount of pores and cracks when treated at 500°C and 600°C. There were many cracks and pores when the temperature increased to 700°C and 800°C. For FHA prepared with the addition of NH_4PF_6, there were a large amount of pores and cracks although the calcining temperature was only 500°C. Further increase of the calcining temperature led to a significant jump in the amount of crack and further deterioration of the morphology.

In conclusion, with the increase of calcining temperature, the morphology of the films becomes more porous with more cracks. The possible reason is that the stress caused by film densification becomes stronger when the temperature increases, and the release of stress leads to cracking. Thus, the higher the calcining temperature, the more cracks the films have. Moreover, films with NH_4PF_6 addition showed even poorer morphology; that is simply because the existence of NH_4^+ will release gas during film formation, which, in turn leads to the generation of pores and cracks.

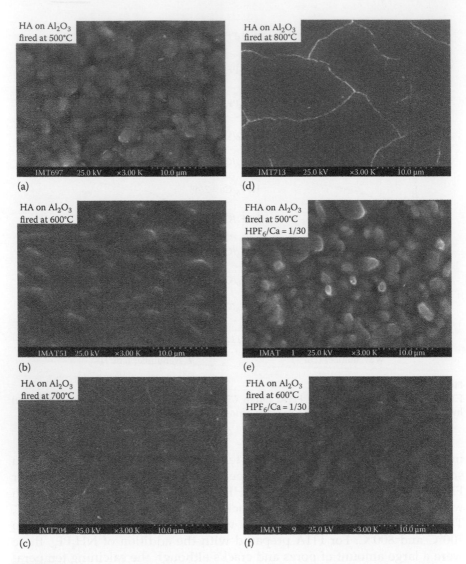

FIGURE 4.5

SEM micrographs of films with and without adding PF_6^{6-}. (a) HA fired at 500°C, (b) HA fired at 600°C, (c) HA fired at 700°C, (d) HA fired at 800°C, (e) FHA fired at 500°C with $HPF_6/Ca = 1/30$, (f) FHA fired at 600°C with $HPF_6/Ca=1/30$.

　　Besides the calcining temperature, the concentration of PF_6^- in the precursor solution has a considerable effect on the crystallinity and morphology of the FHA films. Table 4.1 shows the relative crystallinity of the different films.

　　As shown previously, at the same calcining temperature, the relative crystallinity of pure HA coatings was the lowest. With the increase of fluorine content, the relative crystallinity significantly increased. This phenomenon

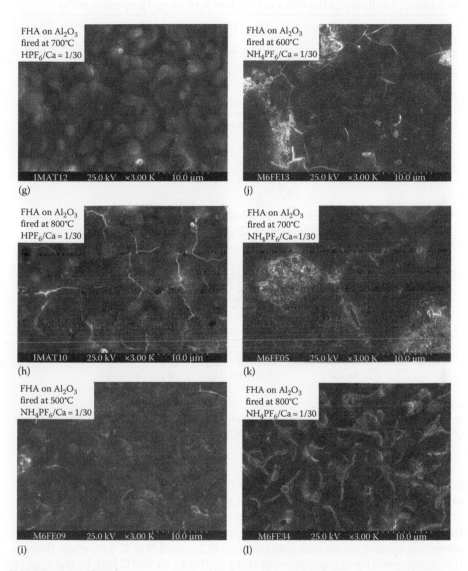

FIGURE 4.5 (continued)

SEM micrographs of films with and without adding PF_6^-. (g) FHA fired at 700°C with $HPF_6/Ca = 1/30$, (h) FHA fired at 800°C with $HPF_6/Ca = 1/30$, (i) FHA fired at 500°C with $NH_4PF_6/Ca = 1/30$, (j) FHA fired at 600°C with $NH_4PF_6/Ca = 1/30$, (k) FHA fired at 700°C with $NH_4PF_6/Ca = 1/30$, (l) FHA fired at 800°C with $NH_4PF_6/Ca = 1/30$. (From Cheng, K., Sol–gel preparation of fluoridated hydroxyapatite films and their in vitro properties, PhD dissertation, 2003.)

TABLE 4.1

Relative Crystallinity of the Films at Different Calcining Temperatures (%)

Sample	$HPF_6/Ca = 0$	$HPF_6/Ca = 1/90$	$HPF_6/Ca = 1/45$	$HPF_6/Ca = 1/30$	$HPF_6/Ca = 2/45$
Crystallinity (%)	51.2	54.1	55.5	60.8	68.0
Sample	$HPF_6/Ca = 1/18$	$HPF_6/Ca = 1/15$	$NH_4PF_6/Ca = 1/90$	$NH_4PF_6/Ca = 1/45$	$NH_4PF_6/Ca = 1/30$
Crystallinity (%)	63.4	61.2	51.4	60.0	63.4

Source: Cheng, K., Sol–gel preparation of fluoridated hydroxyapatite films and their in vitro properties, PhD dissertation, 2003.

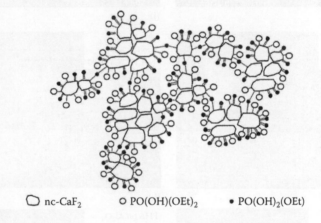

○ nc-CaF$_2$ ○ PO(OH)(OEt)$_2$ ● PO(OH)$_2$(OEt)

FIGURE 4.6
Schematic of gel formation through bonding of F in nc-CaF$_2$ and H in P-containing groups. (From Cheng, K. et al., *J. Sol–Gel Sci. Technol.*, 38, 13, 2006.)

could be explained by the formation of the FHA phase. In the case of no addition of HPF$_6$, most Ca^{2+} ions in the gel coordinated to phosphate groups, while others remained "free" to crystallize into Ca(NO$_3$)$_2$. These Ca(NO$_3$)$_2$ further reacted with PO$_4^{3-}$-containing groups to form HA. In this case, the observed apatite formation temperature is around 500°C [30], even though the decomposition temperature of pure Ca(NO$_3$)$_2$ is 561°C [31]. In the case of HPF$_6$ addition, the formation of nanocrystalline CaF$_2$ consumed Ca^{2+} ions, thus reducing or totally depriving the amount of crystalline Ca(NO$_3$)$_2$ in the gel. This facilitated the formation of apatite due both to the high reactivity of nanoscale CaF$_2$ and to the reduction or even total absence of Ca(NO$_3$)$_2$. As a result, even pure fluorapatite (FA) (Ca$_{10}$(PO$_4$)$_6$F$_2$) appeared at much lower temperature of 400°C in the system contained HPF$_6$ (Figure 4.6).

Therefore, the increase in the amount of HPF$_6$ in the precursor solution is favorable for the formation of the FHA phase with high crystallinity.

F also affects the morphology of the films (Figure 4.7). Pure HA film surface was smooth, with little pores and cracks; with the presence of fluorine, the surface became rough, and many pores were observed; while the surface morphology improved with the increase of fluorine content, some pores still showed up; even higher fluorine content led the film to be more smooth. However, since then, further increase of HPF_6 led to increased pores and

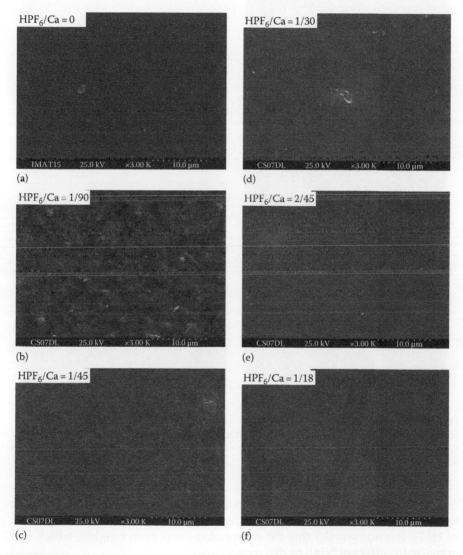

FIGURE 4.7

SEM micrographs of films with different fluorine concentration. (a) $HPF_6/Ca=0$, (b) $HPF_6/Ca=1/90$, (c) $HPF_6/Ca=1/45$, (d) $HPF_6/Ca=1/30$, (e) $HPF_6/Ca=2/45$, (f) $HPF_6/Ca=1/18$.

(continued)

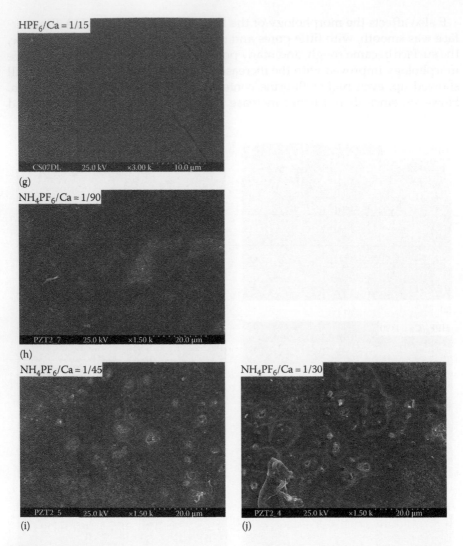

FIGURE 4.7 (continued)
SEM micrographs of films with different fluorine concentration. (g) $HPF_6/Ca = 1/15$, (h) $NH_4PF_6/Ca = 1/90$, (i) $NH_4PF_6/Ca = 1/45$, (j) $NH_4PF_6/Ca = 1/30$. (From Cheng, K. et al., *Biomaterials*, 26, 6288, 2005.)

cracks and thus resulted in a rather rough surface. As for NH_4PF_6 addition, the effects of fluorine on the morphology of films were rather similar to that of HPF_6. It is noteworthy that the morphology was even rougher, which was actually the effect of NH_4^+, as discussed earlier.

How does the amount of HPF_6 content affect the morphology of the films? When HPF_6 is added into the precursor solution continuously, the transparency of the solution reduces, proving that the size of CaF_2 particles increases. The increase of CaF_2 particle size will lead to the reduction of particle number and finally affects gel formation. As a result, the quality of the film obtained deteriorates, and the amount of pores and cracks also increases. On the other hand, however, with the increase of HPF_6 content, $Ca(NO_3)_2$ in the gel decreases, resulting in the decrease of gas released in the subsequent calcining process, which then improves the microstructure. Considering these two factors, the microstructure of films initially improves with increasing content of F and then deteriorates with even higher F content.

For films in the case of NH_4PF_6 addition in preparation, the corresponding precursor solution did not possess excellent ability of gelation. In addition, a relatively large amount of gas would be released due to the existence of NH_4^+, making the film morphology even worse.

4.2.4 Solubility of FHA Coatings

As F is smaller than the OH group in size, the corresponding lattice of FHA is smaller, compared with pure HA. The lattice constant of FA (a = 9.3684A, c = 6.8841A) is smaller than that of HA (a = 9.418A, c = 6.884A). As a solid solution of FA and HA, the structure of FHA is more compact, and the solubility of FHA is relatively lower. That would be very important for HA films showing lower solubility and longer lifetime.

Many scholars have studied the dissolution process of FHA films at different pH value and proposed many mechanisms of dissolution. Most of them have admitted that the introduction of F is an important factor for the stability of FHA.

The solubility of the films was usually measured by immersing them in covered containers with Tris solution buffered at pH 7.25, held in a 37°C water bath. Tris buffer solution can provide the same pH as the body environment and does not contain any additional ions except H^+ and Cl^-, which eliminates the influence of the deposition of other minerals. Therefore, the appearance of metal ions and anions after film immersion and the variation of their concentrations can be considered as the result of the dissolution of the thin film and thus can be utilized to evaluate the relative stability.

Figure 4.8 depicts how the fluorine content affects the solubility of the films. As shown in the figure, the greater the fluorine content, the more slowly the films dissolve. On the other hand, calcination had an effect on the solubility of the film as well, even though the F content did not alter, as shown in Figure 4.9. Besides, calcination also affected the dissolution of pure HA (Figure 4.10).

In order to explain this phenomenon, it is necessary to clarify the dissolution process of the FHA films. Generally [34], the solubility of calcium phosphate varies in the following order: ACP > HA > FA. Since FA is less soluble than HA, the substitution of F in the structure of HA, which makes FHA, would make the solubility somewhere between HA and FA. Therefore, the order

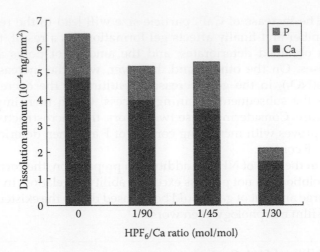

FIGURE 4.8
Dissolution amount of films with different fluorine content after 2 weeks. (From Cheng, K. et al., *J. Biomed. Mater. Res. B Appl. Biomater.*, 69B, 33, 2004.)

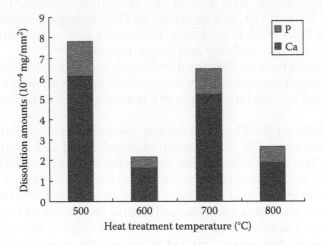

FIGURE 4.9
Dissolution amount of FHA calcined at different temperatures. (From Cheng, K., Sol–gel preparation of fluoridated hydroxyapatite films and their in vitro properties, PhD dissertation, 2003.)

of solubility should be ACP > HA > FHA > FA. In the immersion test, due to the films that usually contain some poorly crystallized calcium phosphate, which is rather similar to ACP, these calcium phosphate will dissolve first, after which the concentration of Ca, PO_4^{3-} increases and dissolution–precipitation equilibrium is achieved. As shown in Figure 4.11, the Ca and P concentrations in Tris solution varied with immersion time. After a rapid increase at the beginning, the concentrations tended to be stable.

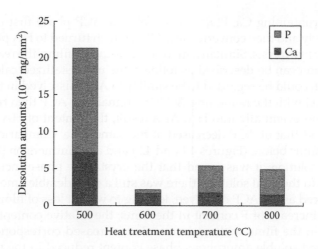

FIGURE 4.10
Dissolution amount of pure HA films calcined at different temperatures. (From Cheng, K., Sol–gel preparation of fluoridated hydroxyapatite films and their in vitro properties, PhD dissertation, 2003.)

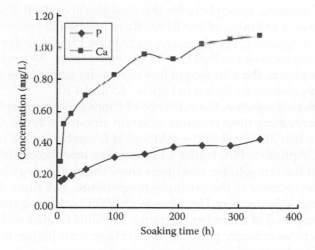

FIGURE 4.11
Ca and P concentration in the Tris solution versus different soaking time. (From Cheng, K., Sol–gel preparation of fluoridated hydroxyapatite films and their in vitro properties, PhD dissertation, 2003.)

The solubility product constant (K_{sp}) in the Tris solution after soaking was calculated, and it was found that the concentration of Ca^{2+}, PO_4^{3-}, and OH^- was much greater than the saturation concentration of HA, which meant that HA had saturated in the Tris solution. Due to the much slower growth process of HA compared with other phosphate, however, it is unlikely to produce HA directly in the HA crystal surface. Previous study has found that the HA deposition in solution containing Ca, PO_4^{3-} was as follows: ACP phase first precipitated spontaneously and then converted into OCP, which turned to HA phase finally after further hydrolysis. Similarly, in the case of apatite films, the reactions during immersion can be described as follows: the uncrystallized calcium phosphate, which could be regarded to be similar to ACP, dissolved in the solution and saturated with the remaining ACP; the remaining ACP then transformed into OCP and eventually into HA. As a result, the content of HA in the film increased and that of ACP decreased at the same time. Comparing the XRD patterns of films before (Figures 4.1 and 4.2) and after immersion (Figure 4.12) in the Tris solution, it was found that the crystalline phase increased after immersion. In the final solution, there was still a considerable amount of Ca^{2+}, PO_4^{3-} produced as the ACP dissolved, leading to weight loss of films.

With the increase of F content in the films, the relative content of crystalline phase in the films without immersion increased correspondingly (see Table 4.1), and soluble amorphous phase content reduced in the meanwhile. Moreover, some fluorine might be present in ACP, which would improve the stability of ACP [34,35]. As a result, less Ca^{2+} and PO_4^{3-} dissolution, as well as less weight loss of the films were observed.

Defects (such as cracks and pores) will increase the actual contact area of films and solution, which benefits the dissolution of ACP. Therefore, for high-temperature calcining of the films, there are two opposite effects: the amount of amorphous phase decreases so that the films dissolve less, while the contact area increases so that the films dissolve more. As a result of these two opposite effects, the film weight loss may differ against the variation of calcining temperature, as shown in Figures 4.9 and 4.10.

In the immersed solution, the existence of F ions (as the amorphous phase of fluorine-containing films contains a certain amount of F, which may produce F ions when dissolved in the solution) is favorable for the transformation from OCP phase to HA. Figure 4.13 shows the morphology of films after immersion in the Tris solution (the insets show the morphology before soaking). With the increase of the calcining temperature, HA films showed significant changes. Surfaces of HA films with 500°C calcining turned rough and were composed of a number of grains. For films treated with high temperature, there were many pores on the surface; with higher temperature, the film surface became very rough and even discontinuous.

The morphology variation of FHA films after immersion is basically the same as that of HA films (Figure 4.13). Compared with the HA films, the FHA film after 800°C calcining was relatively flat but with larger pores. With different fluorine content, the morphology of films after immersion also had some changes. Films with lower fluoride content became rougher, and more grains were observed on the surface, while films with higher fluorine content appeared rather smooth.

In conclusion, fluoride-containing films are much more advantageous with respect to stability. This advantage is due to the fact that the introduction of F not only reduces the dissolution of the crystalline phase but also increases

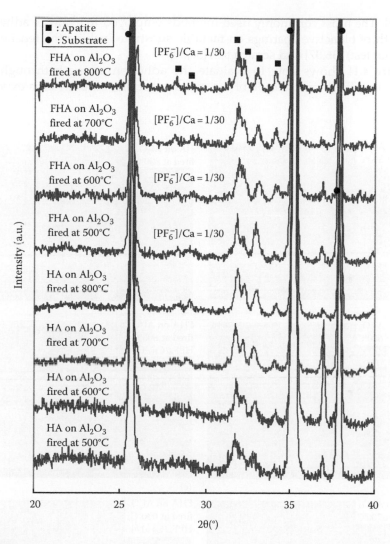

FIGURE 4.12
XRD patterns after soaking in Tris solution. (From Cheng, K., Sol–gel preparation of fluoridated hydroxyapatite films and their in vitro properties, PhD dissertation, 2003.)

the content of crystalline phase in the films. As a result, the overall stability of the bioactive films has been improved.

4.2.5 Adhesion Strength of FHA Coatings

Adhesion strength between the film and substrate is a critical factor in successful implantation and long-term stability of any coated implant. The introduction of F into HA films has some influence on the adhesion strength as well. The most commonly used methods employed to evaluate adhesion strength of bioactive coatings on metallic substrate include the tensile test (pull-out test) [36,37] and scratch test [38,39].

Figure 4.14 shows how to evaluate the adhesion strength through the tensile test. First, on one side of the metal substrate, bare metal is exposed

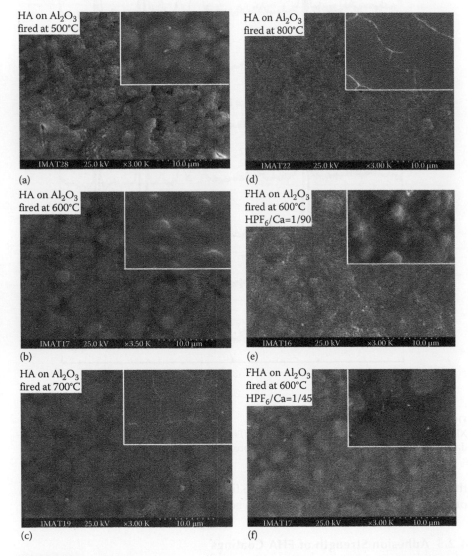

FIGURE 4.13
SEM micrographs of the films after 2 week immersion in Tris solution. (a) HA fired at 500°C, (b) HA fired at 600°C, (c) HA fired at 700°C, (d) HA fired at 800°C, (e) FHA fired at 600°C with $HPF_6/Ca = 1/90$, (f) FHA fired at 600°C with $HPF_6/Ca = 1/45$.

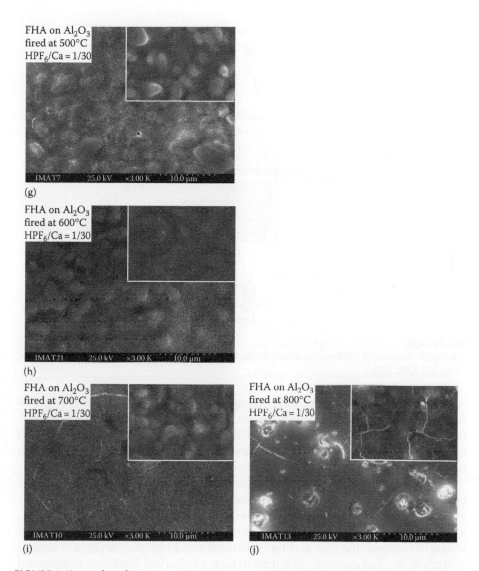

FIGURE 4.13 (continued)
SEM micrographs of the films after 2 week immersion in Tris solution. (g) FHA fired at 500°C with $HPF_6/Ca = 1/30$, (h) FHA fired at 600°C with $HPF_6/Ca = 1/30$, (i) FHA fired at 700°C with $HPF_6/Ca = 1/30$, (j) FHA fired at 800°C with $HPF_6/Ca = 1/30$. (From Cheng, K., Sol–gel preparation of fluoridated hydroxyapatite films and their in vitro properties, PhD dissertation, 2003.)

through polishing, and then two aluminum rods with the same diameter and length are stuck to both sides coaxially, and the tensile strength test is carried out with a universal testing machine. The force at which fracture occurs as the film side is divided by the contact area, and the result is considered as the bonding strength between film and substrate.

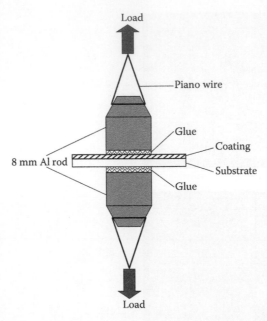

FIGURE 4.14
Schematic diagram of the bonding strength characterization. (From Cheng, K. et al., *Thin Solid Films*, 517, 5361, 2009.)

Figure 4.15 shows the bonding strength of films with different content of fluorine. Obviously, FHA films showed stronger adhesion than pure HA films. When F content was relatively low, adhesion increased rapidly with increasing F content. With even higher F content, the adhesion strength tapered off gradually.

However, as shown in the morphology of the films, there are pores and cracks in the film. That will lead to inevitable glue infiltration, which may eventually result in inaccuracy in the pull-out test [41,42]. Comparatively, a scratch test may avoid such adverse influences. In this test, the adhesion strength between the film and the substrate was evaluated using a scanning scratch tester. For example, in a typical test, a spherical Rockwell C diamond stylus of 15 μm radius with a progressive load from 0 to 1000 mN was used. The stylus scanned the film surface at a speed of 2 μm/s while maintained scanning amplitude of 50 μm perpendicular to the scratching direction. The load at which the first damage is made in the film is called the "lower critical load," and the load at which the total peeling-off of the film from the substrate occurs is referred to as the "upper critical load." In hard films, usually the "lower critical load" is used to measure the cohesion of the film. To evaluate the film/substrate adhesion, the "upper critical load" was chosen as an indication of the adhesion strength.

A typical scratch track of the film is shown in Figure 4.16, and the corresponding friction response was given in terms of relative voltage as a

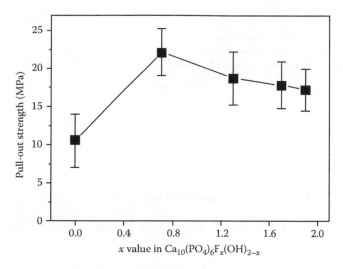

FIGURE 4.15
Bonding strength between the film and substrate. (From Cheng, K. et al., *Thin Solid Films*, 517, 5361, 2009.)

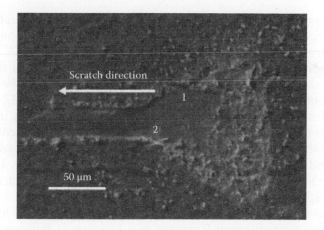

FIGURE 4.16
Scratch track of an HA coating. (From Zhang, S. et al., *Surf. Coat. Technol.*, 200, 6350, 2006.)

function of load. Increasing voltage corresponded to increasing coefficient of friction. Figure 4.17 is that for HA and FHA fired at 600°C. At the beginning of the scratch, the coefficient of friction increased as the load increased due to cohesive failure because of the "soft" nature of the film. Before point 1 (Figures 4.16 and 4.17), there were fluctuations as a result of the surface roughness. After point 1, the indenter started to plow into the film, resulting in a steeper increase in the coefficient of friction. As the load increased to point 2, or 370 mN for pure HA (curve a), the indenter

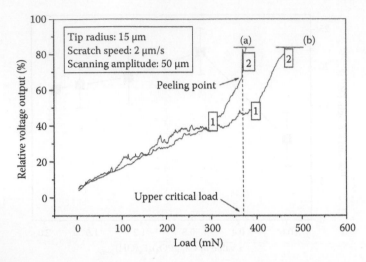

FIGURE 4.17
Coefficient of friction in terms of relative voltage as a function of normal load while scratching (a) pure HA coatings and (b) FHA coating on Ti6Al4V substrate. (From Zhang, S. et al., *Surf. Coat. Technol.*, 200, 6350, 2006.)

completely peeled off the film and scratched onto the substrate causing an abrupt increase in friction. This load was taken as an indication of the adhesion strength for the film. Curve b records that for FHA. After the indenter digs into the film (note the change of the slope in the curve), there is no abrupt increase in friction until the indenter reaches the substrate at about 470 mN. Comparing curves a and b in Figure 4.17, b appears smoother (less fluctuation in friction before the indenter digs in) and the film adheres to the substrate better (slower slope after the indenter digs in and lack of abrupt increase). The abrupt increase of friction signals meant a sudden and brittle peeling-off of the film from the substrate. The lack of this abrupt change demonstrated that b has more ductile interface and thus better film–substrate bonding [43].

Figure 4.18 summarizes the adhesion strength ("upper critical load") of all the FHA films with different F content as a function of fluorine and firing temperature. Both fluorine content and firing temperatures had significant effect on the adhesion strength: with increasing fluorine concentration or firing temperature, there was an obvious increase of the critical load. For the same fluorine content films, higher firing temperatures gave rise to jumps in adhesion strength. At the same firing temperature, the general trend was that adhesion increases with fluorine content. However, the increase was more profound at higher temperatures. At 500°C, the critical load only increased marginally with fluorine content, but at 600°C and 700°C, the curves went up rapidly with fluorine content. In addition, only when the F content was relatively low, the adhesion strength increased rapidly. If the F content exceeded a certain value, the increase of strength was not so obvious.

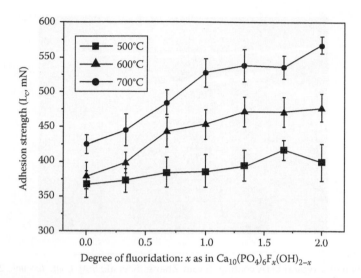

FIGURE 4.18
Adhesion strength of pure HA and fluoridated HA coatings on Ti6Al4V substrate as indicated by upper critical load in scratch test. Firing temperatures are indicated. (From Zhang, S. et al., *Surf. Coat. Technol.*, 200, 6350, 2006.)

Films adhere to substrates through either mechanical interlocking or chemical bonding or both. Since all substrates have the same finishing, mechanical interlocking is deemed identical; thus, the increase in adhesion strength is attributed to stronger chemical bonds developed at the film–substrate interface during firing.

Figure 4.19 shows the interface microstructure of an FHA film fired at 600°C. The cross section micrograph showed dense and uniform structure, and no delamination and/or cracks at the interface were observed.

Figure 4.20 shows a typical EDS cross-sectional elemental distribution of Ca, O, F, and P in an FHA fired at 600°C. Three regions were marked: the substrate (Rs), the film (Rc), and in between was the transition (Rt). Within the transitional region, the Ca and P concentrations decreased significantly toward the Ti substrate, while the F and O decreased slowly and gradually from the film to the substrate. In other words, there existed relatively large amount of F and O in the transition region. It showed that, along the cross section, certain Ti–P–Ca–O–F chemical bonds formed at the interface [37,44,45] that could contribute to the increase in adhesion. The formation of chemical bonds at the interface may be also responsible for the change in failure mode (from brittle to more ductile).

Figure 4.21 shows the cross-sectional HRTEM image of an FHA film. There are multiple sets of fringes, representing the different lattices of apatite crystals. As seen from the image, the grain size is about several nanometers to several tens of nanometers. Some typical apatite lattices, such as [211], [112], and [111] could be found in the high-resolution images.

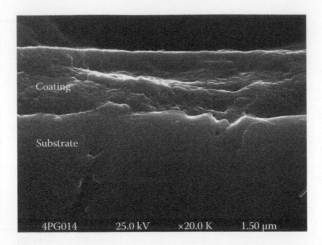

FIGURE 4.19
Cross section of a typical FHA coating. (From Zhang, S. et al., *Surf. Coat. Technol.*, 200, 6350, 2006.)

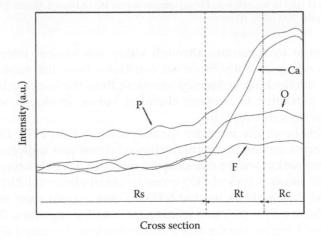

Cross section

FIGURE 4.20
EDS element analysis of the cross section of FHA fired at 600°C. (From Zhang, S. et al., *Surf. Coat. Technol.*, 200, 6350, 2006.)

Figure 4.22 is the electron diffraction diagram (r = 0.29 μm) of the cross section. The value of $R_1^2:R_2^2:R_3^2:R_4^2:R_5^2$ basically equaled to 3:4:7:9:12, which was consistent with the increase of R^2 of {hk0} lattices in the hexagonal system. That proves that the grains belonged to the hexagonal system. As FHA is hexagonal, the result is in well agreement with the presence of lattice fringes shown in Figure 4.21. In addition, the presence of diffraction rings indicates that there are a considerable number of grains present, i.e., a polycrystalline structure.

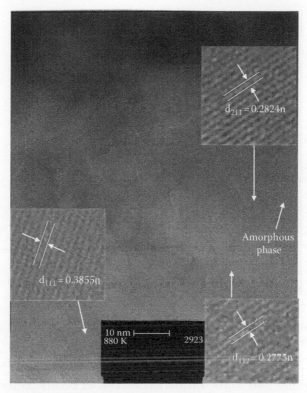

FIGURE 4.21
Cross-sectional HRTEM photo of the film. (From Cheng, K. et al., *Surf. Coat. Technol.*, 198, 242, 2005.)

As discussed previously, there might be some chemical bonds formed at the interface; the HRTEM image of the interface between the film and the substrate in Figure 4.23 gives more direct evidence. The interface showed that the film and substrate bond together closely without any gap. Part of the lattice fringes of the apatite can be found in Figure 4.23, such as [002]. More important, from the microdiffraction photo of a typical grain, $R_1^2 : R_2^2 : R_3^2 : R_4^2 : R_5^2 : R_6 2 : R_7^2 = 1:2:4:5:8:9:10$ could be calculated. That indicated such grains were tetragonal. Considering the previous XRD and EDS results, the grains were rutile TiO_2.

According to the previous analysis, the structure of the films can be described as follows, as shown in Figure 4.24: crystalline FHA layer with thickness of about 1.5 μm, FHA with low fluoride content, and the transition layer containing fluorine and titanium dioxide.

The formation of this structure may be explained as follows: First, chemi- and physisorption process occurred during dipping and drying. Since native oxide (TiO_y, $y \leq 2$) forms spontaneously on titanium and titanium alloy surface upon exposure to air, hydroxide ions and water molecules are absorbed

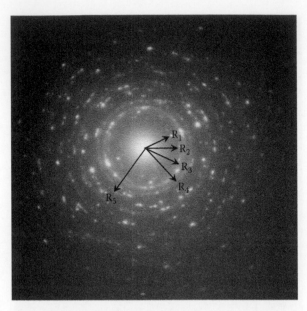

FIGURE 4.22
Selected area electron diffraction pattern of the film at cross section. (From Cheng, K. et al., *Surf. Coat. Technol.*, 198, 242, 2005.)

by Ti cations that lead to the formation of Ti–OH bonds on the outermost surface [46]. XPS depth profiling confirmed a 5–8 nm titanium oxide layer as soon as the Ti6Al4V substrates were prepared. In dipping sols, fluorine ions released by F-precursor react with Ca^{2+} to form nanocrystalline CaF_2 [29]. The F in nc-CaF_2 easily forms hydrogen bond with H in the OH group, i.e., O–H$\cdots$F–Ca–F [44]. Therefore, when the Ti6Al4V substrate is immersed into the dipping sol, nc-CaF_2 could be easily adsorbed onto the substrate surface through the formation of hydrogen bond. The higher the fluorine concentration in the dipping sol, the more nc-CaF_2 forms and adsorbs on the substrate surface. It has been reported that the adsorbed nc-CaF_2 on the titanium alloy substrate surface attracts more O near the interface [44]. On the other hand, due to the amphoteric property of Ti–OH, the substrate surface is positively charged in current dipping sol (pH < 1). Thus, surface adsorption of electronegative groups, such as NO_3^-, PO_4^{3-} groups, will follow. Following the adsorption, an important step is the diffusion during firing when the surface-adsorbed nc-CaF_2, PO_4^{3-} groups, etc., react with titanium to form certain chemical bonds if the temperature is high enough to activate the formation process. The diffusion of fluorine and oxygen from the film into the transition region encourages this reaction and results in the accumulation of O and F in this region (Figure 4.20). Since higher fluorine content in the dipping sol attracts more O near the interface [44], firing in an open atmosphere provides inexhaustible oxygen for the process. As a result,

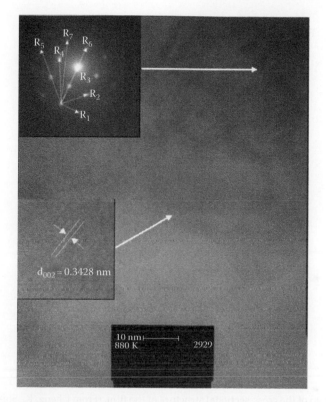

FIGURE 4.23
HRTEM photo of the film–substrate interface. (From Cheng, K. et al., *Surf. Coat. Technol.*, 198, 242, 2005.)

TiO$_2$ or even complex Ti–P–O–F–Ca bonds may be formed in the transitional region that helps improve the adhesion strength. Higher firing temperature obviously promotes diffusion of fluorine and oxygen as well as provides activation for the chemical bonding process, thus leading to higher adhesion strength. Previous studies by Filiaggi et al. [47] also indicated that similar interactions at the interface of the substrates and the dipping sol have great influence on the ultimate adhesion property.

Residual stress at the interface resulting from the deposition process may exert adverse effect on adhesion strength. The difference in the thermal expansion coefficient for the film and substrate results in thermal mismatch that in turn contributes negatively to adhesion strength via increased residual stress [48]. The incorporation of fluorine ion into apatite structure decreases the coefficient of thermal expansion (CTE) from $15 \times 10^{-6}/°C$ to $9.1 \times 10^{-6}/°C$ when HA is changed to FA. The latter is much closer to that of the Ti-alloy substrate ($8.9 \times 10^{-6}/°C$). In fact, the mismatch in CTE between HA and Ti-alloy is 68.5% while only 2.2% between FA and Ti6Al4V. As a result, the incorporation of fluorine in HA reduces the thermal mismatch and

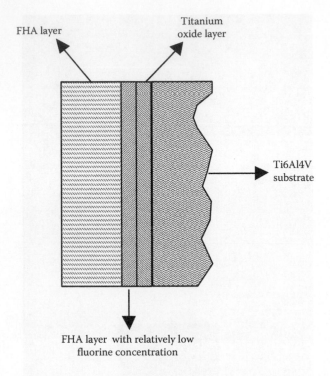

FIGURE 4.24
Schematic photo of the cross-sectional structure of the film. (From Cheng, K., Sol–gel preparation of fluoridated hydroxyapatite films and their in vitro properties, PhD dissertation, 2003.)

thus residual stress and consequently improves the adhesion strength. The reduction in residual stress also contributes to better ductility [49].

Both tensile test and scratch test reveal that the film adheres to Ti-alloy substrate better as the fluorine concentration increases in the film. This adhesion enhancement is much more profound at higher firing temperature. The increase in adhesion likely comes from the formation of chemical bonding at the interface and the relief of thermal mismatch stress due to the incorporation of F into the HA structure. As a result, the film and substrate interface becomes more ductile.

4.2.6 In Vitro and In Vivo Evaluations

From cell proliferation point of view, pure HA is preferred because large dissolution of HA leads to increases of Ca and P ion concentrations that favor cell activities [50,51]. Although long-term stability (dissolution resistance) and enhanced cell attachment are achieved on FHA films, cell proliferation is compromised because release of Ca and P ions is impaired [33]. To strike a balance between these two polarities, FHA films with moderate F content

TABLE 4.2

Sample Annotation and F Content in the Coatings

	F0	F067	F148	F190	F200
HPF_6/Ca ratio	0	1/90	2/90	3/90	4/90
Molecular formula	$Ca_{10}(PO_4)_6$ $(OH)_2$	$Ca_{10}(PO_4)_6$ $(OH)_{1.33}F_{0.67}$	$Ca_{10}(PO_4)_6$ $(OH)_{0.52}F_{1.48}$	$Ca_{10}(PO_4)_6$ $(OH)_{0.10}F_{1.90}$	$Ca_{10}(PO_4)_6F_{2.00}$

Source: Cheng, K. et al., *Biomaterials,* 26, 6288, 2005.

are tested and they do demonstrate good performances both in vitro and in vivo [52].

Here cell attachment and proliferation on different FHA films (the molecular formula of some typical FHA films were obtained and tabulated in Table 4.2) and cell culture in leaching out solutions were measured and assayed. The experiment process is conducted as follows.

The cells used in the culture test were the osteoblast progenitor cells derived from adult rabbit bone marrow, extracted from adult rabbit back shank. The cells were washed twice in phosphate buffer solution (PBS) and collected via centrifugation. In the standard incubation condition (5% CO_2, 37°C), they were incubated for about 1 week in different cultural media containing Dulbecco's Modified Eagle's Medium (DMEM), dexamethasone, vitamin C, and β-sodium glycerol-phosphate. Finally, the cells developed into osteoblast-like cells. After sterilization in 120°C water steam for 20 min, the samples were placed in 24-well plates for osteoblast-like cell implantation at a set density of 1×10^4 cells/cm^2.

After implantation, the films were taken out and the osteoblast-like cells attached on the films were fixed and observed with an SEM. As shown in the typical SEM micrographs (Figure 4.25), the cells attached well on the film surface. They spread well, and the filopodia were clearly seen, indicating good

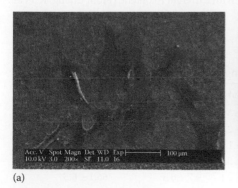
(a)

(b)

FIGURE 4.25
Typical morphology of cells attached on the films: (a) F148 and (b) F0. (From Cheng, K. et al., *Biomaterials,* 26, 6288, 2005.)

livelihood. The cell numbers were evaluated by hemocytometer. Figure 4.26 plots the cell density as a function of culture time for films with different F content but the same initial cell implanting density of 1×10^4 cell/cm². In the pure HA film (F0), the cell concentration reached about 5×10^4 cell/cm²; for F067 and F148, only about $(3–4) \times 10^4$ cell/cm²; for F190, it also reached about 5×10^4 cell/cm², which is similar to pure HA; as for F200, the concentration reached 11×10^4 cell/cm².

Flow cytometry analysis indicated (Table 4.3) the percentage of cells in different cell cycles (G0G1: pre-DNA synthesis or resting; S: DNA synthesis phase; G2M: post-DNA synthesis and mitosis). The percentage of the cells in every cycle is different: for all the films and culture time involved, more cells

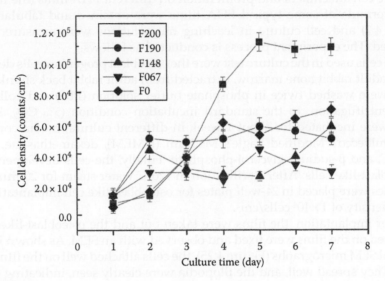

FIGURE 4.26
Cell growth curve of cells attached on coatings with different F content. (From Cheng, K. et al., *Biomaterials*, 26, 6288, 2005.)

TABLE 4.3

Cell Cycles of Cells on Different Coatings at Different Time Periods

Period of Cell Cycle	G0G1 (%)	S (%)	G2M (%)	Proliferation Index (%)
F190, 4 day	50.2 ± 4.8	49.8 ± 3.2	0	50.2 ± 4.5
F148, 4 day	60.0 ± 5.2	34.4 ± 2.6	5.6 ± 0.3	65.6 ± 5.4
F0, 4 day	86.5 ± 6.8	11.2 ± 1.3	2.3 ± 0.2	88.8 ± 5.7
F190, 7 day	32.9 ± 2.1	67.1 ± 5.3	0	32.9 ± 2.9
F067, 7 day	56.4 ± 4.3	34.8 ± 2.6	9.7 ± 0.6	66.1 ± 6.5
F0, 7 day	88.2 ± 5.6	11.8 ± 1.3	0	88.2 ± 5.1
Control	99.5 ± 6.3	0	0.5 ± 0.1	100 ± 6.4

Source: Cheng, K. et al., *Biomaterials*, 26, 6288, 2005.

were found in the S period when they are cultured on the films than those on the control. Compared to the control, the difference is significant.

For effects of leaching out of the film, the samples were immersed in DMEM solution at 37°C for 4 days at a ratio of 10 mL/cm^2 (DMEM volume per unit film area). The immersed solution was used for the incubation of osteoblast-like cells. Untreated DMEM solution was also processed as the control. The MTT statistics (at $p < 0.05$) of the cells cultured for 1–5 days are shown in Figure 4.27. At first, all the cells showed relatively low absorbance, indicating the cells were adapting to the new environment. Thus, the number of living ones fell below that of the control. After 3 days, the absorbance increased drastically on all the films, meaning the number of living cells increased, i.e., the cells proliferated. At this moment, leaching out from films with different F content showed significant difference (considering at $p < 0.05$) in cell proliferation: pure HA and FHA with less F content (F0, F067, and F148) showed even better cell proliferation than control (higher absorbance). On film with very high F content or pure FA (F190 and F200), less cell proliferation was seen than the control.

When the osteoblast cells are cultured on the FHA film, there are two different ways in which F affects the behavior of the cultured cells. First, surface property variations due to F existence in FHA films could have a direct influence on interactions between the cell and the film; second, changes in the leaching outs of the films due to the F incorporation could have an influence on cell growth by means of ion concentration changes in the culture medium. As such, the cell behavior will be affected. Considering these two factors, it is obvious that only the surface layer of the film is involved.

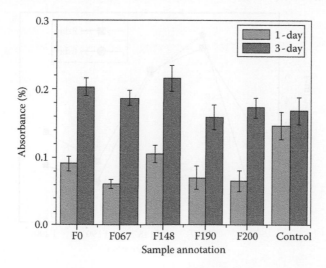

FIGURE 4.27
MTT results of cells cultured in leaching-out solutions of the coatings (considered at $p < 0.05$). (From Cheng, K. et al., *Biomaterials*, 26, 6288, 2005.)

Based on the dissolution results obtained in a pH condition similar to that of the cell culture media, the maximum depth involved can be estimated through calculation based on weight loss. The depth is calculated to range from about 170 to 68 nm, inversely proportional to the F content.

Apparently, all the films showed good biocompatibility because osteoblast-like cells could attach and spread well on the films, as shown in the SEM results (Figure 4.25). In the cell growth curve (Figure 4.26), when the films had low F content (F067 and F148), less cell attachment was observed as compared to pure HA. Only when the concentration was high (to the extent of F190), did it show positive effect on cell attachment: F190 had a cell attachment similar to that of pure HA, while F200 produced the maximum. Further cell cycle analysis results are tabulated in Table 4.3. With the increasing F content in the film, the percentage of cells in S period increased, indicating that the cells attached on the film seemed to become more active in the DNA replication process. From both the cell attachment and cell cycle results, it can be concluded that it is quite clear that the variations of cell behaviors are greatly affected by the F content in the films.

As widely accepted, rough surfaces promote cell attachment. However, even F067 and F148 have rougher surfaces than all the other films and the number of cells attached is still less (Figure 4.26). The same situation happened to F190 and F200, where F190 seems to have rougher surface than F200, but the cells attached on F190 are significantly less than F200. Obviously, the cell attachment differences here are not mainly induced by the roughness variations.

The aforementioned cell behavior variations could have resulted from the surface property change owing to F incorporation. Figure 4.28 gives the zeta

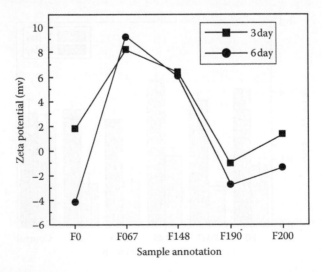

FIGURE 4.28
Zeta potential of FHA coatings after different times of immersion in Tris solution. (From Cheng, K. et al., *Biomaterials*, 26, 6288, 2005.)

potential results for 3-day and 6-day soaking. The tendency of the zeta potential against the F content coincided with the trend of the cell attachment as a function of F content: roughly lower potential corresponded to higher number of attachment. The variation of zeta potential may influence the adsorption of Ca^{2+} ions, which, in turn, affects the adsorption of extracellular matrix (ECM, including osteocalcin and osteopontin) [53]. Consequently, the attachment of the cells is affected because osteoblast-like cells actually attach on the ECM preadsorbed on the surface of the sample.

The influence of F on surface potentials can be attributed to the FHA film's different dissolution behavior. According to the apatite self-inhibition dissolution model [54,55], after the initial dissolution of a small portion of apatite, a steady calcium-rich layer consisting of Ca^{2+} ions and their anionic counterparts will first form and then control the further dissolution of apatite. If this "calcium-rich layer" is regarded as the "compact layer" in the double layer theory of colloids, the exact content of the "calcium-rich layer" will affect the zeta potential of apatite. With the F existence in the surface layer of apatite, more Ca^{2+} ions will be attracted in the "calcium-rich layer" due to the strong electronegativity of F, leading to increase of the zeta potential (F067 and F140). However, even higher F content will result in solubility decrease, which will in turn produce a low Ca^{2+} concentration in the solution, leading to decrease of the zeta potential (F190 and F200).

When a film exists in the body, the dissolution of the film will affect the surface potential as mentioned earlier, as well as the long-term stability [5,6]. The dissolution also produces ion concentration variations around the vicinity of the film in the initial time after implantation. This could influence new bone formation. One approach to getting insight into the influence could be an in vitro test through observing behavior of osteoblast-like cells in a culture medium with leaching out, which is prepared by soaking the film in the medium.

The MTT measurement demonstrates that the leaching outs of the different films do have different effects on cell proliferation. The number of living cells (or percentage of absorbance) for pure HA is rather similar to that of FHA with low F concentration, while for film with high F concentration, the number of living cells is significantly lower than pure HA.

Based on the relation between F content and dissolution, the maximum concentration of F ions is estimated to be $\sim 6.3 \times 10^{-6}$ mol/L, which falls below the determination limit of normal measurement methods such as ion-selective potentiometry. In such a low concentration, it is impossible for F ions in the culture medium to play a significant role in affecting the behavior of osteoblast-like cells [56]. Comparing the Ca^{2+} dissolution with that of the MTT result (Figure 4.27), it is seen that as the concentration of Ca^{2+} increases, the number of living cells increases in cell culture experiments. This is in agreement with the report where the existence of Ca^{2+} benefits osteoblast cell proliferations [57]. The aforementioned variation of the F content with cell proliferation could be attributed to Ca^{2+} ion concentration changes that depend on the F content in FHA film, i.e., an increase in F content in the film leads to a decrease in Ca^{2+} release, which, in turn, decreases cell proliferation.

Although the situation for the cell growth in the body is different from that in the culture media, the aforementioned analysis may help in understanding influences of the films as comprehensive as possible. The cell culture results demonstrate that the F content in FHA films affects the behaviors of cultured cells in different ways: high F content produces a low surface potential, which favors cell attachment, but at the same time, a reduction in Ca^{2+} release in a culture medium could inhibit cell proliferation. Previous work shows that high F content in FHA films has low solubility in SBF, which is good to guarantee long-term stability; also, a certain F content promotes apatite deposition in SBF [11]. From the film design point of view, all these effects of F content do not necessarily function in line with one another. However, all these effects are important in clinical applications and thus need to be taken into consideration simultaneously and holistically. Based on the experimental data, FHA films with moderate F content, $Ca_{10}(PO_4)_6F_{0.67-1.48}(OH)_{0.52-1.33}$, may have a more promising potential. Savarino [52] also showed that the $Ca_{10}(PO_4)_6F_{1.0}(OH)_{1.0}$ film in the in vivo test was better than both $Ca_{10}(PO_4)_6(OH)_{2.0}$ and $Ca_{10}(PO_4)_6F_{2.0}$ films. The aforementioned analysis on F influences and optimized F content serves well in explaining Savarino's results.

In general, FHA films with different F content prepared by sol–gel method show good cytocompatibility. F in the FHA films results in low zeta potential that promotes cell attachment. High F content also results in decreased Ca^{2+} release in culture media due to reduced solubility, and, as a result, cell proliferation is inhibited. For clinical application, it is suggested that a moderate content of F, such as $Ca_{10}(PO_4)_6F_{0.67-1.48}(OH)_{0.52-1.33}$, is most suitable as a compromise among cell attachment, cell proliferation, apatite deposition, and dissolution resistance.

4.3 Sol–Gel-Derived Zinc- and Fluorine-Doped Hydroxyapatite Coatings

In vitro and in vivo studies revealed that reactions such as dissolution, precipitation, ion exchange, and structural rearrangement occurred at the material–tissue interface [58], where the released ions can create solution-mediated effects on cellular activity. As an essential trace element in the body, zinc is found to have positive roles in promoting bone formation and inhibiting bone resorption [59–61]. Several studies have been carried out on bulk zinc-containing materials based on coprecipitated HA [62–64], and the dissolution or biodegradation of zinc-containing HA could release zinc ions and show beneficial effects on bone formation. Therefore, based on the previous discussion, the incorporation of Zn into FHA film may show even better effect on cellular responses.

4.3.1 Direct or Composite Doping

The easiest way to prepare Zn-doped FHA films on Ti-alloy is to incorporate Zn-containing compounds into the sol, so that the stimulating effect of Zn and the low solubility of FHA could be imparted into one film [65]. It is found that Zn can easily be enriched on the surface of the films or exists in grain boundaries. Although such Zn distribution favors the initial Zn release in vivo to act as an agent for enhancing bone formation, large amount of Zn incorporation renders undesirable porosity. Moreover, the fast release of Zn is rather uncontrollable and therefore would not have the desired pharmacological effect.

In contrast, Zn ions can directly locate within the crystal structure of β-tricalcium phosphate (β-TCP), which can then serve as an ideal slow release carrier of Zn. Ito and colleagues [66,67] have reported that low levels of Zn can be incorporated into β-TCP (ZnTCP) ceramics to enhance bone formation. However, these ZnTCP ceramics lack sufficient mechanical and chemical stability needed for osseointegration and therefore cannot be directly used as coating materials.

In order to compromise Zn release, bioactivity, and long-term stability in a coating, a bilayer coating that consists of a chemically stable FHA bottom layer coupled with a top layer with Zn-releasing ability was proposed. The top layer, Zn-releasing fluoridated hydroxyapatite (ZTFHA) coating, was prepared by sol–gel method, in which nanosized ZnTCP particles were dispersed as Zn carriers within an FHA matrix [68].

In such coating, nanosized ZnTCP particles acted as Zn carriers and Zn mainly existed in the β-TCP phase dispersed among FHA matrix. ZTFHA coatings demonstrated slow and sustained release ability because soluble ZnTCP particles were involved. Preliminary cell culture results showed an obvious increase in cell viability on the ZTFHA coatings compared with an FHA coating without Zn, indicating that capacity for Zn release ability clearly had a stimulatory effect on osteoblast cells. In vivo testing showed that the osseointegration was obviously improved with the ZTFHA coating compared with an FHA coating. This novel ZTFHA coating shows promise for establishing early cementless fixation of orthopedic implants.

4.3.2 Sol–Gel Preparation of Zinc-Containing FHA Coatings

In a typical sol–gel preparation of ZnFHA films [69], $Zn(NO_3)_2 \cdot 6H_2O$ was dissolved in absolute ethanol to form 0.1 M/L solution, and then designed amount of $Zn(NO_3)_2$ solution was added into the typical sol of FHA to form a series of dipping sols. Extra ethanol was added to adjust the Ca ionic concentration to be the same in different dipping sols and was used for film preparation. As for ZTFHA films, the sol–gel preparation is even simpler, just add designed amount of nano-ZnTCP powder into the typical sol of FHA and then use the obtained sol for film preparation. It is noteworthy that after nano-ZnTCP addition, the sol transforms into a much thicker "colloid sol," and as a result, the times of coatings could be reduced to once or twice to a film thickness of 1–2 μm.

4.3.3 Microstructures of Zinc-Containing FHA Coatings

Figure 4.29 shows the SEM micrographs of the ZnFHA films [65]. The morphology (Figure 4.29a) of the FHA film was dense. As Zn was incorporated, the film became porous and more porous with increasing Zn content. As for the films with composite Zn doping, all of the ZTFHA films (Figure 4.30) exhibited heterogeneous and rough morphology [69]. Most parts of the coating were covered by island-like ZnTCP agglomerates, which were embedded in the FHA matrix and existed stably. The morphology of the resulting coating showed a more heterogeneous and rougher surface than that of an FHA

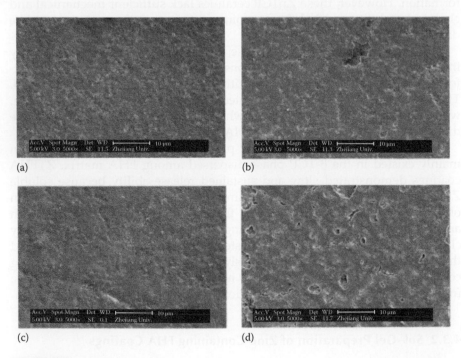

(a)   (b)

(c)   (d)

FIGURE 4.29
SEM micrographs of the 600°C fired films with different Zn content. (a) Zn/Ca = 0, (b) Zn/Ca = 0.0046, (c) Zn/Ca = 0.0077, and (d) Zn/Ca = 0.0093. (From Miao, S. et al., *Surf. Coat. Technol.*, 198, 223, 2005.)

FIGURE 4.30
Scanning electron microscopic micrographs of ZTFHA coatings: (a) ZTFHA-1, Zn/(Zn+Ca) = 0.03; (b) ZTFHA-2, Zn/(Zn+Ca) = 0.06; and (c) ZTFHA-3, Zn/(Zn+Ca) = 0.09. Zn/(Zn+Ca) means the amount of Zn in ZnTCP; the amount of ZnTCP added was fixed at 67.2 mg/mL. (From Miao, S. et al., *J. Am. Ceram. Soc.*, 94, 255, 2011.)

coating. The surface tension driving force during the solidification of the liquid layer may cause the accumulation of the particles on the surface. The accumulation degree is closely related to the particle size of ZnTCP, which decreases with increasing Zn content in ZnTCP. This leads to a relatively smooth surface morphology and a uniform Zn distribution in the ZTFHA-3 coating (Figure 4.30c).

4.3.4 Zinc Release Behavior

Zn has been shown to be capable of increasing osteoblast proliferation, biomineralization, and bone formation [59,70]. However, the pharmacological performance of zinc ions is dependent on the release behavior. Generally, the initial burst release of ions in the body may induce adverse reactions. Yamamoto evaluated the cytotoxicity of zinc salts using the colony formation method and found that the 50% inhibitive concentration of zinc for

MC3T3-E1 cells cultured in a modified Eagle's medium + 10% fetal bovine serum (FBS) was about 5.85 mg/L [71]. Thus, the release process should be well adjusted to a slow and sustained rate, which is required during bone tissue regeneration.

The soaking test results (Figure 4.31) proved that the ZnFHA films are able to release zinc [72]. The zinc release profile of the ZnFHA films showed that Zn initially released fast within the first 16 h and then slowed down. The zinc concentration of ZnFHA-2 in Tris solution was higher than that of ZnFHA-1, which was consistent with the designed amount. This indicates that the release of zinc can be controlled by zinc amount in the films. After soaking for 168 h, the zinc of ZnFHA-1 was almost released out, while the ZnFHA-2 could still release zinc.

Zn release from ZTFHA coatings was time and load dependent, according to the Zn concentration variation in the Tris-buffered solution (Figure 4.32) [69]. The Zn release process took place slowly and was sustained up to 275 h. The release process could be regulated by changing the Zn content of the ZnTCP particles. In the ZTFHA-3 coating, the Zn content was about 1.7 times that of the ZTFHA-2 coating, while Zn release was only 1.1 times higher than that from the ZTFHA-2 coating. Hence, the ZTFHA-3 coating had a slower and more sustained Zn-releasing capacity.

For ZTFHA films, Zn release was realized through the dissolution of the ZnTCP phase. Thus, the variation in Zn content of the ZnTCP can modify the Zn release behavior. Compared with that realized through

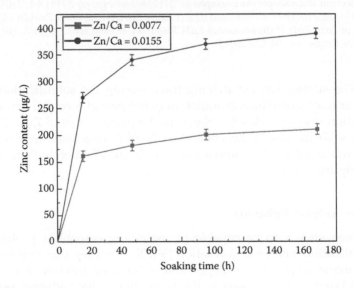

FIGURE 4.31
In vitro zinc release profile of ZnFHA coatings. (From Miao, S. et al., *J. Mater. Sci. Mater. Med.*, 18, 2101, 2007.)

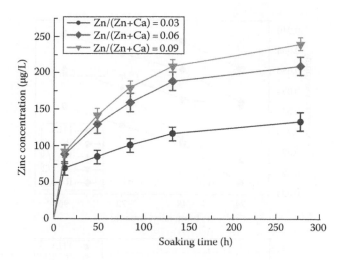

FIGURE 4.32
In vitro Zn release profiles for ZTFHA coatings. (From Miao, S. et al., *J. Am. Ceram. Soc.*, 94, 255, 2011.)

direct incorporation of Zn into FHA coatings [59,73], the Zn release process of ZTFHA coating is slower and more sustained and therefore more desirable.

In a more detailed comparative immersion test, ZTFHA film showed quite different Zn-releasing behavior from the control ZnFHA film [74]. In Figure 4.33a, during the first 96 h, for ZnFHA film, the Zn concentration increased rapidly during the initial 48 h and then tapered off. For ZTFHA films, however, the 24 h initial release of Zn was rather slow for films with Ca_{ZnTCP}/Ca_{FHA} (R) of 1/8 and 1/4, but became similar to ZnFHA when R reached 1/2 and 3/4. During the first 96 h, ZTFHA films showed almost linear increases of Zn content. The end concentrations of Zn after each 96 h cycle were also compared in Figure 4.33b. Obviously, ZnFHA film released most of the Zn during the first 96 h, and after that, during the rest cycles, very few Zn ions were leached. While ZTFHA films showed quite different behaviors: for those with R equal to 1/8, 1/4, and 1/2, almost unchanged Zn concentrations were observed at the end of each cycle. But an obvious concentration decrease was recorded for that with R equal to 3/4.

The corresponding Ca/Zn ratios are also plotted in Figure 4.33c; for ZnFHA film, the Ca/Zn ratios kept almost unchanged at the ends of each cycle. While for ZTFHA films, the Ca/Zn ratios varied greatly; for coatings with R of 1/8, 1/4, and 1/2, the ratios after the first 96 h were rather similar, while they increased greatly after the second cycle and dipped a little after the third cycle. Particularly, for that with R of 3/4, Ca/Zn ratios remained steady after each cycle, and the values were similar to those of ZnFHA.

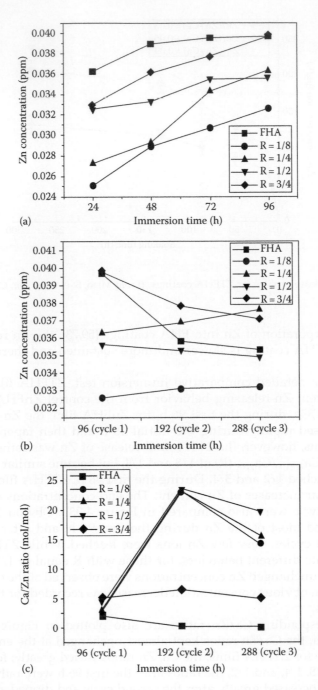

FIGURE 4.33
Immersion results of ZTFHA coatings. (a) Zn concentration variations during the first 96 h, (b) Zn concentrations after different immersion cycles, and (c) Ca/Zn ratio variations after different immersion cycles. (From Cheng, K. et al., *Thin Solid Films*, 519, 4647, 2011.)

TABLE 4.4

Designed Zn Content in the Final Coatings

	Zn/(Zn+Ca) in Sol (%)	Zn/(Zn+Ca) in ZnTCP (%)	Ca_{ZnTCP}/Ca_{sol} (R)	Zn/(Zn+Ca) in Coatings (%)
FHA	1.0	N/A	0	1.0
ZTFHA, R=1/8	0	6.0	1/8	0.7
ZTFHA, R=1/4	0	6.0	1/4	1.3
ZTFHA, R=1/2	0	6.0	1/2	2.1
ZTFHA, R=3/4	0	6.0	3/4	2.7

Source: Cheng, K. et al., *Thin Solid Films*, 519, 4647, 2011.

Comparing with ZnFHA, ZTFHA films (except that R=3/4) show quite different Zn release behavior: the concentrations of Zn increase slowly but steadily in the first 96 h (Figure 4.33a), and the concentrations after each cycle remain almost unchanged, while that with R=3/4 shows similar Zn release behavior to ZnFHA, which is a rather fast release during the first 96 h and significantly decreased release in the remaining 96 h cycles.

In fact, as shown in Table 4.4, the Zn contents in ZTFHA films with R of 1/4 and 1/2 are actually higher than that in ZnFHA films, and their Zn-releasing behaviors actually mean that a slower and more sustained Zn release is achieved.

The reason for such release is attributed to the different Zn chemical state: both the surface state and interfacial state of Zn are rather instable and thus more soluble, which makes the majority contribution to the initial rapid release of Zn. After that, since β-TCP is more soluble than FHA, Zn in ZnTCP releases at a medium rate. When there is a proper amount of ZnTCP (e.g., R equals 1/8, 1/4, or 1/2) in the composite coatings, the release of Zn is actually controlled by β-TCP dissolution since most of Zn exists in that phase (Figure 4.34). Further increased ZnTCP amount (R equals to 3/4) may lead to more interfacial Zn, which results from chemical reaction between ZnTCP and FHA precursor sol. These interfacial Zn ions are released rather quickly during the first immersion cycle (Figure 4.33a). Such analyses coincide well with the Ca/Zn ratio results in Figure 4.33c: in the first cycle, all the coatings show relatively low Ca/Zn ratio since the Zn is actually released from those soluble surface or interface phase; after that, the increased Ca/Zn ratios are in well agreement with Zn content in the ZnTCP phase; particularly, for biphasic coating with R=3/4, since there is much interfacial phase, the Ca/Zn ratios remain low during the whole process.

Obviously, a composite Zn incorporation route could easily realize a slow and sustained Zn release.

The morphological evolution of the film also confirms that Zn is released through ZnTCP dissolution. In the typical SEM micrographs in Figure 4.35a

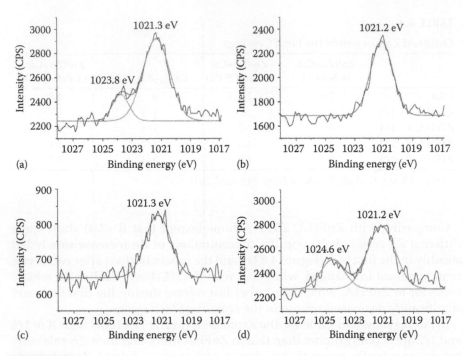

FIGURE 4.34
Detail scan spectra of Zn2p3/2 peaks. (a) ZnTCP/FHA coating with R=1/4, (b) ZnTCP/FHA coating with R=3/4, (c) FHA coating, and (d) ZnTCP powder. (From Cheng, K. et al., *Thin Solid Films*, 519, 4647, 2011.)

[74], the original rough surface, which contained many microscale islands (ZnTCP particles), transformed into one with many microscale pores; such morphological change meant that many ZnTCP particles were dissolved. As for the ZnFHA coatings, except sporadic pores, few changes were observed. Obviously, for the ZnFHA coating, very few parts were dissolved, since this phase is much more dissolution resistant [73].

Obviously, Zn release behavior of ZTFHA film was slow and sustained for up to tens of days. Moreover, the zinc release process could be regulated by varying both the zinc content in ZnTCP and the amount of ZnTCP coating. The controllable release behavior of Zn ions therefore could be a promising candidate for orthopedic implants, as it is well accepted that Zn ions could influence the surrounding bone tissue in a sustained way. This will be confirmed in the in vivo test.

4.3.5 In Vitro and In Vivo Evaluations

As discussed earlier, slow and sustained Zn release, which is desired for osteointegration, is realized through a composite Zn incorporation route.

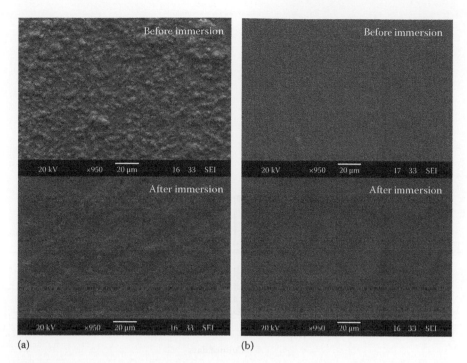

FIGURE 4.35
SEM morphological comparisons of the coatings before and after immersion. (a) ZTFHA coating with R = 3/4 and (b) ZnFHA coatings. (From Cheng, K. et al., *Thin Solid Films*, 519, 4647, 2011.)

Therefore, it is reasonable to consider such coatings could show good performance in the in vitro test and in vivo test.

The MTT results of cell cultured on ZTFHA films are shown in Figure 4.36. Based on the MTT results (Figure 4.36) [69], ZTFHA coating showed a statistically significant increase in cell viability compared with the control FHA coating at days 1 and 5. Zn has been shown to have stimulatory effects on bone formation and the stimulatory effects when the Zn concentration reached 65.3 µg/L in vitro. Nevertheless, the adverse effects caused by increased Zn ion concentration should also be considered. A 50% inhibition of MC3T3-E1 cell culture growth has been reported at Zn concentrations of 5.85 mg/L. The results shown in Figure 4.32 suggest that Zn released from ZTFHA films is below the toxic range. Therefore, the enhanced cell viability on ZTFHA films is attributed to the pharmacological effects resulting from proper Zn release. Osseointegration is the essential and pivotal requirement for the long-term success of the implant to maintain the mechanical stability with the surrounding bone tissue. Some researches [75,76] have demonstrated that pure Ti implant may sometimes lead to the direct bonding to the bone tissue through its native oxide surface under

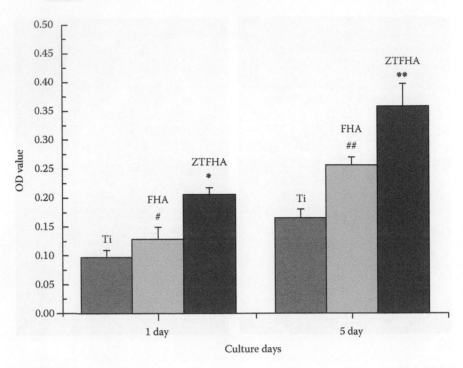

FIGURE 4.36
MTT results of different coatings (#significantly different from the controls [ANOVA, LSD
Po0.05]; ZTFHA sample was that with Zn/(Zn+Ca)=0.09). (From Miao, S. et al., *J. Am. Ceram.
Soc.*, 94, 255, 2011.)

press-fit condition. However, Takatsuka et al. [77] and Yan et al. [78] also
reported that Ti alloys did not have the ability of bonding to bone directly.
That means in many cases, the implant surface of Ti alloy was in poor
contact with bone tissue, and a typical layer of fibrous tissue was often
observed at the interface.

To further investigate the effect of composite route of Zn incorporation,
a comparison of cell growth between β-TCP/FHA and ZTFHA films were
carried out (Figure 4.37) [68]. Both of these layers were able to support
cell growth as the number of viable cells on the layers increased with culture time. A significant difference could be seen at days 1 and 5 when
comparing the β-TCP/FHA and ZTFHA layers. When the cell responses
were compared at each time point, the ZTFHA layer showed a statistically
significant increase in cell viability in comparison with the β-TCP/FHA
layer. In addition, cells growing on the ZnTCP/FHA layer showed a higher
growth rate with culture time. Considering the similar dissolution of
β-TCP and ZnTCP, it is reasonable to ascribe the increased cell activity to
the existence of Zn.

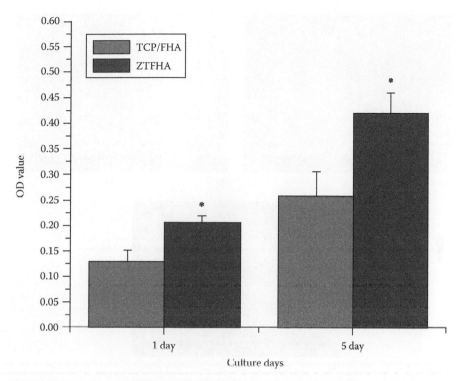

FIGURE 4.37
MTT results of cells cultured on the TCP/FHA and ZTFHA coatings at days 1 and 5. (Values represent the mean + SD of triplicate measurements. *Significantly different from the TCP/FHA coating [$p < 0.05$].) (From Miao, S. et al., *Acta Biomater.*, 4, 441, 2008.)

After 5-day culture [68,69], the cells showed a spherical morphology on a Ti alloy surface (Figure 4.38a). In contrast, irregularly spread cells were observed on both the FHA and ZTFHA films (Figure 4.38b and c), and the cells appeared to be confluent and overlapped with each other. Cells were more widely spread and extended on ZTFHA films, while on FHA films, some relatively spherical cells could often be seen. The cells on ZTFHA coatings therefore appeared more viable than those on Ti or the FHA coating.

Figure 4.39 shows the toluidine blue–stained histological images of the bone at the implant interface [69]. There were few contact areas of new bone with the implant surface in the Ti group. In some cases, a typical layer of fibrous tissue could be identified between the bone and Ti implant interface, suggesting poor long-term osseointegration of titanium (Figure 4.39a and d). Both FHA and ZTFHA coating groups showed mature cortical or trabecular bone growth in contact with the implants, while fibrous tissue at the bone–implant interface was absent (Figure 4.39b and c). A higher amount of bone

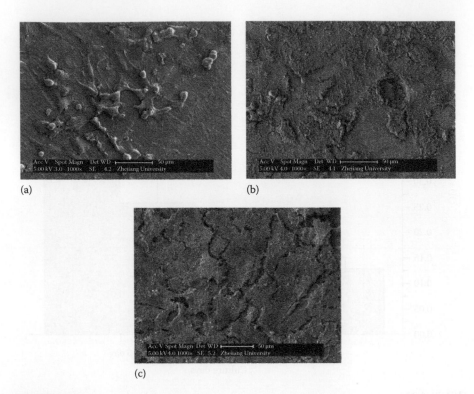

(a)   (b)

(c)

FIGURE 4.38
In vitro cell biocompatibility of different coatings: (a) Ti, (b) FHA, and (c) ZTFHA. (From Miao, S. et al., *J. Am. Ceram. Soc.*, 94, 255, 2011.)

tissue and direct bone contact occurred in the ZTFHA group compared with that seen in the FHA group. The latter consistently showed a gap between the implant and the bone tissue. The newly formed bone tissue in contact with the ZTFHA implant appeared highly mineralized and had a similar structure to that of the lamellar bone with organized osteon groups and a Haversian canal (Figure 4.39f).

The white arrow refers to the tetracycline fluorescent band, indicating active new bone formation [69]. Results of fluorescence microscopy are depicted in Figure 4.40. Tetracycline labels, which were administered 2 and 1 weeks before animal sacrifice, were clearly observed as thin yellow lines, indicating the newly formed bone tissue. An intensive new bone formation was clearly observable at the bone–implant interface in both the FHA and ZTFHA groups, indicating active bone formation near the implants.

In conclusion, ZTFHA films with a composite Zn doping route exhibit excellent osteointegration performance. It is believed that the Zn existence plays an important role in osteointegration.

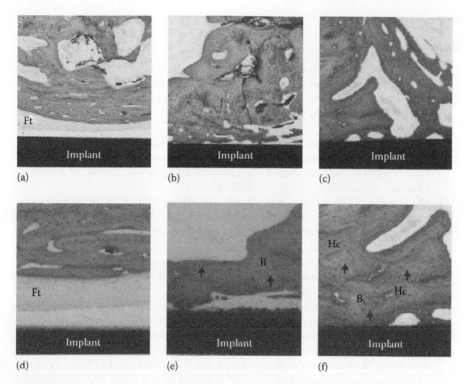

(a)   (b)   (c)

(d)   (e)   (f)

FIGURE 4.39
Light microphotographs of the bone–implant interface at 12 weeks after implantation in Ti alloy group (a and d), FHA group (b and e), and ZTFHA group (c and f). The black arrows indicate osteocytes (toluidine blue staining; original magnification: (a through c) ×4, (d, e, and f) ×10). Hc, Haversian canal; B, bone tissue; Ft, fibrous tissue. (From Miao, S. et al., *J. Am. Ceram. Soc.*, 94, 255, 2011.)

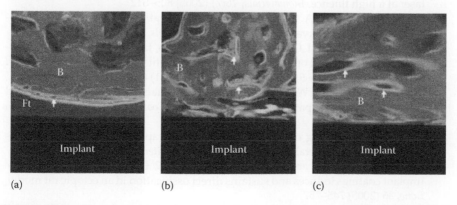

(a)   (b)   (c)

FIGURE 4.40
Fluorescence microphotographs of the bone–implant interface at 12 weeks after implantation in the Ti alloy group (a), FHA group (b) and ZTFHA group (c). (From Miao, S. et al., *J. Am. Ceram. Soc.*, 94, 255, 2011.)

References

1. K. De Groot, R. Geesink, C. P. A. T. Klein, and P. Serekian, Plasma sprayed coatings of hydroxylapatite, *Journal of Biomedical Materials Research*, 21(12) (1987) 1375–1381.
2. J. L. Ong, D. L. Carnes, and K. Bessho, Evaluation of titanium plasma-sprayed and plasma-sprayed hydroxyapatite implants in vivo, *Biomaterials*, 25(19) (2004) 4601–4606.
3. K. Ogata, S. Imazato, A. Ehara, S. Ebisu, Y. Kinomoto, T. Nakano, and Y. Umakoshi, Comparison of osteoblast responses to hydroxyapatite and hydroxyapatite/soluble calcium phosphate composites, *Journal of Biomedical Materials Research Part A*, 72A(2) (2005) 127–135.
4. M. Stewart, J. F. Welter, and V. M. Goldberg, Effect of hydroxyapatite/tricalcium-phosphate coating on osseointegration of plasma-sprayed titanium alloy implants, *Journal of Biomedical Materials Research Part A*, 69A(1) (2004) 1–10.
5. L. Gineste, M. Gineste, X. Ranz, A. Ellefterion, A. Guilhem, N. Rouquet, and P. Frayssinet, Degradation of hydroxylapatite, fluorapatite, and fluorhydroxyapatite coatings of dental implants in dogs, *Journal of Biomedical Materials Research*, 48(3) (1999) 224–234.
6. S. Overgaard, M. Lind, K. Josephsen, A. B. Maunsbach, C. Boger, and K. Soballe, Resorption of hydroxyapatite and fluorapatite ceramic coatings on weight-bearing implants: A quantitative and morphological study in dogs, *Journal of Biomedical Materials Research*, 39(1) (1998) 141–152.
7. K. Cheng, G. Shen, W. Weng, G. Han, J. M. F. Ferreira, and J. Yang, Synthesis of hydroxyapatite/fluoroapatite solid solution by a sol–gel method, *Materials Letters*, 51(1) (2001) 37–41.
8. H. W. Kim, H. E. Kim, and J. C. Knowles, Fluor-hydroxyapatite sol–gel coating on titanium substrate for hard tissue implants, *Biomaterials*, 25(17) (2004) 3351–3358.
9. D. Ferro, S. M. Barinov, J. V. Rau, R. Teghil, and A. Latini, Calcium phosphate and fluorinated calcium phosphate coatings on titanium deposited by Nd: YAG laser at a high fluence, *Biomaterials*, 26(7) (2005) 805–812.
10. K. A. Bhadang and K. A. Gross, Influence of fluorapatite on the properties of thermally sprayed hydroxyapatite coatings, *Biomaterials*, 25(20) (2005) 4935–4945.
11. K. Cheng, W. Weng, H. Qu, P. Du, G. Shen, G. Han, J. Yang et al., Sol–gel preparation and in vitro test of fluorapatite/hydroxyapatite films, *Journal of Biomedical Materials Research Part B: Applied Biomaterials*, 69B(1) (2004) 33–37.
12. K. Cheng, S. Zhang, W. Weng, and X. Zeng, The interfacial study of sol–gel-derived fluoridated hydroxyapatite coatings, *Surface and Coatings Technology*, 198 (2005) 242–246.
13. G. Willmann, Coating of implants with hydroxyapatite material connections between bone and metal, *Advanced Engineering Materials*, 2 (1995) 95–105.
14. Y. Liu, K. de Groot, and E. B. Hunziker, BMP-2 liberated from biomimetic implant coatings induces and sustains direct ossification in an ectopic rat model, *Bone*, 36 (2005) 745–757.
15. Y. Liu, L. Enggist, A. F. Kuffer, D. Buser, and E. B. Hunziker, The influence of BMP-2 and its mode of delivery on the osteoconductivity of implant surfaces during the early phase of osseointegration, *Biomaterials*, 28 (2007) 2677–2686.

16. K. L. Wong, C. T. Wong, W. C. Liu, H. B. Pan, M. K. Fong, W. M. Lam, W. L. Cheung et al., Mechanical properties and in vitro response of strontium-containing hydroxyapatite/polyetheretherketone composites, *Biomaterials*, 30 (2009) 810–817.

17. E. S. Thian, J. Huang, S. M. Best, Z. H. Barber, R. A. Brooks, N. Rushton, and W. Bonfield, The response of osteoblasts to nanocrystalline silicon-substituted hydroxyapatite thin films, *Biomaterials*, 27 (2006) 2692–2698.

18. D. Guo, K. Xu, X. Zhao, and Y. Han, Development of a strontium-containing hydroxyapatite bone cement, *Biomaterials*, 26 (2005) 4073–4078.

19. W. C. Xue, J. L. Moore, H. L. Hosick, S. Bose, A. Bandyopadhyay, W. W. Lu, K. M. C. Cheung et al., Osteoprecursor cell response to strontium-containing hydroxyapatite ceramics, *Journal of Biomedical Materials Research*, 79A (2006) 804–814.

20. A. Grandjean-Laquerrier, P. Laquerriere, E. Jallot, J. M. Nedelec, M. Guenounou, D. Laurent-Maquin, and T. M. Phillips, Influence of the zinc concentration of sol–gel derived zinc substituted hydroxyapatite on cytokine production by human monocytes in vitro, *Biomaterials*, 27 (2006) 3195–3200.

21. S. Storrie and S. I. Stupp, Cellular response to zinc-containing organoapatite: An in vitro study of proliferation, alkaline phosphate activity and biomineralization, *Biomaterials*, 26 (2005) 5492–5499.

22. M. Yamaguchi, Role of zinc in bone formation and bone resorption, *Journal of Trace Elements in Experimental Medicine*, 11 (1998) 119–135.

23. M. Otsuka, Y. Ohshita, S. Marunaka, Y. Matsuda, A. Ito, N. Ichinose, K. Otsuka et al., Effect of controlled zinc release on bone mineral density from injectable Zn-containing b-tricalcium phosphate suspension in zinc-deficient diseased rats, *Journal of Biomedical Materials Research*, 69A (2004) 552–560.

24. L. L. Hench, Sol–gel materials for bioceramic application, *Current Opinion in Solid State and Materials Science*, 2 (1997) 604–610(7).

25. B. Mavis and A. C. Tas, Dip coating of calcium hydroxyapatite on Ti–6Al–4V substrates, *Journal of the American Ceramic Society*, 83(4) (2000) 989–991.

26. C. J. Brinker, A. J. Hurd, P. R. Schunk, G. C. Frye, and C. S. Ashley, Review of sol–gel thin film formation, *Journal of Non-Crystalline Solids*, 147–148 (1992) 424–436.

27. G. W. Scherer, Recent progress in drying of gels, *Journal of Non-Crystalline Solids*, 147–148 (1992) 363–374.

28. K. Cheng, Sol–gel preparation of fluoridated hydroxyapatite films and their in vitro properties, PhD dissertation, Zhejiang University, Hangzhou, China, 2003.

29. K. Cheng, G. Han, W. Weng, H. Qu, P. Du, G. Shen, J. Yang et al., Sol–gel derived fluorinated hydroxyapatite films, *Materials Research Bulletin*, 38(1) (2003) 89–97.

30. W. Weng and J. L. Baptista, Alkoxide route for preparing hydroxyapatite and its coatings, *Biomaterials*, 19(1–3) (1998) 125–131.

31. D. R. Lide, *Handbook of Chemistry and Physics*, 85th edn. (CRC Press, Boca Raton, FL, 2004), pp. 4–49.

32. K. Cheng, S. Zhang, and W. J. Weng, Sol–gel preparation of fluoridated hydroxyapatite in $Ca(NO_3)_2$-$PO(OH)_{3-x}(OEt)_x$-HPF_6 system, *Journal of Sol–Gel Science And Technology*, 38 (2006) 13–17.

33. K. Cheng, W. Weng, H. Wang, and S. Zhang, In vitro behavior of osteoblast-like cells on fluoridated hydroxypatite coating, *Biomaterials*, 26(32) (2005) 6288–6295.

34. J. C. Elliott, Structure and chemistry of the apatites and other calcium ortho-phosphates, *Studies in Inorganic Chemistry*, Vol. 18 (Elsevier, Amsterdam, the Netherland, 1994), pp. 4–6.
35. Z. Amjad, *Calcium Phosphates in Biological and Industrial Systems* (Kluwer, Boston, MA, 1998), pp. 27–30.
36. D. M. Liu, Q. Yang, and T. Troczynski, Sol–gel hydroxyapatite coatings on stainless steel substrates, *Biomaterials*, 23(3) (2002) 691–698.
37. M. F. Hsieh, L. H. Perng, and T. S. Chin, Hydroxyapatite coating on Ti6Al4V alloy using a sol–gel derived precursor, *Materials Chemistry and Physics*, 74(3) (2002) 245–250.
38. J. M. Fernandez-Pradas, M. V. Garcia-Cuenca, L. Cleries, G. Sardin, and J. L. Morenza, Influence of the interface layer on the adhesion of pulsed laser deposited hydroxyapatite coatings, *Applied Surface Science*, 195(1–4) (2002) 31–37.
39. K. Takahashi, T. Hayakawa, M. Yoshinari, H. Hara, C. Mochizuki, M. Sato, and K. Nemoto, Molecular precursor method for thin calcium phosphate coating on titanium, *Thin Solid Films*, 484(1–2) (2005) 1–9.
40. K. Cheng, C. B. Ren, W. J. Weng, P. Y. Du, G. Shen, G. R. Han, and S. Zhang, Bonding strength of fluoridated hydroxyapatite coatings: A comparative study on pull-out and scratch analysis, *Thin Solid Films*, 517(17) (2009) 5361–5364.
41. J. Wang, P. Layrolle, M. Stigter, and K. De Groot, Biomimetic and electrolytic calcium phosphate coatings on titanium alloy: Physicochemical characteristics and cell attachment, *Biomaterials*, 25(4) (2004) 583–592.
42. S. Zhang, X. T. Zeng, Y. S. Wang, K. Cheng, and W. J. Weng, Adhesion strength of sol–gel derived fluoridated hydroxyapatite coatings, *Surface and Coatings Technology*, 200 (2006) 6350–6354.
43. S. Zhang and X. L. Bui, In: K. L. Mittal (Ed.), *Adhesion Aspects of Thin Films*, Vol. 2 (VSP, Utrecht, the Netherlands, 2005), p. 37.
44. A. Montenero, G. Gnappi, F. Ferrari, M. Cesari, E. Salvioli, L. Mattogno, S. Kaciulis et al., Sol–gel derived hydroxyapatite coatings on titanium substrate, *Journal of Materials Science*, 35 (2000) 2791–2797.
45. K. Cheng, S. Zhang, and W. Weng, The F content in sol–gel derived FHA coat-ings: An XPS study, *Surface and Coatings Technology*, 198(1–3) (2005) 237–241.
46. D. M. Brunette, P. Tengvall, M. Textor, and P. Thomsen, *Titanium in Medicine* (Springer, Berlin, Germany, 2001), pp. 171.
47. M. J. Filiaggi, R. M. Piliar, and D. Abdulla, Evaluating sol–gel ceramic thin films for metal implant applications II. Adhesion and fatigue properties of zir-conia films on Ti–6Al–4V, *Journal of Biomedical Materials Research*, 33(4) (1996) 239–256.
48. W. Weng, S. Zhang, K. Cheng, H. Qu, P. Du, G. Shen, J. Yuan, and G. Han, Sol–gel preparation of bioactive apatite films, *Surface and Coatings Technology*, 167(2–3) (2003) 292–296.
49. M. A. Stranick and M. J. Root, Influence of strontium on monofluorophosphate uptake by hydroxyapatite XPS characterization of the hydroxyapatite surface, *Colloids and Surfaces*, 55 (1991) 137–147.
50. Y. L. Chang, C. M. Stanford, and J. C. Keller, Calcium and phosphate supple-mentation promotes bone cell mineralization: Implications for hydroxyapatite (HA)-enhanced bone formation, *Journal of Biomedical Materials Research*, 52(2) (2000) 270–278.

51. S. Maeno, Y. Niki, H. Matsumoto, H. Morioka, T. Yatabe, A. Funayama, Y. Toyama et al., The effect of calcium ion concentration on osteoblast viability, proliferation and differentiation in monolayer and 3D culture, *Biomaterials*, 26(23) (2005) 4847–4855.

52. L. Savarino, M. Fini, G. Ciapetti, E. Cenni, D. Granchi, N. Baldini, M. Greco et al., Biologic effects of surface roughness and fluorhydroxyapatite coating on osteointegration in external fixation systems: An in vivo experimental study, *Journal of Biomedical Materials Research Part A*, 66A(3) (2003) 652–661.

53. C. Chenu, S. Colucci, M. Grano, P. Zigrino, R. Barattolo, G. Zambonin, N. Baldini et al., Osteocalcin induces chemotaxis, secretion of matrix proteins and calcium mediated intracellular signaling in human osteoclast-like cells, *Journal of Cell Biology*, 127(4) (1994) 1149–1158.

54. S. V. Dorozhkin, A review on the dissolution models of calcium apatites, *Progress in Crystal Growth and Characterization of Materials*, 44(1) (2002) 45–61.

55. P. Schaad, F. Poumier, J. C. Voegel, and Ph. Gramain, Analysis of calcium hydroxyapatite dissolution in non-stoichiometric solutions, *Colloids and Surfaces A: Physicochemical and Engineering Aspects*, 121(2–3) (1997) 217–228.

56. Y. Zhang, G. Sun, Y. Jin, and Y. Wang, Effects of fluoride on cell cycle and apoptosis in cultured osteoblasts of rats, *Wei Sheng Yan Jiu*, 32(5) (2003) 432–433.

57. M. Kanatani, T. Sugimoto, M. Fukase, and T. Fujita, Effect of elevated extracellular calcium on the proliferation of osteoblastic MC3T3-E1 cells: Its direct and indirect effects via monocytes, *Biochemical and Biophysical Research Communications*, 181(3) (1991) 1425–1430.

58. P. Ducheyne and Q. Qiu, Bioactive ceramics: The effect of surface reactivity on bone formation and bone cell function, *Biomaterials*, 20(23–24) (1999) 2287–2303.

59. J. Ovesen, B. Moller-Madsen, J. S. Thomsen, G. Danscher, and L. I. Mosekilde, The positive effects of zinc on skeletal strength in growing rats, *Bone*, 29(6) (2001) 565–570.

60. H. Kawamura, A. Ito, S. Miyakawa, P. Layrolle, K. Ojima, N. Ichinose, and T. Tateushi, Stimulatory effect of zinc-releasing calcium phosphate implant on bone formation in rabbit femora, *Journal of Biomedical Materials Research*, 50(2) (2000) 184–190.

61. A. Ito, H. Kawamura, M. Otsuka, M. Ikeuchi, H. Ohgushi, K. Ishikawa, K. Onuma et al., Zinc-releasing calcium phosphate for stimulating bone formation, *Materials Science and Engineering: C*, 22(1) (2002) 21–25.

62. R. A. Barrea, C. A. Perez, A. Y. Ramos, H. J. Sanchez, and M. Grenon, Distribution and incorporation of zinc in biological calcium phosphates, *X-Ray Spectrometry*, 32(5) (2003) 387–395.

63. C. Ergun, T. J. Webster, R. Bizios, and R. H. Doremus, Hydroxylapatite with substituted magnesium, zinc, cadmium, and yttrium. I. Structure and microstructure, *Journal of Biomedical Materials Research*, 59(2) (2002) 305–311.

64. I. Mayera and J. D. B. Featherstone, Dissolution studies of Zn-containing carbonated hydroxyapatites, *Journal of Crystal Growth*, 219(1–2) (2000) 98–101.

65. S. Miao, W. Weng, K. Cheng, P. Du, G. Shen, G. Han, and S. Zhang, Sol–gel preparation of Zn-doped fluoridated hydroxyapatite films, *Surface and Coating Technology*, 198(1–3) (2005) 223–226.

66. A. Ito, K. Ojima, H. Naito, N. Ichinose, and T. Tateishi, Preparation, solubility, and cytocompatibility of zinc-releasing calcium phosphate ceramics, *Journal of Biomedical Materials Research*, 50(2) (2000) 178–183.

67. M. Ikeuchi, A. Ito, Y. Dohi, H. Ohgushi, H. Shimaoka, K. Yonemasu, and T. Tateishi, Osteogenic differentiation of cultured rat and human bone marrow cells on the surface of zinc-releasing calcium phosphate ceramics, *Journal of Biomedical Materials Research*, 67A(4) (2003) 1115–1122.

68. S. Miao, W. Weng, K. Cheng, P. Du, G. Shen, and G. Han, Fabrication and evaluation of Zn containing fluoridated hydroxyapatite layer with Zn release ability, *Acta Biomaterialia*, 4(2) (2008) 441–446.

69. S. Miao, N. Lin, K. Cheng, D. Yang, X. Huang, G. Han, W. Weng et al., Zn-releasing FHA coating and its enhanced osseointegration ability, *Journal of the American Ceramic Society*, 94(1) (2011) 255–260.

70. S. L. Hall, H. P. Dimai, and J. R. Farley, Effects of zinc on human skeletal alkaline phosphatase activity in vitro, *Calcified Tissue International*, 64 (1999) 163–172.

71. A. Yamamoto, R. Honma, and M. Sumita, Cytotoxicity evaluation of 43 metal salts using murine fibroblasts and osteoblastic cells, *Journal of Biomedical Materials Research*, 39(2) (1998) 331–340.

72. S. Miao, W. Weng, K. Cheng, P. Du, G. Shen, and G. Han, In vitro bioactivity and osteoblast-like cell test of zinc containing fluoridated hydroxyapatite films, *Journal Materials Science: Materials in Medicine*, 18 (2007) 2101–2105.

73. S. V. Dorozhkin and M. Epple, Biological and medical significance of calcium phosphates, *Angewandte Chemie International Edition*, 41(17) (2002) 3130–3146.

74. K. Cheng, J. Zhou, W. Weng, S. Zhang, G. Shen, P. Du, and G. Han, Composite calcium phosphate coatings with sustained Zn release, *Thin Solid Films*, 519(15) (2011) 4647–4651.

75. S. L. Carlson, T. R. Rostlunt, B. Abrektsson, T. Abrektsson, and P. I. Branemark, Osseointegration of titanium implants, *Acta Orthopaedica*, 57 (1986) 285–289.

76. C. L. B. Lavelle, D. Wedgwood, and W. B. Love, Some advances in endosseous implants, *Journal of Oral Rehabilitation*, 8(4) (1981) 319–331.

77. K. Takatsuka, T. Yamamuro, T. Nakamura, and T. Kokubo, Bone-bonding behavior of titanium alloy evaluated mechanically with detaching failure load, *Journal of Biomedical Materials Research*, 29(2) (1995) 157–163.

78. W. Q. Yan, T. Nakamura, M. Kobayashi, H. M. Kim, F. Miyaji, and T. Kokubo, Bonding of chemically treated titanium implants to bone, *Journal of Biomedical Materials Research*, 37(2) (1997) 267–275.

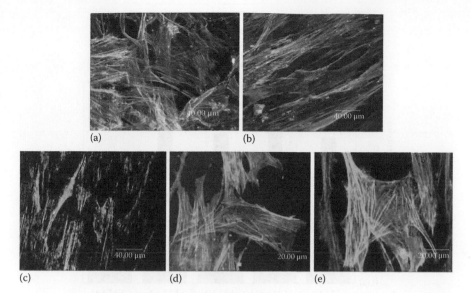

(a)　　　　　　　　　(b)

(c)　　　　　　　(d)　　　　　　　(e)

FIGURE 1.12

Confocal fluorescence microscopy of nuclear DNA (stained blue), actin cytoskeleton (stained green), and vinculin plaque (stained red) in HOB cell as revealed with multiple labeling using TOTO-3, FITC conjugated phalloidin, and Texas red conjugated streptavidin on (a) S1, (b) S2, (c) uncoated Ti substrate (control), (d) S3, and (e) S4. (Reprinted from *Biomaterials*, 26, Thian, E. S., Huang, J., Best, S. M., Barber, Z. H., and Bonfield, W., Magnetron co-sputtered silicon-containing hydroxyapatite thin films—An in vitro study, 2947–2956, Copyright 2005, with permission from Elsevier; *Biomaterials*, 27, Thian, E. S. et al., The response of osteoblasts to nanocrystalline silicon-substituted hydroxyapatite thin films, 2692–2698, Copyright 2006, with permission from Elsevier.)

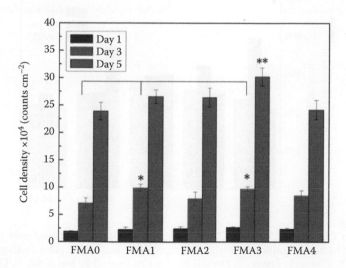

FIGURE 3.37

Cell proliferation on Mg_xFHA coatings in terms of cell density: *At day 3, FMA1 and FMA3 have a significantly higher cell number than that on FMA0 ($p < 0.05$). **At day 5, cell numbers on FMA3 are significantly higher than all the other coatings ($p < 0.05$). (From Cai, Y., Zhang, J., Zhang, S., Mondal, D., Venkatraman, S.S., and Zeng, X., Osteoblastic cell response on fluoridated hydroxyapatite coatings: The effect of magnesium incorporation, *Biomed. Mater.*, 5, 054114, 2010. With permission from Institute of Physics.)

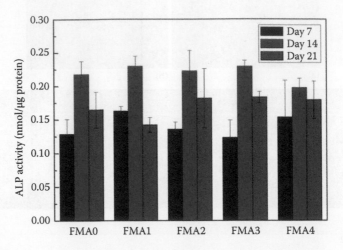

FIGURE 3.38
Intracellular ALP activities of MG63 cells cultured on Mg$_x$FHA coatings over the culture period: no significant difference between Mg-incorporated coatings (from FMA1 to FMA4) and Mg-free one (FMA0). (From Cai, Y., Zhang, J., Zhang, S., Mondal, D., Venkatraman, S.S., and Zeng, X., Osteoblastic cell response on fluoridated hydroxyapatite coatings: The effect of magnesium incorporation, *Biomed. Mater.*, 5, 054114, 2010. With permission from Institute of Physics.)

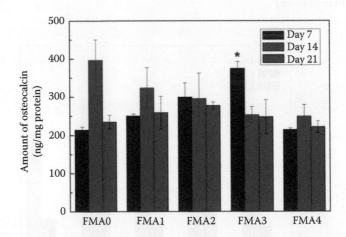

FIGURE 3.39
Intracellular OC expression of MG63 cells cultured on Mg$_x$FHA coatings over the culture period: *FMA3 has a significantly higher OC expression than all the other coatings ($p < 0.05$). (From Cai, Y., Zhang, J., Zhang, S., Mondal, D., Venkatraman, S.S., and Zeng, X., Osteoblastic cell response on fluoridated hydroxyapatite coatings: The effect of magnesium incorporation, *Biomed. Mater.*, 5, 054114, 2010. With permission from Institute of Physics.)

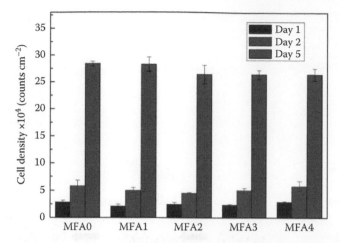

FIGURE 3.54
Cell proliferation on MgF_yHA coatings in terms of cell density: no significant difference is observed on different coatings ($p < 0.05$).

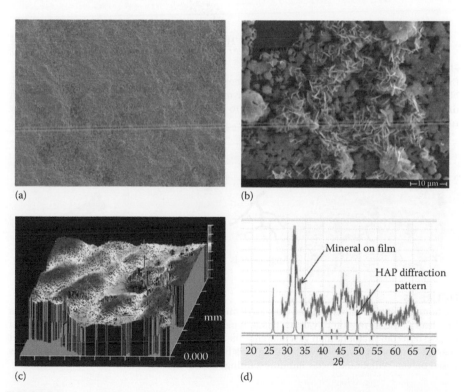

(a)　　(b)　　(c)　　(d)

FIGURE 5.2
Scanning electron microscopy of bioresorbable polymer films (a) before incubation and (b) after biomimetic incubation in a modified simulated body fluid (mSBF). Films were composed of the bioresorbable polymer poly(ε-caprolactone) (PCL). Incubation of films in mSBF resulted in the formation of a continuous mineral layer that displayed a nanoporous, plate-like morphology. (c) A profilometric analysis of the films demonstrated as well the formation of a continuous mineral coating. (d) Characterization of mineral phase by x-ray diffraction confirmed the formation of hydroxyapatite (HAP). The blue trace represents the diffraction patterns associated with pure HAP.

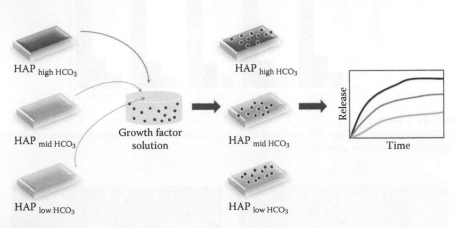

FIGURE 5.3
Biomimetic HAP coatings with different extent of carbonate incorporation (represented in black, blue, and green) allow for control over mineral dissolution and associated release of drugs. Higher carbonate incorporation leads to higher dissolution rates and therefore faster release kinetics.

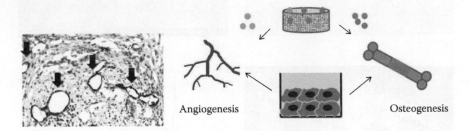

FIGURE 5.5
Sustained release of growth factors from HAP mineral coatings can influence cellular activities that lead to the formation of various tissue types, such as blood vessels and bone. Sustained release of recombinant human vascular endothelial growth factor (VEGF) induced new blood vessel formation in vivo. Arrows pointing dark brown stained lumens represent blood vessels based on von Willebrand factor immunostaining.

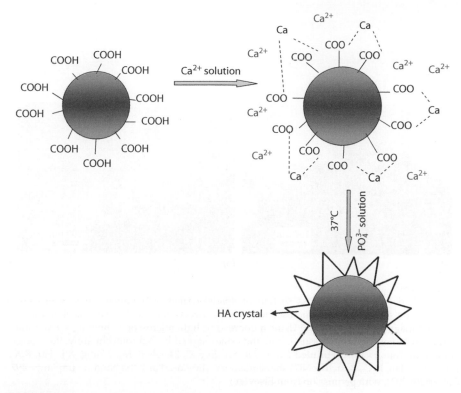

FIGURE 6.26
Schematics of HA coating formation on PLGA microspheres. (From Zhang, Z. et al., *Nanosci. Nanotechnol. Lett.*, 3(4), 472, 2011.)

(a)   (b)

FIGURE 6.28
Photographs for Ca and P mixed solution with different concentrations: (a) Solution1 and (b) Solution2.

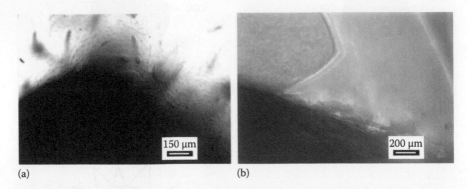

(a)   (b)

FIGURE 7.24
Nondecalcified histological cross section of apatite-coated NiTi implant at 6 weeks after implantation. (a) Using a light microscope, showing direct contact of bone and coated implant (original magnification × 100). (b) Using a fluorescent light microscope, showing active transformation of bone (osseous tissue is red and zones signed by Achromycin are yellow, original magnification × 75). (Reprinted from *Mat. Sci. Eng. C*, 24, Chen, M.F., Yang, X.J., Hu, R.X., Cui, Z.D., Man, H.C., Bioactive NiTi shape memory alloy used as bone bonding implants, 497, Copyright 2004, with permission from Elsevier.)

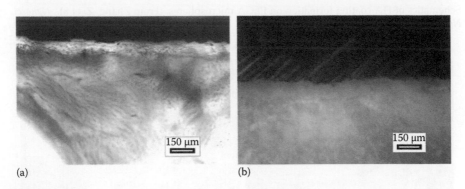

(a)   (b)

FIGURE 7.25
Nondecalcified histological cross section of apatite-coated NiTi implant at 13 weeks after implantation. (a) Using a light microscope, showing a new bone layer contacting with coated implant (original magnification × 100). (b) Using a fluorescent light microscope, showing a green bone layer between the marrow and the coated implant (original magnification × 75). (Reprinted from *Mat. Sci. Eng. C*, 24, Chen, M.F., Yang, X.J., Hu, R.X., Cui, Z.D., Man, H.C., Bioactive NiTi shape memory alloy used as bone bonding implants, 497, Copyright 2004, with permission from Elsevier.)

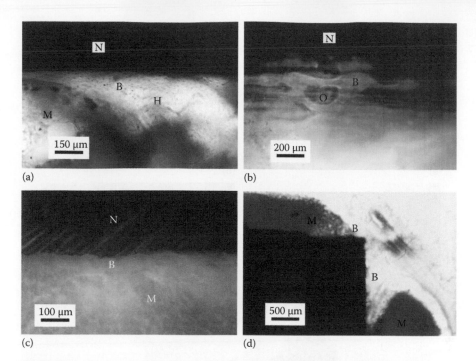

FIGURE 7.28
Nondecalcified histological cross section of apatite-coated NiTi implant at 26 weeks after implantation. (a) Using a light microscope, showing a mature bone layer. (b) Using a fluorescent light microscope, showing pink osteoid on apatite-coated NiTi implant. (c) Formation of new bone layer in medullary cavity. (d) A special specimen. N, NiTi; B, Bone; M, Marrow; O, Osteoid; H, Haversian.

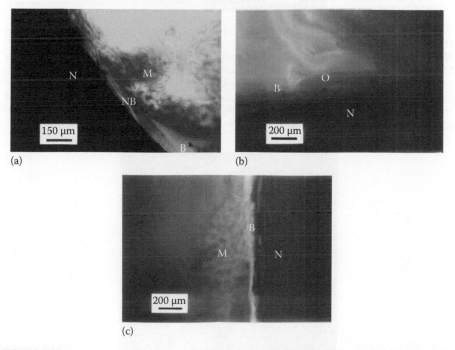

FIGURE 7.29
Nondecalcified histological cross section of uncoated NiTi implant at 26 weeks after implantation. (a) Using a light microscope, showing a new bone layer between bone and NiTi implant. (b) Using a fluorescent light microscope, showing new bone and mineralizing osteoid. (c) Noncontinuous new bone in medullary cavity. N, NiTi; B, bone; M, marrow; O, osteoid.

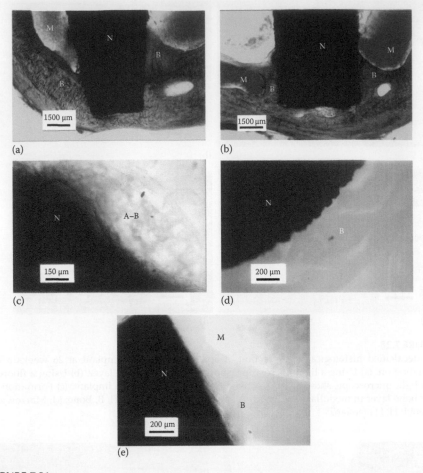

FIGURE 7.31
Nondecalcified histological cross section of apatite-coated NiTi and uncoated NiTi implants at 53 weeks after implantation. Using a light microscope, showing a mature bone layer encompasses apatite-coated NiTi implant (a), and absorbable bone at one end of uncoated NiTi implant (b and c); using a fluorescent light microscope, showing rough interface between host bone and apatite-coated NiTi implant (d), and smooth interface between host bone and uncoated-NiTi (e). N, NiTi; B, bone; M, marrow; A-B, absorbable bone.

FIGURE 7.59
Collagen distribution in the coating on NiTi alloy after soaking in 0.5 g/L 1.5CSBF for 3 days.

5

Biomimetic Hydroxyapatite Materials for Therapeutic Delivery

Travelle W. Franklin-Ford, Darilis Suarez-Gonzalez, Jae Sung Lee, and William L. Murphy

CONTENTS

5.1 Introduction: Apatites in Biology and Medicine

Biomimetic principles seek to imitate or reproduce key elements of biological systems using synthetic techniques. In biomineralization, inorganic materials are often designed, structured, and optimized to replicate functions of nature's minerals for a desired application. Biomimetic approaches can also provide a simple, deconstructed surrogate for biological function in the case of complex biological archetypes that are difficult to replicate ex vivo [1]. For more complicated biological systems, staged addition of features may be used to increase model sophistication as needed. Thus, biomimetic principles can potentially illustrate a biological system, explore its functionality, and exploit benefits to improve upon natural limitations.

Biomineralization is a classic example of a process that can be mimicked for multiple technological applications. Biomineralization is a natural process in which organisms synthesize and regulate growth of over 60 minerals for a number of critical functions [2]. Hydroxyapatite (HAP), a naturally occurring, hexagonal mineral crystal with the formula $Ca_{10}(PO_4)_6OH_2$ is synthesized by a myriad of organisms and is an important constituent of hard tissues such as vertebrate teeth and bones [3–5]. HAP comprises roughly 70% of the material found in cancellous bone by weight, with the other 30% coming from organic materials including collagen. In nature, HAP is the most stable of the calcium phosphate minerals at physiological pH [6], and other calcium phosphate mineral phases can transition to HAP in physiological buffer conditions [7,8]. Because bones and teeth are hard materials that degrade and wear over time, development of biomimetic substitutes for these materials has been prevalent during the past three decades.

In nature, HAP provides strength, support, and external coverage for critical body parts such as shells and bones and also facilitates an organism's basic functions by maintaining calcium concentrations in the blood [9]. Therefore, there is a strong motivation for the use of HAP as a high-strength biomaterial and as a bone implant. Interestingly, pure synthetic HAP is not typically attractive as a stand-alone orthopedic implant, as it demonstrates brittle mechanical properties and low material strength. This distinction between natural HAP-based hard tissues (e.g., bone) and pure synthetic HAP provides a simple illustration of the value of biomimetic approaches. One specific distinction is that biological apatites found in bones, teeth, and antlers are more calcium deficient and contain carbonate ions in lieu of phosphates when compared to pure synthetic HAP [10]. Therefore, perhaps the simplest example of biomimetic mineral synthesis involves creating calcium-deficient or otherwise "substituted" versions of HAP. This type of substitution renders HAP more soluble by influencing the crystal size and crystallinity, making the material slower to grow and faster to dissolve and release mineral ions. A number of ions can substitute within the basic calcium phosphate mineral structure at either the hydroxyl or phosphate ion sites $Ca_{10}(A_X)B_Y$ (A, AB, or B substitution), resulting in a change in crystal structure and mineral stability. Examples of variations of stoichiometric HAP include carbonated apatite, fluorapatite, and chlorapatite [11,12]. Each of these variations in HAP synthesis has functional consequences that can be useful for particular technological applications.

Another fundamental property of natural HAP exploited in technological applications is its well-characterized ability to bind to biological molecules (e.g., proteins, DNA). As a result, HAP materials have made strong contributions to the fields of chromatography and therapeutic drug delivery. HAP has a high affinity for proteins and DNA because of its elemental composition, which can be used to selectively separate proteins or DNA passed through a column [13–15]. Further, HAP-based drug delivery applications have exploited the controllable electrostatic interactions of the inorganic mineral with a variety of drug molecules, including DNA, peptides, proteins, and small molecules for subsequent delivery in vivo.

In essence, the fundamental properties of HAP and its prevalence in nature's hard tissues have led to the widespread development of HAP-based biomaterials with biomimetic functions. These biomaterials have continued to gain widespread use as components of metallic and polymeric medical devices to improve osteoconduction and deliver proteins and other drug molecules, thereby improving osteoinduction (Figure 5.1) [16–18]. This chapter will discuss the contemporary use of HAP-based biomaterials for therapeutic drug delivery, with a particular emphasis on their fabrication, stability, dissolution, and interaction with drugs (e.g., proteins). The focus will be on introducing key HAP properties of relevance to drug delivery and describing examples in which HAP biomaterials have been used in biomedical applications.

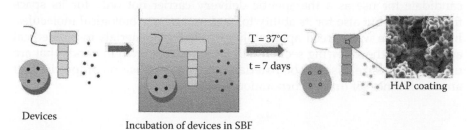

Devices

Incubation of devices in SBF

T = 37°C
t = 7 days

HAP coating

FIGURE 5.1

Biomimetic coating of medical devices. Hydroxyapatite (HAP) nucleation from aqueous solution (e.g., simulated body fluid [SBF]) can be used to coat a wide array of devices. The process is highly adaptable and can be used with complex geometries and a variety of template materials. Devices such as bone screws, sutures, scaffolds, and injectable particles have been successfully coated for localized drug delivery applications.

5.2 Introduction: Localized Therapeutic Delivery

Local therapeutic delivery vehicles have emerged as an alternative to systemic drug delivery in some biomedical applications. Generally, the minimal requirements for successful localized drug carriers include (1) biocompatibility—avoidance of adverse biological reactions after in vivo implantation; (2) ability to chemically or physically incorporate drug molecules; (3) biodegradability into byproducts that are not detrimental to the patient locally or systemically; and (4) implantability or injectability into desired in vivo locations. In addition to these base requirements, emerging localized release approaches must either improve upon efficacy of a particular drug molecule or decrease side effects caused by systemically administered therapy. Calcium phosphate biomaterials such as HAP-based biomaterials are particularly interesting vehicles for localized therapeutic delivery, as they have a long tenure as osteoconductive bone graft substitutes in orthopedic, dental, and plastic/reconstructive surgery applications [19–21] and are therefore known to be biocompatible and implantable. The ability of biomimetic HAP biomaterials to degrade over time and incorporate/release various drug molecules has extended their use beyond bone graft substitutes into a variety of controlled delivery applications for proteins, peptides, and small molecules (Figure 5.1).

A wide variety of drug carrier materials have been used for localized therapeutic delivery, and they can be generally classified into naturally derived organic biomaterials, synthetic organic biomaterials, and inorganic biomaterials. Common examples in clinical applications include demineralized, inactivated, or insoluble bone matrix, autolyzed antigen-extracted allogeneic bone, collagen types I and IV (e.g., sponges, gels, or membranes), fibrin, synthetic poly (α-hydroxyacids) (e.g., polylactic acid, polyglycolic acid) [22], titanium, and calcium phosphates [23]. HAP is a particularly attractive candidate for use as a therapeutic delivery carrier not only for its space filling ability but also for its ability to bind and release biological molecules. In view of the widespread applicability of HAP biomaterials in biomedical applications, the following section will introduce common methods that are used to synthesize this class of materials, with an emphasis on properties that are of relevance for drug incorporation and release.

5.3 Common Methods to Form Hydroxyapatite Biomaterials

As a group, calcium phosphate minerals vary individually based on the number of calcium ions present, the presence of water, and the fabrication method. The presence of impurities and the spatial arrangement or ratio of elements during fabrication influence the mineral properties, which can in turn influence degradability and drug release [24]. Pure HAP has a calcium/phosphorous (Ca/P) ratio of 1.67 and a low level of degradability over time, while biomimetic HAP retains the $Ca_{10}(PO_4)_6OH_2$ formula but is calcium deficient and its Ca/P ratio ranges from 1.55 to 1.67 [25]. The degradation characteristics of HAP are important, as they substantially influence release of drugs from HAP biomaterials. Matusmoto et al. were among the first to demonstrate that changing the degradation of pure HAP via changes in sintering temperature influences the release rate of bound proteins, with faster degradation kinetics leading to more rapid protein release [26]. HAP properties (e.g., degradability, osteoconductivity) can also be influenced by porosity, which is affected, for example, by the presence of water molecules during HAP sintering. Pores of a sufficient size can allow for bone ingrowth, creating a macroporous, osteoconductive interface with a large surface area to promote new tissue formation [27,28]. Likewise, pores can be used to increase drug binding capacity and facilitate drug release. Microporous and macroporous HAP each show promise in their ability to deliver proteins and other drug molecules. The purpose of this section is to introduce common fabrication methods used to form the types of HAP that have been used to achieve sustained, localized drug delivery in therapeutic applications.

HAP is fabricated via a variety of methods, including solid state reactions, precipitation from solution, hydrothermal processing, electrospraying [29], electrospinning [30–32], flux cooling [33], rapid combustion [34], flame spray pyrolysis [35], ultrasound [36], mechanochemical processing [37], microwave-assisted precipitation [38], microemulsion/surfactant-assisted precipitation [39–41], hydrolysis of metal organic precursors [42–44], chemical vapor deposition, and plasma deposition [45]. Though varied, these methods can be categorized based on the temperature ranges at which they form materials: physiological (low temperature) and elevated (high temperature) [6]. The following subsections briefly introduce common types of HAP fabrication methods.

5.3.1 Fabrication of Calcium Phosphate Minerals: Effects of Temperature

Temperature is a key element of calcium phosphate mineral formation and strongly influences the mineral phase produced. Precipitated HAP or "low-temperature HAP" is obtained by aqueous precipitation of calcium and phosphate ions homogeneously (i.e., precipitation via nucleation in solution) or heterogeneously (i.e., precipitation via nucleation upon a solid phase substrate). High-temperature HAP is produced by thermal reactions greater than 800°. Sintering, which is the process of removing air spaces from CaP materials, typically heats minerals to temperatures in excess of 1200°C, producing highly crystalline pure HAP with a Ca/P molar ratio of 1.67 and little to no degradability in aqueous solution [46]. Low-temperature calcium phosphate minerals can be decompensated or calcinated [47] to change the phase of the material to form stable α- or β-tricalcium phosphate (TCP) or high-temperature HAP. Calcium phosphate cements are a mix of calcium phosphate formulations—with mixing at 40°C–80°C, the CaP formulations dissolve and mix into a less soluble, rigid CaP cement that sets to form precipitated HAP at pH > 4.2 or brushite at pH ≤ 4.2 [48]. Cement processing varies based on the volume of the cement and the ratio of particles to liquid, and these parameters dictate cement rigidity and toughness.

5.3.2 Sintered Calcium Phosphate Ceramics

The initial calcium phosphate crystals introduced into a sintering process can be nucleated from a supersaturated solution or grown on an organic platform before being sintered at high, supra-physiological temperatures. Sintering heats HAP particle suspensions of low density to form more dense masses, causing a decrease in specific surface area and porosity. Sintering can also combine two different calcium phosphate phases, such as HAP and TCP, and progress across three morphologic stages to form a new material. The characteristic setup of the initial phase allows for particles to be in close contact to one another before they merge, allowing numerous small particles to become fewer larger ones. At the conclusion of the intermediate stage, the contact area between particles grows larger, leading to the final stage, when the pore spaces are distributed between isolated closed pores that begin to shrink over time [47]. Sintering is a critical element of preparing ceramic HAP, as it imparts the material with strength and resistance to erosion. It is greatly influenced by the initial CaP phases present and the ratio of ions. For example, too much TCP and too little HAP can cause a coarser surface and altered morphology of the resulting sintered ceramic.

5.3.3 Calcium-Deficient Hydroxyapatite

HAP phases can also be prepared at lower temperature. For example, HAP can be produced by combining exact stoichiometric quantities of calcium and phosphate salts at basic pH, then boiling for several days with CO_2 extraction, filtration, and drying. This process, and other similar aqueous processes, is extremely sensitive to the presence of additives, and often results in the formation of calcium deficient hydroxyapatite (CDHA). CDHA is nonstoichiometric in its Ca/P ratio and known to have a large number of defect sites. It is often highly substituted and poorly crystalline, for example, ion substitutions such as carbonate substitution for phosphate can influence the crystal lattice parameters, morphology, crystallinity, solubility, and thermal stability [6]. Heating CDHA with a Ca/P ratio between 1.5 and 1.67 above 700°C creates a biphasic calcium phosphate, which contains both β-TCP and HAP phases. This biphasic calcium phosphate has been used in a variety of bone void filling applications discussed elsewhere [49,50], but will not be discussed in detail here.

5.3.4 Calcium Phosphate Particles: Toward Drug Incorporation and Release

A variety of applications have also led to the development of HAP microparticles and nanoparticles. Solid HAP particles have a limited surface area for drug binding and therefore have poor drug incorporation efficiency. Incorporating porogens can increase surface area and facilitate better drug integration. Cosijns et al. fabricated porous HAP tablets by eccentrically pressing a mixture of milled HAP and a 50% (w/w) pore former, then sintering the final product at 1250°C. The investigators discovered the upper limit of pore incorporation and the sintering temperature at which tablets maintained their stability. Interestingly, these tablets could be used to efficiently incorporate proteins, and the rate of protein release was proportional to the pore size. Indeed, protein release kinetics could be accelerated by up to 26-fold via an increase in pore size [51].

Hollow HAP particles in spherical or rod-like structures provide an alternative vehicle to encapsulate therapeutic agents [52]. Hollow HAP microspheres spray dried at 180°C or calcined at 500°C for 1 h resulted in porous surface microspheres ranging from 1 to 10 μm. Immersed in a simulated body fluid (SBF), the mass of the particles decreased initially until day 8, suggesting particle dissolution, but then increased over time, suggesting new apatite precipitation on the microsphere surface. The slight changes in mass between days 8 and 21 were attributed to a shift from amorphous HAP to crystalline HAP, and after 21 days a dissolved, porous structure remained. As a delivery vehicle, hollow spheres are able to incorporate and release insulin into phosphate-buffered saline (PBS) at physiological temperature and pH over a 7 h timeframe. The hormone release kinetics followed a burst release trajectory over the first 3 h, followed by a linear release. Because the insulin was released prior to significant HAP dissolution, the authors proposed that the release was dependent on sphere porosity and the insulin–microsphere binding affinity [53]. This study is representative of a series of studies that have used micro- or nanoscale HAP materials for protein incorporation and release.

5.3.5 Calcium Phosphate Cements

Self-hardening in situ calcium phosphate cements are attractive for in vivo applications because they form HAP adjacent to existing bone, contouring to the geometry of a defect more closely than premolded implants. The setting process is often a result of HAP dissolution and precipitation, and the cement hardens as the precipitated HAP particles interlock. Cements are advantageous over other preparations because they are injectable and microporous and can set at low temperatures allowing for incorporation of biologically active drugs such as protein growth factors [54]. Cement base materials are composed of a variety of minerals, including tetracalcium phosphate and anhydrous calcium phosphate. When they form crystalline apatite, CaP cements are osteoconductive with respect to the surrounding bone [21]. However, cements have a lower compressive modulus and failure strength than typical sintered HAP ceramics, which substantially limits their applicability in load-bearing sites. Some studies have used composite cements to increase the load-bearing capabilities with some success. Regardless of their material properties, these cements are potentially good candidates for use as drug carriers, provided that the drug remains active within the hardened cement.

5.4 "Bioinspired" Calcium Phosphate Synthesis

Biomineralization constitutes the formation of inorganic minerals within biological systems, which typically involves induction, nucleation, and mineral growth facilitated by the presence of an organic matrix and cellular activity [55]. Organisms including bacteria, protozoa, and mollusks produce minerals with complex architectures in extracellular spaces, directed by an organic extracellular matrix which is critical in dictating the structural framework [55]. The intricate organization and adaptability of natural biomineralization processes has led to widespread interest in biomimetic or "bioinspired" mineralization processes ex vivo, which seek to replicate some of nature's mineral properties. In particular, bioinspired HAP synthesis proposes to simplify and deconstruct apatite biomineralization and thereby create a material that retains critical functions of native HAP. Critical HAP functions, as suggested by Legeros et al., include protein adhesion, interconnecting porosity, biodegradability, bioactivity, osteoconduction, and the ability to elicit a pro-osteogenic cellular response [56].

5.4.1 Heterogeneous Mineral Nucleation in Simulated Body Fluid

There are myriad of ways in which HAP has been produced in simulated physiological conditions to form a crystalline mineral. The most common is by incubating a substrate material (e.g., an orthopedic implant) in an SBF to encourage heterogeneous nucleation. Using this approach, Kokubo et al. initially demonstrated that an HAP film can be formed on a silicate bioglass immersed in a circulating SBF containing mineral ions present in the blood [57,58]. The resultant mineral coating has a carbonate-substituted, calcium-deficient HAP phase, nanoscale porosity, and a lower stability than sintered, stoichiometric HAP. We and many others have extended the Kokubo approach to enable synthesis of numerous calcium phosphate mineral compositions on a broad spectrum of materials, and the mineral properties are generally dictated by the characteristics of the SBF (Figure 5.2). For example, Lu and Leng describe the formulation of different HAP phases

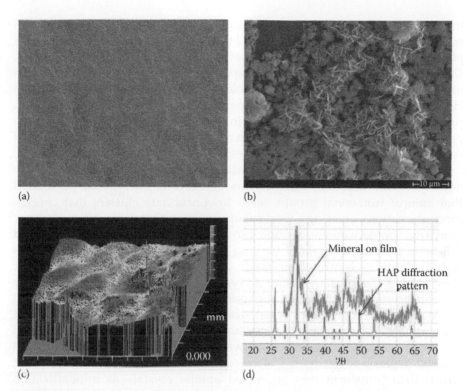

(a)　　(b)

(c)　　(d)

FIGURE 5.2
(See color insert.) Scanning electron microscopy of bioresorbable polymer films (a) before incubation and (b) after biomimetic incubation in a modified simulated body fluid (mSBF). Films were composed of the bioresorbable polymer poly(ε-caprolactone) (PCL). Incubation of films in mSBF resulted in the formation of a continuous mineral layer that displayed a nanoporous, plate-like morphology. (c) A profilometric analysis of the films demonstrated as well the formation of a continuous mineral coating. (d) Characterization of mineral phase by x-ray diffraction confirmed the formation of hydroxyapatite (HAP). The blue trace represents the diffraction patterns associated with pure HAP.

based on the composition of the SBF and its relationship to the ions found in human blood plasma [59]. Interestingly, crystallinity of HAP depends on its synthesis temperature, and low-temperature synthesis yields low crystallinity. For example, Matsumoto et al. demonstrated that a difference of 20°C has a direct influence on crystal size and an inverse relationship with crystal surface area [26].

One of the common themes in biomineralization, and by extension in biomimetic mineralization, is heterogeneous nucleation. This type of nucleation occurs when nuclei are energetically stabilized on an organic surface. The saturation of mineral ions in the surrounding solution serves as the driving force for the rate of mineral formation, as well as dissolution. The greater the concentration of ions in a supersaturated solution, the higher the driving force for mineral precipitation. Mineral precipitation is delayed in solutions with insufficient supersaturation, as less saturated solutions can exist in metastable states that may need to be seeded or "doped" with calcium phosphate seeds or other charged moieties to induce mineral nucleation [60]. Undersaturated solutions are counterproductive to the mineral nucleation process and encourage mineral dissolution, by decreasing the size of the HAP crystals. Because there is no energy barrier limiting the dissolution process, it can occur frequently even in the midst of the mineral nucleation process. This dissolution/precipitation dynamic is less volatile as the level of supersaturation remains high, but does still typically occur.

The precise mechanism by which calcium phosphate minerals nucleate in nature has been debated for many years, and several studies have postulated that anionic functional groups serve to concentrate clusters that contain inorganic cations, creating a local supersaturation that promotes nucleation. Similarly, spontaneous nucleation can occur in vitro from supersaturated solutions, forming crystals of varying shapes and sizes. Weiner describes two methods to control biomimetic growth of large crystals by either slow cooling the substance onto a crystal seed or using a crystal seed to nucleate crystal formation in a supersaturated solution. In contrast, natural organisms create crystals of equal size along a determined trajectory and orientation and are not able to withstand the temperature increases needed by the slow cooling approach. Moreover, it has been determined that solid mineral deposits in nature are often organized randomly or amorphously, which then transform into organized apatite crystals as mineralization progresses [61]. Mann et al. determined that the key to controlling the crystal product is the ability to create an inorganic–organic interface that recognizes molecules and incorporates them into the growing crystal lattice. Proteins that recognize specific crystal surfaces can also control mineral growth and dissolution by determining the surface topology of the crystal face [62].

Mineral formation is also often achieved within or upon an extracellular matrix, including self-assembled matrices and planar substrates (e.g., Figure 5.2). Cui et al. used self-assembly to prepare mineralized collagen nanofibrils, noting that the self-assembled collagen nanofibrils served as a template for HAP precipitation. The fibrils were manipulated by changing the composition of collagen monomers and placing them into solutions with varying calcium and phosphate ion concentrations. The crystal lattice formed on the surface of the fibrils had a higher density than the interior mineral. Importantly, HAP crystals were formed in the gap zones between collagen fibrils, and the hexagonal crystals were oriented with their c-axis along the longitudinal axis of the fibrils. These data suggested that the fibrils served as a template for oriented HAP crystal growth, and further data from Cui et al. showed that negatively charged side groups on the collagen molecules served as nucleation sites for HAP [63]. Importantly, this type of mineralized collagen matrix bears key similarities to natural bone, which is a composite of organic collagen fibrils and oriented inorganic apatite, and this composite leads to bone's unique structural and mechanical properties.

5.4.2 Biomolecule Incorporation during Bioinspired HAP Formation

Apatites formed via heterogeneous nucleation in simulated physiological conditions have several desirable features, as they can allow for mineral formation on a variety of geometries, with mineral properties that can be readily modified during formation by varying solution characteristics. Therefore, biomimetic HAP formation can create a biocompatible mineral interface on the surface of a variety of material types fashioned in variable geometries. HAP formation in SBFs also allows for protein incorporation during mineral growth, which can add biological activity to a mineral interface. HAP coatings on materials can then be used for delivery of "encapsulated" proteins, small molecules, DNA, and other drugs to a particular region of interest.

In one approach, investigators have examined incorporation of biological molecules (e.g., proteins) into a growing mineral coating. For example, bovine serum albumin (BSA) has been incorporated into an SBF solution during the formation of an HAP coating on the surface of a poly(lactic-co-glycolic acid) material. This study explored a variety of processing variables, including co-precipitation time and subsequent protein adsorption [64]. The results showed that co-precipitation leads to higher incorporation of protein than adsorption to a preformed mineral layer, as well as seemingly higher protein retention after scaffold rinsing. The authors also observed that protein co-precipitation changes mineral morphology, leading to decreased crystal size and crystallinity. This study also demonstrated that BSA incorporation influences mineral properties, as there was a correlation between BSA incorporation and mechanical strength. Although this study did not focus on the mechanism of protein incorporation or discuss biological activity of the protein, it did demonstrate that protein can incorporate into a growing mineral and influence resultant mineral properties. Other studies have co-precipitated HAP and BSA onto titanium surfaces to achieve efficient protein loading. They have found that loading efficiency varies with protein concentration, as expected, and that subsequent protein release kinetics depends on the strength and number of bonds that form between the mineral and BSA as the coating increases its porosity.

Importantly, the presence of protein during crystal growth can affect the nucleation and growth of the mineral in several ways. It is important to consider the potential influence of proteins, as there is the potential for undesirable trade-offs between efficient protein incorporation into a growing mineral and soluble protein effects on the growing mineral. The presence of proteins in SBF can inhibit or promote mineral nucleation and growth depending on the concentration of protein included. In the presence of albumin, the interfacial energy of mineral nuclei is decreased, stabilizing the smaller nuclei that grow at a slower rate. As a result, the higher the albumin concentration, the longer it takes for the crystal to form. Likewise, a high protein concentration also limits diffusion of ions to the crystal surface. In contrast, if there is not sufficient protein concentration to coat the surface of mineral nuclei, then crystal growth occurs at a faster rate. Over time, as more protein adsorbs to the surface, the solution protein concentration decreases and eventually less stable mineral nuclei form. Importantly, since the protein can incorporate into the crystal lattice, the mineral may have a protein-dependent morphology, composition, or crystallinity. Each of these parameters is critical, as they can alter subsequent mineral dissolution and associated protein release kinetics.

5.5 Therapeutic Delivery from Hydroxyapatite Biomaterials

HAP typically serves as an inexpensive, nontoxic, minimally resorbed material that is able to facilitate appositional bone growth. Thus, HAP has been extensively used as a bone substitute and a coating material on orthopedic and dental implants. It has more recently been appreciated that HAP also serves as an outstanding carrier for biological molecules, and it is widely applicable to a number of small molecules, proteins, and nucleic acids. A number of therapeutic proteins (e.g., growth factors) have been shown to enhance tissue regeneration and bone graft healing, but their effects are short lived when directly injected into open defects. Since HAP biomaterials are inserted directly into a defect site, they can provide local therapeutic delivery and circumvent some limitations of systemic therapies. Contemporary therapeutic delivery systems require compatibility with clinical settings as well as the capacity to efficiently incorporate and release therapeutic agents with controllable dosage and timing. The high binding affinity of HAP for a broad range of proteins and nucleic acids is well documented and provides an adaptable mechanism for drug incorporation and release. The following sections will introduce a series of specific therapeutic applications that have been explored to date using HAP biomaterials as drug carriers.

5.5.1 Antibiotics against Bacterial Infection

Because of the high incidence of HAP in bone implant coatings, a large amount of work has focused on incorporating antibiotics onto HAP at the time of implantation to prevent infection. Implant-associated infections pose a serious impediment in adult orthopedic implant procedures including total joint replacements and fracture fixation. The number of implant-associated infections has continuously increased due to growing demands for surgical implantation and remains a primary cause of postoperative complications. Without proper treatment, implant-associated infections often necessitate additional surgical procedures such as simple debridement and implant removal, followed by reimplantation. It is during this window between removal and reimplantation that antibiotic delivery carriers have often been used to improve therapy.

Osteomyelitis is a particularly severe infection found mostly in the metaphysis of long bones. It may be secondary to injury or joint replacement. Common bacterial strains involved in osteomyelitis include *Staphylococcus aureus*, *Pseudomonas aeruginosa*, and *Escherichia coli*. A majority of *S. aureus* pathogens are methicillin sensitive (60%–80%) and 10% are methicillin resistant (termed methicillin resistant *S. aureus* or "MRSA"), but in the United States, resistance is as high as 40%–50%. In many cases, antibiotic treatment must be initiated before susceptibility testing is obtained, treatment includes surgical debridement and systemic antibiotics for several weeks. Acute osteomyelitis can transform to chronic osteomyelitis if left untreated, causing changes to the bone architecture. Bone grafts are often necessary to reconstruct the bone loss caused by chronic osteomyelitis and prevent substantial limb deformity and risk of amputation. Thus, there is substantial motivation to develop therapeutic approaches to treat implant-associated infections and osteomyelitis.

Current approaches for systemic, intravenous antibiotic therapy can lead to significant complications, including systemic toxicity, low efficiency, and need for hospitalization. Localized antibiotic treatment has the potential to decrease systemic side effects and offers the potential to ease the administration of dosing (less parenteral administration), decrease the number of hospitalizations, and provide significant cost savings. Localized antibiotic delivery from HAP-based implants has been proposed as a treatment of skeletal infection or a prophylactic strategy. Local delivery provides a higher antibiotic concentration at the site of infection without systemic toxicity and better dead area management [65].

HAP deposition on implants has a long and well-established history of osteoconductivity, combined with HAP coatings that can be used to deliver antibiotics locally will potentially provide great advantages in orthopedic applications. Traditional HAP or β-TCP ceramics can also be readily processed into blocks, scaffolds, or granules and used as antibiotic carriers. In addition to overcoming systemic side effects, various geometries of HAP-based bone void fillers such as microspheres [66–68], nanoparticles [52], porous scaffolds [69,70], and self-setting cements [71–73] have been adopted to incorporate and release antibiotic drugs. For example, self-setting HAP bone cement was used as an antibiotic carrier to replace poly(methylmethacrylate) (PMMA) beads, which required a secondary surgery to remove the nondegradable PMMA increasing the propensity to induce a local foreign body reaction. HAP cement designed to release vancomycin in a sustained manner reduced infection by *S. aureus*, methicillin-resistant *S. aureus* (MRSA), and *S. aureus* with small colony variant (SCV) in a rabbit tibia model of chronic bone infection. HAP cylinders designed to release vancomycin in a continuous fashion also disrupted a biofilm of *S. aureus* tested in vitro. Because of the superior bactericidal activity of vancomycin, its biologically active release from cylindrical HAP scaffolds is promising for bone void filling applications [74].

Interestingly, several strategies to control the antibiotic release rate have been proposed, including encapsulating the antibiotic inside an HAP block, incorporating it into biomimetic HAP coatings, or creating protective polymeric coatings on HAP implant surfaces. A clinical study incorporated antibiotic within HAP blocks and showed that the local antibiotic level in the wound was significantly higher than normal serum levels and was able to treat MRSA. Seven osteomyelitis patients were free of infection without complication [75]. Block HAPs may be more advantageous when compared to bone cements as they are simpler to remove if complications occur. However, they are limited by their brittle mechanics and their inability to conform to defect geometry.

Stigter and coworkers successfully incorporated various antibiotics into carbonated HAP coatings via co-precipitation during coating growth. For all released antibiotics tested, antibacterial activity was confirmed against *S. aureus*. This method allowed for a greater incorporation of antibiotics and a longer time period of release when compared to the adsorption of antibiotics on plasma-sprayed HAP coatings. Interestingly, the incorporation efficiency and release kinetics were dependent on the molecular characteristics of the antibiotic. Antibiotics containing carboxylic acid groups showed higher incorporation efficiency and slower release, likely due to their interaction with calcium in the HAP coatings [76,77].

Shinto et al. incorporated three different antibiotics—gentamicin, cefoperazone, and flomoxef—into the cavity of porous HAP blocks and observed the most rapid release from flomoxef and the slowest from gentamicin. Gentamicin released from the cavity of porous HAP blocks was still effective against *S. aureus* at 12 weeks in vitro and, when implanted in rat proximal tibia, a similar release profile was detected [78]. In subsequent clinical studies, these HAP blocks have demonstrated efficacy in preventing infection recurrence after treatment in chronic osteomyelitis patients [65,79].

Some studies have generated multicomponent materials to further control antibiotic release. For example, HAP combined with calcium sulfate and chitosan provided a stable, sustained release of gentamicin. In this case, the advantages of each of the three materials can be leveraged to provide optimal antibiotic delivery. Calcium sulfate dissolution creates an environment that neutralizes cytotoxicity and slows down antibiotic release, while chitosan/HAP provides an additional structural support and scaffolding that alleviates the need for additional bone graft material. The result yields a biodegradable composite bone graft substitute that gradually releases gentimycin and limits cytotoxicity [80].

5.5.2 Silver Particles

Silver particles are often used as a natural antibiotic against a broad range of bacterial strains and provide an alternative method to decrease nonspecific or antibiotic-resistant bacterial infections commonly observed in osteomyelitis. Silver is well-known as a reliable antimicrobial agent for wound healing [81], and silver agents have been shown to be minimally toxic to human cells on biomedical implants [82], including HAP implants. For example, investigators have efficiently incorporated silver into brushite and HAP minerals. HAP-bound silver particles showed minimal release and reduced antibacterial activity when compared to the brushite materials that showed sustained release. However, the addition of silver to HAP cement decreased the compressive strength by 50% [83]. In another study, silver-doped HAP coatings on titanium disks disrupted *S. aureus* and *S. epidermidis* biofilms initially after device implantation. Fortunately, this coating did not alter the behavior of osteoblast precursor cells when compared to the HAP-coated samples without silver [43]. A similar study conducted by the same authors sputter coated silver and HA onto titanium targets that were then heated to 400°C. The authors reported antibiotic effects against *S. aureus* and *S. epidermidis* and no significant difference in the water contact angle of the coating surface with the addition of silver, supporting its benign use as an antimicrobial [30].

5.5.3 Growth Factors

Healing of substantial skeletal defects can take place on the order of months and is often complicated by the formation of reactive fibrous tissue (i.e., scar tissue) instead of the desired tissue. After bone injuries, multiple endogenous growth factors are expressed and work in concert to orchestrate proper bone repair. Exogenous growth factors have been administered during the healing process to accelerate and actively guide tissue regeneration. To date, collagen, chitosan, and other naturally derived materials, as well as synthetic poly(α-hydroxyesters) and other synthetic polymer carriers, have been used to deliver recombinant human bone morphogenetic protein-2 (rhBMP-2), rhBMP-7 (OP-1), and other members of the transforming growth factor beta (TGF-β) superfamily of growth factors for bone and soft tissue regeneration [22,84–88]. Collagen carriers are plagued by rapid growth factor release at supra-physiological dosage, which has raised safety concerns, including potential for heterotopic bone formation, edema, and increased cancer risk. In this context, HAP biomaterials represent a promising candidate for growth factor delivery, as they offer excellent protein binding and retention while maintaining protein biological activity. In addition, they provide the potential to control the rate of growth factor release by controlling the dissolution rate of the mineral (Figure 5.3). The following paragraphs introduce a subset of the studies that have used HAP-based materials to deliver growth factors, and paragraphs are separated based on the specific growth factors delivered.

Bone morphogenetic proteins (BMPs) can induce osteogenic differentiation of mesenchymal stem cells and other osteoprogenitor cells to direct bone ingrowth and new bone formation in bone defects. Thus, there has been widespread

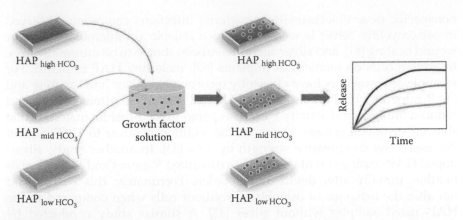

FIGURE 5.3
(See color insert.) Biomimetic HAP coatings with different extent of carbonate incorporation (represented in black, blue, and green) allow for control over mineral dissolution and associated release of drugs. Higher carbonate incorporation leads to higher dissolution rates and therefore faster release kinetics.

interest in delivering BMPs from bone void filling materials such as HAP-based materials. Liu and coworkers incorporated rhBMP-2 either on biomimetic HAP coatings by surface adsorption or into biomimetic HAP coatings by co-precipitation and found that in each condition the released rhBMP-2 induced ectopic bone formation in vivo [89]. Similarly, Schnettler et al. demonstrated ectopic bone formation and also accelerated bone ingrowth throughout an HAP-coated titanium implant with rhBMP-2 [90]. Levine et al. showed that when HAP implants were placed in an anterior muscle, BMP-treated implants induced greater vascularity and bone formation than untreated controls [91]. Ripamonti and coworkers have reported that BMP-7, also referred to as human osteogenic protein-1 (hOP-1), loaded on sintered porous HAP promoted heterotopic and orthotopic bone formation in adult primates [92].

Bone regeneration in large bone defects requires not only bone induction but also other growth factor-dependent processes such as angiogenesis, which is the formation of blood vessels from a preexisting vascular supply. Blood vessels are a source of key osteoprogenitor cells and mesenchymal stem cells, as well as the nutrients and oxygen to feed the developing tissue. Lack of functional vasculature often leads to delays in the healing process and can result in fibrous nonunions. Therefore, local delivery of pro-angiogenic growth factors has been a subject of recent research interest from our group and others (e.g., Figures 5.4 and 5.5). Vascular endothelial growth factor (VEGF) is among the most potent pro-angiogenic growth factors, and it is known to be expressed at elevated levels for nearly a week after bone fracture. It was previously shown that rhVEGF efficiently bound to various HAP-based bone implant materials, including bone cement, porous blocks, biomimetic coatings, and solvent dehydrated human bone grafts. The bound VEGF subsequently released in a sustained manner and retained its

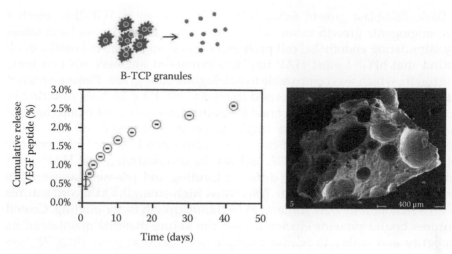

FIGURE 5.4

Release of vascular endothelial growth factor (VEGF) mimetic peptide from β-TCP granules coated with an HAP mineral layer. HAP coatings were formed by incubation of granules in a modified simulated body fluid (mSBF) for 7 days. Release of the VEGF mimetic peptide was sustained for over 40 days.

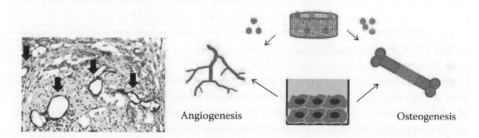

FIGURE 5.5

(See color insert.) Sustained release of growth factors from HAP mineral coatings can influence cellular activities that lead to the formation of various tissue types, such as blood vessels and bone. Sustained release of recombinant human vascular endothelial growth factor (VEGF) induced new blood vessel formation in vivo. Arrows pointing dark brown stained lumens represent blood vessels based on von Willebrand factor immunostaining.

biological activity [93,94]. Another study reported that rhVEGF released from polymer scaffolds coated with a biomimetic HAP layer increased both blood vessel density and bone formation in a rat cranium critical defect [95,96]. More recent studies from our group have demonstrated that VEGF can be efficiently bound to HAP coatings (Figure 5.4) and released in a controllable manner to induce angiogenesis within porous scaffolds in vivo (Figure 5.5). These studies emphasize the potential use of HAP as a component in tissue engineering strategies, in which combinations of biomaterials, molecules, and cells are used to repair, restore, or improve upon tissue function.

Basic fibroblast growth factor (bFGF, also known as FGF-2) is another pro-angiogenic growth factor, which can promote blood vessel formation by stimulating endothelial cell proliferation and migration. Schnettler et al. found that bFGF-loaded HAP implants enhanced angiogenesis and bone ingrowth, which was comparable to autologous bone grafts. Plasma-sprayed HAP coatings have also been used to release bFGF for 10 days, resulting in a significant increase in new bone formation and improved biomechanical properties three months after implantation in ovariectomized rats. Midy et al. compared the ability of carbonate substituted HAP or stoichiometric HAP to bind and release bFGF, and results demonstrated that HAPs with similar structure can exhibit different binding and release characteristics [97]. In another study, Lee et al. [98] coated high-strength Orthocord sutures with a nanoporous HAP mineral to facilitate growth factor binding. Coated sutures bound proteins efficiently, and the suture material maintained its integrity and ability to release multiple proteins over more than 30 days in a sustained fashion in a sheep infraspinatus tendon. This approach was then used to release biologically active bFGF in a sheep model of rotator cuff repair. Six weeks postoperatively, histological analysis demonstrated that bFGF-releasing sutures promoted higher levels of angiogenesis and increased tissue formation in the healing rotator cuff. This result supports the use of HAP coatings as growth factor delivery platforms not only for bone void filling applications but also in soft tissue healing applications.

Transforming growth factor-β1 (TGF-β1) and platelet-derived growth factor (PDGF) are multifunctional regulators for a broad range of cellular and biological processes. In particular, TGF-β1 plays an important role in bone formation by stimulating angiogenesis, osteogenic cell proliferation and differentiation, and extracellular matrix production. TGF-β1 mixed with HAP bone cement has been shown to stimulate osteogenic differentiation of rat pre-osteoblastic cells in vitro and increase the bone volume in rat calvarial bone defects in vivo [99]. Arm et al. released PDGF from the surface of HAP in the presence of albumin and measured its effects on bone ingrowth. Although the released PDGF was found to be biologically active, release from the surface was minimal over the course of three weeks [100,101], and there was no significant effect on bone ingrowth.

5.5.4 Hormones

Insulin injections provide life-sustaining medications for type 1 diabetics and a growing number of type 2 diabetics. Limited by low patient compliance for repeated dosing, investigators continue to seek new delivery approaches for sustained insulin release, including oral administration, subcutaneous pumping, and extended release injections. Insulin encapsulated into poly(ethylene glycol)-modified calcium phosphate nanoparticles has been shown to prevent release at gastric pH and sustain release over eight hours in the intestine to facilitate drug absorption [102]. Circular dichroism spectra of insulin conformation before and after release were not changed during the nanoparticle fabrication or functionalization process [91]. This work is an extension of the previous work from the same group, in which insulin was encapsulated into porous HAP microspheres via solvent evaporation and coated with poly(ethylene vinyl acetate) for sustained release. Their results demonstrated a 2 day release of 92% of the encapsulated insulin, including a burst release and a gradual plateau [103] The released, biologically active insulin had a relatively low duration of action due to low initial incorporation efficiency, suggesting the need to repeat studies in other animal models or with other CaP biomaterials as a better model for scale-up to human use. HAP materials have also demonstrated controlled release of other biologically active hormones that regulate bone formation, including growth hormone and parathyroid hormone. For example, Yu et al. demonstrated a bioactive release of parathyroid hormone related protein (PTHrP). PTHrP was incorporated into an HAP-coated titanium surface via a co-precipitation in modified SBF and released in a sustained manner [104].

5.5.5 Bisphosphonates

Bisphosphonate, an osteoclast inhibitor, is clinically used for the treatment of peri-prosthetic osteolysis, osteoporosis, and similar degenerative bone diseases. Current treatment regimens involve oral administration or intravenous injection, which can each lead to side effects including fever [105], bone pain, rigors [106], gastrointestinal ulcer [107], osteonecrosis, and ocular inflammation. Bisphosphonates incorporated within HAP materials exploit the strong affinity between the Ca^{2+} ions in HAP and the two phosphonate groups attached to a single carbon atom [108]. This mechanism ideally provides two modes of therapy: (1) inhibiting bone resorption activity of osteoclasts to prevent the loss of bone mass via the effects of the bisphosphonate and (2) promoting implant osteointegration to native bone tissue via the osteoconductive HAP coating. A HAP coating on an implant surface can connect the bisphosphonate to the implant and plausibly decrease implant loosening as the appositional bone may be more substantial [109]. Taken together, HAP-bisphosphonate delivery provides potential to improve in vivo osteointegration while inhibiting natural bone resorption.

Peter et al. adsorbed the bisphosphonate, zoledronate, onto HAP-coated titanium implants and sought to determine the optimal concentration to increase mechanical fixation. Results showed a decrease in osteoblast function and impaired mineralization with elevated zoledronate concentration, as evidenced by a decrease in implant–bone pullout force [110]. Gao et al. delivered the bisphosphonates, zoledronate, ibandronate, and pamidronate, and demonstrated positive effects on promoting bone–implant fixation in demineralized, osteoporotic bone [111]. Although the mechanism of action is not yet clear, Gao et al. hypothesized that the bisphosphonates are bound directly to HAP and are released slowly into the new bone mineral, causing osteoclast apoptosis and supporting bone formation and associated bone–implant fixation. Due to their success in these and other bone healing applications, local HAP/bisphosphonate delivery has been hypothesized for use in more rare etiologies, including Gorham Stout syndrome and childhood hemangioma. Gorham Stout syndrome is thought to be a debilitating degenerative disease with abnormal osteoclast activity, analogous to osteoporosis. Current treatments include vitamin D, sodium fluoride, and systemic bisphosphonates. Sun et al. hypothesized that zoledronate released from HAP at a precise time and concentration could enhance bone reconstruction [112]. In another study, Yu et al. hypothesized that bisphosphonates bound to HAP and released locally could promote anti-angiogenic effects for children with unresolved hemangiomas. Benign hemangiomas can have severe side effects in some cases, including cartilage damage, congestive heart failure, or airway obstruction. Slow released zoledronate is known to inhibit serum bFGF and VEGF, but systemic or toxic levels of administration may impair skeletal maturity. The researchers hypothesized that sustained drug release from a bioresorbable carrier locally implanted could treat congenital disease with minimal side effects and high efficacy [113]. These emerging applications of HAP/bisphosphonate therapies may ultimately show substantial clinical impact.

5.5.6 Anticancer and Anti-Inflammatory Drugs

Several types of antineoplastic agents have been incorporated within HAP carriers with demonstrated release. Barroug et al. demonstrated that cisplatin adsorbed on the surface of HAP nanoparticles was released gradually over 20 days and the released cisplatin exhibited dose-dependent cytotoxic effects against a murine osteosarcoma cell line [114]. When mixed with HAP bone cement, 6-mercaptopurine was released over 4–25 days with release kinetics dependent on the porosity of the HAP cement [115,116]. Paclitaxel adsorbed on an HAP disc also proved effective against osteosarcoma cells and metastatic breast cancer cells [117].

Cisplatin, a widely used antitumor agent, binds more favorably to HAP as the crystallinity of the material decreases, resulting in slow drug release during HAP dissolution. For example, Palazzo et al. compared binding and release of cisplatin, alendronate, and di(ethylenediamineplatinum)medronate (DPM) to HAP surfaces. Specifically, they created HAP in either plate-like or needle-shaped morphologies, and the needle-like mineral showed a lower Ca/P ratio and lower crystallinity than the plate-like mineral. Additionally, the plate-like mineral had a slightly larger surface area and potentially higher drug binding capacity relative to the needle-shaped mineral. However, in this case, the less crystalline needle-shaped surface resulted in a higher amount of cisplatin and alendronate binding. This result may be partially explained by the chemistry of drug–HAP binding. In some cases, drug binding to a HAP surface can lead to ion exchange and an associated decrease in the Ca/P ratio, and in the case of the Palazzo et al. study, the negatively charged alendronate likely elicited the exchange of two moles of phosphate for each mole of alendronate. Cisplatin is positively charged and could therefore bind to the phosphate-rich HAP surface. In contrast, DPM is neutral and was likely influenced less by the Ca/P ratio of the mineral. In general, the drug concentration and the Ca/P ratio and crystallinity likely had a substantial effect on drug binding by varying the presentation of drug binding sites on the surface. Within 16 days, 45% and 21% of the cisplatin and alendronate were released, while nearly 85% of the DPM was released over the same timeframe. Taken together, the Palazzo et al. results demonstrated that the HAP surface area and Ca/P ratio influenced drug delivery from HAP more strongly than the specific morphology of the HAP crystals [118].

Nonsteroidal anti-inflammatory drugs (NSAIDS) are widely used in clinical applications but can have side effects on the gastric mucosa, kidneys, and platelets when therapeutic doses are delivered systemically. Therefore, there is a motivation to develop localized delivery systems for NSAIDS, including HAP-based systems. Tvala et al. demonstrated that ibuprofen bound to calcium phosphate granules could be released gradually into a simulated physiological environment over 2 days. Melville et al. revealed that ibuprofen adsorbed on HAP particles had a release rate dependent on the crystal size and surface area, which were readily controlled during HAP particle preparation [119]. Otsuka et al. loaded the NSAID indomethacin homogenously within HAP bone cement, and the drug was released gradually over 3 weeks. They also found that in vivo release was slower than in vitro release, likely because the proteins and cells present in vivo altered the indomethacin–HAP interaction or the properties of the HAP cement surface.

5.5.7 Other Drug Classes

In addition to proteins and small molecules, HAP can also be used as a vehicle to deliver nucleic acids such as DNA and glycosaminoglycans (GAGs) such as heparin. Although these drug classes are not discussed in detail here, several previous studies have detailed the well-characterized polyanion–HAP interaction (e.g., DNA–HAP interaction) and its use for nonviral transfection, siRNA delivery, and shRNA delivery. In one recent example, Choi et al. demonstrated sustained, localized release of plasmid DNA from CaP mineral coatings and the DNA release controlled by carefully tailoring the coating properties [120]. The results demonstrated uniquely high levels of substrate-mediated transfection of stem cells, restricted to the cells in contact with the mineral coating. This approach may find widespread use in clinical applications that call for efficient gene delivery. For GAG delivery, Bajpai et al. developed a calcium phosphate system to continuously release heparin to rats, resulting in a dose-dependent increase in blood clotting time [121]. The carrier material was optimized to provide an effective dose without heparin overdose, which would have substantial clinical consequences.

5.5.8 Cells

Mesenchymal stem cells (MSCs) are multipotent progenitor cells that can differentiate into multiple mature cell types, including osteoblasts, chondrocytes, and adipocytes. Natural bone healing processes involve recruiting osteoprogenitor cells such as MSCs from the periosteum and bone marrow. These cells then differentiate into bone-forming osteoblasts. Based on their well-characterized ability to form calcified bone tissue, MSCs are being extensively developed for bone repair and regeneration. Bruder and coworkers isolated, cultured, and implanted autologous MSCs within porous biphasic calcium phosphate blocks (HAP:β-TCP=65:35) in canine critical-sized segmental defects. The MSC-loaded groups formed rapid integration at the bone–implant interface and produced a larger total quantity of bone, while the untreated groups had nonunion with limited bone formation [122]. Using the same HAP carriers, they also tested culture-expanded human MSCs by implanting in critical-sized segmental defects in the femurs of adult athymic rats [122]. The results indicate that the groups treated with cells had increased bone formation with improved biomechanical strength. Quarto et al. reported the human clinical use of autologous bone marrow–derived, culture-expanded osteoprogenitor cells combined with macroporous HAP scaffolds to treat large bone defects [123]. All three patients treated showed good bone–implant integration and bone formation throughout the scaffold, including a fracture callus without any complications up to at least 15 months. This study and others suggest a strong potential for use of MSCs combined with HAP-based implants in large bone defect healing applications. It is interesting to note that the osteogenic activity of MSCs has also been shown to be dependent on the geometric characteristics of HAP carriers, such as pore size and porosity [124].

5.6 Foundational Properties of Hydroxyapatite

The previous sections have described the ways in which HAP has been used in clinical applications and introduced scenarios in which HAP has been used as a versatile carrier for drug delivery. A series of foundational material properties of HAP have enabled the aforementioned clinical applications, as well as its tremendous potential in drug delivery. Indeed, HAP materials provide inherent benefits not only as orthopedic implants and implant coatings but also as chromatographic agents for purification of biological molecules and bulking agents in some clinical applications (e.g., vocal fold healing). This section will introduce some fundamental properties of HAP that are of relevance to its clinical and scientific applications, particularly therapeutic drug delivery.

5.6.1 Background: The Need for HAP Implants/Coatings in Orthopedics

One of the particular advantages of HAP as a drug carrier is its already widespread use in orthopedic applications. Thus, it is worthwhile to consider the manner in which HAP is used in current orthopedic applications.

Large bone defects caused by trauma, cancer, or congenital defects are most often replaced with autologous or allogeneic bone grafts [125]. Autologous cancellous bone is readily revascularized and can be rapidly incorporated into the recipient site. However, while autologous cancellous bone grafts fill the defect, they do not provide substantial structural support and may be limited by tissue availability. Autologous cortical bone grafts provide better structural support but are similarly limited by tissue availability and associated donor site morbidity. Allogeneic and xenogeneic grafts are less limited in terms of the tissue availability, but may be capable of transmitting bacterial or viral agents that lead to infection or disease, and can also be immunogenic. Demineralized bone matrix represents an attractive alternative to fit well-circumscribed defect areas that do not require load-bearing structural support, but these matrices are limited by batch-to-batch variability in properties.

Creating an osteoconductive surface on a durable, implantable material represents an alternative to autologous and allogeneic bone grafts. A sufficiently osteoconductive material can promote healing via osteoprogenitor cell adhesion, proliferation, and differentiation, which can improve biocompatibility and biological activity. Perhaps the most widely used material to confer medical implants with osteoconductivity is the HAP mineral. HAP was first introduced into orthopedic implants as the major component by weight of natural bone tissue, with a surface area that facilitates strong cell/molecular interactions [9]. HAP has a well-known biological interaction with bony tissues and can be secured into a bone interface when osteoblasts colonize the biocompatible surface [126]. Initially, plasma-sprayed HAP coatings were commonly used to accelerate peri-implant bone formation and to thereby decrease the time needed for implant fixation. After extensive investigations, HAP-coated implants have been shown to not only improve implant stability and bone–implant interface strength but also facilitate mineralized tissue formation, improve bone ingrowth, and decrease implant migration [127, 128]. HAP coatings also increase fatigue resistance, which improves long-term fixation of metal implants such as titanium implants [129]. Unmodified HAP is a brittle material [130] and has suboptimal mechanical properties that include poor tensile strength and low resistance to impact [131]. Therefore, coating structurally superior materials with HAP and keeping the HAP coating thickness low reduces the likelihood of interfacial fracture.

HAP on its own does not support load bearing and therefore does not serve as a structural implant in large bone defects. Instead, HAP biomaterials typically serve as bone spacers or fillers in non-load-bearing sites [20] or are used to encourage appositional bone formation as coatings on titanium alloys and other structural implants [132]. Plasma spraying, high-velocity oxygen fuel spraying, electrophoretic deposition, sol–gel deposition, hot isostatic pressing, frit enameling, ion-assisted deposition, pulsed laser deposition, electrochemical deposition, and sputter coating are all common methods by which HAP is coated onto implants or fabricated into stand-alone implants. These methods do not allow for protein incorporation during processing due to harsh material processing conditions. Thus, the amount of drug incorporated within common HAP-based implants is limited, and the drug release is often rapid since the drug is only bound to the nonporous outer surface.

5.6.2 HAP Chromatography and HAP–Protein Binding

Therapeutic drug delivery from HAP-based materials is critically dependent on the characteristics of drug–HAP interactions, and this concept is perhaps most directly illustrated by HAP–protein binding. In general terms, protein adsorption occurs almost instantaneously when a protein in an aqueous solution is exposed to solid materials such as HAP. Due to what is referred to as the "Vroman effect," proteins circulating in the plasma adsorb as a monolayer based on their size and affinity, with larger proteins often displacing smaller proteins from the surface. As proteins relax on a surface, they can undergo conformational changes and may not regain their original conformation upon desorption. In particular, an adsorbed protein may undergo conformational changes in which an increase in net surrounding charge decreases protein conformational stability, yielding unfolding and an extended conformation on the surface [133]. The typical stages of protein adsorption include transport of the protein from the bulk solution to the interface, attachment at the sorbent surface, and finally relaxation of the protein on the surface. Once the protein is bound to the solid surface, it will not typically desorb via simple dilution but can be displaced if a protein with more favorable binding energy on the surface is introduced. In this process, the protein cycles between the adsorbed and dissolved conformations. Adsorption of proteins to the surface also depends on several aspects of the surrounding environment, including the buffer osmolarity, ionic strength, ion composition, temperature, pH, and functional groups present on the protein and the substrate.

Proteins bind to pure, stoichiometric HAP via a mechanism that is based on the polar or charged side chains of the protein molecule interacting with calcium and phosphate ions in the HAP crystal. Amine and carboxylate groups are the major side groups implicated during protein binding to the HAP surface. This mechanism of protein binding has been used for protein purification for decades, with HAP and the more stable mineral fluorapatite used as resins for protein or DNA purification. Described by Tiselius et al. in 1956, HAP chromatography provides separation of protein mixtures containing medium to high molecular weight proteins [14]. At a constant pH,

BSA, lysozyme, ovalbumin, hemocyanin, and other proteins (MW range: 14×10^3 to 8.9×10^6 Da) can be successfully separated by eluting protein fractions in solutions with a linear increase in soluble phosphate ion concentrations. This ion exchange mechanism allows for versatile purification of a broad range of proteins. HAP chromatography differs from common separations based on ion exchange, as separation of the protein from the resin particles depends primarily on the phosphate buffer concentration used for elution rather than the electrostatic properties of the molecules themselves.

Protein–HAP binding is also significantly influenced by the HAP crystal structure. Most proteins bind to calcium phosphate minerals in solution, and the binding strength depends in part on the isoelectric point of the protein and the solution pH. Ca^{2+} ions in the HAP crystal interact with carboxylate or other negatively charged (acidic) groups on the protein, whereas PO_4^- ions interact with positively charged (basic) amino acid side chains. Protein binding to HAP can be complex, as it can be influenced by both anionic and cationic exchange processes that vary according to the organization of the protein's amino acid side chains and the environmental pH. The hexagonal HAP crystal also has two types of adsorbing surfaces—the a-axis and the c-axis. Typical needle-like crystals of HAP are thought to be primarily a-axis surfaces. In an aqueous solution, calcium ions on the a-axis surface present the hydroxyl group of HAP to the exterior particle surface, facilitating suspension of the mineral in aqueous solution [134]. These exposed hydroxyls are then flanked by calcium ions, creating a putative adsorption site for acidic molecules, termed as "C" site. In contrast, "P" sites are deficient in calcium ions and bind to basic molecules. Because the HAP surface is heterogeneous with "C" and "P" sites, its composite surface charge depends on the ratio of calcium to phosphate ions as well as the solution pH—the higher the pH, the more negative the surface charge. Likewise, because pockets of "C" sites or "P" sites can occur, proteins bind to HAP in a variety of conformations and they bind nonuniformly across the surface. Therefore, proteins with multiple binding sites can switch between stable conformations, facilitating intermolecular interactions between proteins.

The isoelectric point of a protein and its regions of hydrophobicity further complicate the HAP–protein binding phenomenon, and proteins may be able to cooperate to maximize binding to the surface. Depending on the solution conditions, cooperative binding or competitive repulsion can be facilitated, depending on the ions in solution and the ability of proteins to bind to the HAP crystal lattice. Within a crystal, C sites are arranged in a rectangular manner while P sites are arranged on c-axis surfaces in a hexagonal manner. Kawasaki et al. systematically varied the HAP form and the protein chemistry to demonstrate that acidic and basic proteins adsorb to different surfaces of HAP [135]. In addition, protein–HAP interactions are a function of the protein's net charge, as single amino acids have a poor affinity to HAP.

Although electrostatic interactions between complementary charges on protein side chains and HAP crystal constituents are commonly thought to dominate protein–HAP interactions, there is evidence of greater complexity in this interaction. For example, amine-terminated side chains can also hydrogen bond to phosphate and hydroxyl groups on the HAP surface. Previous studies have shown that blocking basic amino acids within basic proteins inhibits binding, while some acidic proteins are not affected by this blockage [136]. Changing a carboxylate to a sulfonic acid maintained the charge but significantly reduced the retention of all proteins other than acidic proteins to negligible levels [137]. The simple assertion that basic proteins bind via amino groups and acidic proteins bind via carboxylates is also more generally an incomplete mechanistic description, as basic protein adsorption has been shown to be affected by carboxylate blocking. Gorbunoff et al. support that both carboxylate and amino groups are critical for binding, irrespective of the protein's isoelectric point, because a higher molarity phosphate was needed for elution of all proteins after carboxylate blocking.

In HAP chromatography methods, acidic and basic proteins vary in the eluents used for protein desorption as well as their molar concentrations [136,138]. Elution of both acidic and basic proteins occurs at a relatively low phosphate ion molarity by either competitive displacement or charge screening, respectively. Acidic proteins cannot be eluted from the HAP surface by calcium ions, whereas basic proteins are displaced by divalent cations such as calcium at low molar concentrations. Because of the carboxyl groups found on acidic proteins, there is a strong interaction between the carboxyl and apatite calcium ions. Additional soluble calcium ions can increase the number of bridges of acidic proteins to the apatite surface, as calcium-bound carboxylate groups can interact electrostatically with phosphate sites on HAP [138].

5.6.3 Factors Influencing Mineral Dissolution

Since a variety of studies have demonstrated a correlation between mineral dissolution and the release kinetics of incorporated or adsorbed drugs, it is worthwhile to consider the mineral dissolution process and the parameters that influence this process. Mineral dissolution depends on two principle phenomena: (1) the creation of "pits" also referred to as "active sites" and (2) the rate of step movement from the active site. The dissolution process can be accelerated by the presence of more pits on a mineral surface, which in turn leads to accelerated drug release from HAP. Surface pitting occurs for a variety of reasons and has been attributed to undersaturation of mineral ions in solution, the presence of hydronium ions in solution, and the inherent existence of point defects on the crystal surface. Dissolution proceeds via a similar process to nucleation, but in reverse, without the need for a to form critical nucleus [60].

A series of models have been proposed to describe HAP dissolution behavior. The "polynuclear model" is based on the work by Christofferson et al. and describes nucleation as a function of calcium ion activity, both in the saturated solution and on the surface. It assumes that the linear growth or dissolution perpendicular to a surface has the same rate for all crystal surfaces and is often used to describe dissolution at neutral or slightly acidic pH. According to the polynuclear model, a crystal surface has concave and convex components, and the convex components continue to grow while the concave components are outgrown. During dissolution, the convex components disappear while concave components grow larger. The polynuclear mechanism has been explored experimentally in buffered solutions, and results indicate that the rate-limiting step during mineral dissolution is the dissolution of mineral particles at the surface [139].

Another well-cited model is described by the "self-inhibition" theory, which asserts that once nuclei reach a critical size the calcium and phosphate ions detach from the surface, move into the solution, and dissolve. At some point in time, the ions may readsorb to the apatite surface and form a calcium rich layer. This dissolution phenomenon tends to occur at more acidic pH 3.7–6.9, where the rates of calcium release are coupled with hydrogen ion uptake. The step movement of ions along the crystal surface depends on the size of the mineral nuclei and whether they have reached their critical size. A nucleus smaller than the critical size can diffuse toward or away from the apatite material. When the nucleus is larger than its critical size, it transports in a manner that is dominated by convection rather than diffusion [139].

Mineral dissolution in vivo is significantly influenced by the surrounding environment and the mineral crystal composition. In undersaturated conditions of the surrounding solution, changes in temperature, ionic strength, pH, and other variables can facilitate mineral dissolution. Calcium phosphate mineral dissolution typically increases with a decrease in pH, and the rate of dissolution is mineral phase dependent. In vivo HAP dissolution characteristics are often similar to HAP dissolution observed in vitro at low pH, suggesting that a low pH may drive the in vivo HAP dissolution process [140]. Other environmental factors present in vivo can also influence HAP dissolution. Factors that accelerate HAP dissolution can function by modifying mineral solubility, influencing mineral surface charge, or blocking new crystal formation. For example, citric acid can inhibit HAP dissolution in a concentration-dependent manner by adsorbing to the dissolving surface [141]. Interestingly, this concentration-dependent behavior is not reproduced for other calcium phosphate phases. Binding the crystal surface at the active site not only decreases the rate of dissolution but also has an effect on free edge energy. It is also noteworthy that HAP is not pure in natural tissues and often contains other elements that strongly affect dissolution [4]. For example, natural apatites such as those in vertebrate bones and teeth are calcium deficient, which leads to high specific surface area and an abundance of stoichiometric/crystallographic imperfections that enhance mineral dissolution [142].

5.7 Concluding Remarks

The fundamental properties of HAP and other calcium phosphate minerals are advantageous for a variety of applications, including therapeutic drug delivery. The ability to readily process HAP biomaterials with a range of physical and structural properties has led to widespread use as medical implants in applications that range from bone regeneration to vocal fold healing. Further, HAP biomaterials have emerged as an ideal carrier for drug delivery, as they are adaptable to suit virtually all drug classes and in vivo locations. Although HAP has been used as a drug carrier primarily in orthopedic applications, new HAP processing methods and composite biomaterials are extending its use into a broader range of clinical scenarios. In addition, some specific drug delivery applications not emphasized here (e.g., gene delivery) benefit from not only drug release but also the associated release of mineral ions such as calcium and phosphate. Indeed, previous studies and emerging research directions suggest that HAP biomaterials have only begun to reach their extensive clinical potential in therapeutic drug delivery.

References

1. Weiner S. Biomineralization: A structural perspective. *Journal of Structural Biology* 2008 September;163(3):229–234.
2. Cusack M, Freer A. Biomineralization: Elemental and organic influence in carbonate systems. *Chemical Reviews* 2008 November;108(11):4433–4454.
3. Lowenstam HA. Minerals formed by organisms. *Science* 1981 March; 211(4487):1126–1131.
4. Dorozhkin SV, Epple M. Biological and medical significance of calcium phosphates. *Angewandte Chemie* 2002 September;41(17):3130–3146.
5. Jarcho M. Calcium phosphate ceramics as hard tissue prosthetics. *Clinical Orthopaedics and Related Research* 1981 June;157:259–278.
6. Uskoković V, Uskoković DP. Nanosized hydroxyapatite and other calcium phosphates: Chemistry of formation and application as drug and gene delivery agents. *Journal of Biomedical Materials Research B: Applied Biomaterials* 2011 January;96(1):152–191.
7. Eanes ED, Termine JD, Nylen MU. An electron microscopic study of the formation of amorphous calcium phosphate and its transformation to crystalline apatite. *Calcified Tissue Research* 1973;12:143–158.
8. Meyer JL, Eanes ED. A thermodynamic analysis of the amorphous to crystalline calcium phosphate transformation. *Calcified Tissue Research* 1978;25:59–68.
9. Palmer LC, Newcomb CJ, Kaltz SR, Spoerke ED, Stupp SI. Biomimetic systems for hydroxyapatite mineralization inspired by bone enamel. *Chemical Reviews* 2008;108(11):4754–4783.

10. Brown WE, Chow LC. Chemical properties of bone mineral. *Annual Reviews Material Science* 1976;6:213–236.
11. Barralet J, Best S, Bonfield W. Carbonate substitution in precipitated hydroxyapatite: An investigation into the effects of reaction temperature and bicarbonate ion concentration. *Journal of Biomedical Materials Research* 1998 July;41(1):79–86.
12. LeGeros Z. Two types of carbonate substitution in the apatite structure. *Experentia* 1969;25:5–7.
13. Brunngraber EG, Occomy WG. Fractionation of brain macromolecules. *Biochemistry Journal* 1965;97:689–696.
14. Tiselius A, Hjerten S, Levin O. Protein chromatography on calcium phosphate columns. *Archives of Biochemistry and Biophysics* 1956 November;65(1):132–155.
15. Bernardi G, Kawasaki T. Chromatography of polypeptides and proteins on hydroxyapatite columns. *Biochimica et Biophysica Acta* 1968;160:301–310.
16. Boix T, Gómez-Morales J, Torrent-Burgués J, Monfort A, Puigdomènech P, Rodríguez-Clemente R. Adsorption of recombinant human bone morphogenetic protein rhBMP-2 m onto hydroxyapatite. *Journal of Inorganic Biochemistry* 2005 May;99(5):1043–1050.
17. Iafisco M, Sabatino P, Lesci IG, Prat M, Rimondini L, Roveri N, Lafisco M, Giorgio I. Conformational modifications of serum albumins adsorbed on different kinds of biomimetic hydroxyapatite nanocrystals. *Colloids Surfaces B: Biointerfaces* 2010 November;81(1):274–284.
18. Luo Q, Andrads JD. Cooperative adsorption of proteins onto hydroxyapatite. *Journal of Colloid and Interface Science* 1998;200:104–113.
19. Chan WD, Perinpanayagam H, Goldberg HA, Hunter GK, Dixon SJ, Santos GC, Rizkalla AS. Tissue engineering scaffolds for the regeneration of craniofacial bone. *Journal of the Canadian Dental Association* 2009 June;75(5):373–377.
20. Holmes RE. Bone regeneration within a corraline hydroxyapatite implant. *Plastic and Reconstructive Surgery* 1979;63:626–633.
21. Xu HHK, Weir MD, Simon CG. Injectable and strong nano-apatite scaffolds for cell/growth factor delivery and bone regeneration. *Dental Materials* 2008 September;24(9):1212–1222.
22. Hollinger JO, Leong K. Poly(alpha-hydroxy acids): Carriers for bone morphogenetic proteins. *Biomaterials* 1996 January;17(2):187–194.
23. Liu Y, de Groot K, Hunziker EB. Osteoinductive implants: The mise-en-scène for drug-bearing biomimetic coatings. *Annals of Biomedical Engineering* 2004 March;32(3):398–406.
24. Elhadj S, De Yoreo JJ, Hoyer JR, Dove PM. Role of molecular charge and hydrophilicity in regulating the kinetics of crystal growth. *Proceedings of the National Academy of Sciences of the United States of America* 2006 December;103(51):19237–19242.
25. Maciejewski M, Brunner TJ, Loher SF, Stark WJ, Baiker A. Phase transitions in amorphous calcium phosphates with different Ca/P ratios. *Thermochimica Acta* 2008 February;468(1–2):75–80.
26. Matsumoto T, Okazaki M, Inoue M, Yamaguchi S, Kusunose T, Toyonaga T, Hamada Y, Takahashi J. Hydroxyapatite particles as a controlled release carrier of protein. *Biomaterials* 2004 August;25(17):3807–3812.
27. Eggli PS, Muller W, Schenk RK. Porous hydroxyapatite and tricalcium phosphate cylinders with two different pore size ranges implanted in the cancellous bone of rabbits. *Clinical Orthopaedics and Related Research* 1988;232:127–138.

28. Chan O, Coathup MJ, Nesbitt N, Ho C-Y, Hing KA, Buckland T, Campion C, Blunn GW. The effects of microporosity on osteoinduction of calcium phosphate bone graft substitute biomaterials. *Acta Biomaterialia* 2012 July;8(7):2788–2794.

29. Francis L, Venugopal J, Prabhakaran MP, Thavasi V, Marsano E, Ramakrishna S. Simultaneous electrospin-electrosprayed biocomposite nanofibrous scaffolds for bone tissue regeneration. *Acta Biomaterialia* 2010 October;6(10):4100–4109.

30. Chen J, Chu B, Hsiao BS. Mineralization of hydroxyapatite in electrospun nanofibrous poly(L-lactic acid) scaffolds. *Journal of Biomedical Materials Research, A* 2006 November;79(2):307–317.

31. Seyedjafari E, Soleimani M, Ghaemi N, Shabani I. Nanohydroxyapatite-coated electrospun poly(l-lactide) nanofibers enhance osteogenic differentiation of stem cells and induce ectopic bone formation. *Biomacromolecules* 2010 October;11:3118–3125.

32. Zhang Y, Venugopal JR, El-Turki A, Ramakrishna S, Su B, Lim CT. Electrospun biomimetic nanocomposite nanofibers of hydroxyapatite/chitosan for bone tissue engineering. *Biomaterials* 2008 November;29(32):4314–4322.

33. Teshima K, Lee S, Sakurai M, Kameno Y, Yubuta K, Suzuki T, Shishido T, Endo M, Oishi S. Well-formed one-dimensional hydroxyapatite crystals grown by an environmentally friendly flux method. *Crystal Growth and Design* 2009 June;9(6):2937–2940.

34. Sasikumar S, Vijayaraghavan R. Synthesis and characterization of bioceramic calcium phosphates by rapid combustion synthesis. *Journal of Materials Science and Technology* 2010 December;26(12):1114–1118.

35. Cho JS, Kang YC. Nano-sized hydroxyapatite powders prepared by flame spray pyrolysis. *Journal of Alloys and Compounds* 2008 September;464(1–2):282–287.

36. Poinern GE, Brundavanam RK, Mondinos N, Jiang Z-T. Synthesis and characterization of nanohydroxyapatite using an ultrasound assisted method. *Ultrasonics Sonochemistry* 2009 April;16(4):469–474.

37. Salas J, Benzo Z, Gonzalez G, Marcano E, Gómez C. Effect of Ca/P ratio and milling material on the mechanochemical preparation of hydroxyapatite. *Journal of Materials Science: Materials in Medicine* 2009 July;2249–2257.

38. Poinern GJE, Brundavanam R, Le XT, Djordjevic S, Prokic M, Fawcett D. Thermal and ultrasonic influence in the formation of nanometer scale hydroxyapatite bio-ceramic. *International Journal of Nanomedicine* 2011 January;6:2083–2095.

39. Koumoulidis GC, Katsoulidis AP, Ladavos AK, Pomonis PJ, Trapalis CC, Sdoukos AT, Vaimakis TC. Preparation of hydroxyapatite via microemulsion route. *Journal of Colloid and Interface Science* 2003 March;259(2):254–260.

40. Uota M, Arakawa H, Kitamura N, Yoshimura T, Tanaka J, Kijima T. Synthesis of high surface area hydroxyapatite nanoparticles by mixed surfactant-mediated approach. *Langmuir* 2005 May;21(10):4724–4728.

41. Yan L, Li Y, Deng Z, Zhuang J, Sun X. Surfactant-assisted hydrothermal synthesis of hydroxyapatite nanorods. *International Journal of Inorganic Materials* 2001 February;3:633–637.

42. Choi AH, Ben-Nissan B. Sol-gel production of bioactive nanocoating for medical applications. Part II: Current research and development. *Nanomedicine* 2007;2(1):51–61.

43. Chen W, Oh S, Ong AP, Oh N, Liu Y, Courtney HS, Appleford M, Ong JL, Al CET. Antibacterial and osteogenic properties of silver-containing hydroxyapatite coatings produced using a sol gel process. *Journal of Biomedical Materials Research* 2007;82A:906–906.
44. Vallés Lluch A, Ferrer GG, Pradas MM. Surface modification of P(EMA-co-HEA)/SiO$_2$ nanohybrids for faster hydroxyapatite deposition in simulated body fluid? *Colloids and Surfaces B: Biointerfaces* 2009 May;70(2):218–225.
45. Hacking SA, Zuraw M, Harvey EJ, Tanzer M, Krygier JJ, Bobyn JD. A physical vapor deposition method for controlled evaluation of biological response to biomaterial chemistry and topography. *Journal of Biomedical Materials Research* 2007;82A:179–187.
46. Raynaud S, Champion E, Bernache-Assollant D, Thomas P. Calcium phosphate apatites with variable Ca/P atomic ratio I. Synthesis, characterisation and thermal stability of powders. *Biomaterials* 2002 February;23(4):1065–1072.
47. Bailliez S, Nzihou, A. The kinetics of surface area reduction during isothermal sintering of hydroxyapatite adsorbent. *Chemical Engineering Journal* 2004; 98:141–152.
48. Bohner M. Calcium orthophosphates in medicine: From ceramics to calcium phosphate cements. *Injury* 2000 December;31(Suppl 4):37–47.
49. Ignjatović NL, Ninkov P, Kojic V, Bokurov M, Srdic V, Krnojelac D, Selakovic S, Uskoković DP. Cytotoxicity and fibroblast properties during *in vitro* test of biphasic calcium phosphate/poly-dl-lactide-co-glycolide biocomposites and different phosphate materials. *Microscopy Research and Technique* 2006,69.976–982.
50. Ignjatović NL, Ajdukovic Z, Uskoković DP. New biocomposite [biphasic calcium phosphate/poly-DL-lactide-co-glycolide/biostimulative agent] filler for reconstruction of bone tissue changed by osteoporosis. *Journal Materials Science: Materials in Medicine* 2005;16:621–626.
51. Cosijns A, Vervaet C, Luyten J, Mullens S, Siepmann F, Van Hoorebeke L, Masschaele B, Cnudde V, Remon JP. Porous hydroxyapatite tablets as carriers for low-dosed drugs. *European Journal of Pharmaceutics and Biopharmaceutics* 2007 September;67(2):498–506.
52. Zhou H, Lee J. Nanoscale hydroxyapatite particles for bone tissue engineering. *Acta Biomaterialia* 2011 July;7(7):2769–2781.
53. Sun R, Chen K, Lu Y. Fabrication and dissolution behavior of hollow hydroxy-apatite microspheres intended for controlled drug release. *Materials Research Bulletin* 2009 October;44(10):1939–1942.
54. Ginebra MP, Traykova T, Planell JA. Calcium phosphate cements as bone drug delivery systems: A review. *Journal of Controlled Release* 2006 June;113(2):102–110.
55. Mann S. Mineralization in biological systems. In: *Structure and Bonding.* Springer Verlag, Berlin, Germany, 1983.
56. LeGeros RZ. Calcium phosphate-based osteoinductive materials. *Chemical Reviews* 2008 November;108(11):4742–4753.
57. Kokubo T. Surface chemistry of bioactive glass-ceramics. *Journal of Non-Crystalline Solids* 1990;120:138–151.
58. Kokubo T. Formation of biologically active bone-like apatite on metals and polymers by a biomimetic process. *Thermochimica Acta* 1996;280/281:479–490.
59. Lu X, Leng Y. Theoretical analysis of calcium phosphate precipitation in simulated body fluid. *Biomaterials* 2005 April;26(10):1097–1108.

60. Tang R, Orme CA, Nancollas GH. Dissolution of crystallites: Surface energetic control and size effects. *Chemphyschem* 2004 May;5(5):688–696.
61. Weiner S, Sagi I, Addadi L. Structural biology. Choosing the crystallization path less traveled. *Science* 2005 August;309(5737):1027–1028.
62. Sarikaya M. Biomimetics: Materials fabrication through biology. *Proceedings of the National Academy of Sciences of the United States of America* 1999 December;96(25):14183–14185.
63. Cui W, Li X, Zhou S, Weng J. In situ growth of hydroxyapatite within electrospun poly(DL-lactide) fibers. *Journal of Biomedical Materials Research, A* 2007 September;82(4):831–841.
64. Liu Y, Hunziker EB, Randall NX, de Groot K, Layrolle P. Proteins incorporated into biomimetically prepared calcium phosphate coatings modulate their mechanical strength and dissolution rate. *Biomaterials* 2003 January;24(1):65–70.
65. Yamashita Y, Uchida A, Yamakawa T, Shinto Y, Araki N, Kato K. Treatment of chronic osteomyelitis using calcium hydroxyapatite ceramic implants impregnated with antibiotic. *International Orthopaedics* 1998 January;22(4):247–251.
66. Ferraz MP, Mateus AY, Sousa JC, Monteiro FJ. Nanohydroxyapatite microspheres as delivery system for antibiotics: Release kinetics, antimicrobial activity, and interaction with osteoblasts. *Journal of Biomedical Materials Research* 2007;81A:994–1004.
67. Pham HH, Luo P, Génin F, Dash AK. Synthesis and characterization of hydroxyapatite-ciprofloxacin delivery systems by precipitation and spray drying technique. *AAPS PharmSciTech* 2002 January;3(1):E1.
68. Victor SP, Kumar TSS. BCP ceramic microspheres as drug delivery carriers: Synthesis, characterization and doxycycline release. *Journal of Material Science: Materials in Medicine* 2008 January;19(1):283–290.
69. Hasegawa M, Sudo A, Komlev VS, Barinov SM, Uchida A. High release of antibiotic from a novel hydroxyapatite with bimodal pore size distribution. *Journal of Biomedical Materials Research. Part B, Applied Biomaterials* 2004 August;70(2):332–339.
70. Rogers-Foy JM, Brosnan DA, Barefoot SF, Friedman RJ, LaBerge M. Hydroxyapatite composites designed for antibiotic drug delivery and bone reconstruction: A caprine model. *Journal of Investigative Surgery* 1999;12:263–275.
71. Joosten LAB, Lubberts E, Helsen MM, Saxne T, Coenen-de Roo CJ, Heinegård D, van den Berg WB, Heinegard D. Protection against cartilage and bone destruction by systemic interleukin-4 treatment in established murine type II collagen-induced arthritis. *Arthritis Research* 1999 January;1(1):81–91.
72. Shirtliff ME, Calhoun JH, Mader JT. Experimental osteomyelitis treatment with antibiotic-impregnated hydroxyapatite. *Clinical Orthopaedics and Related Research* 2002 August;401:239–247.
73. Yu D, Wong J, Matsuda Y, Fox L, Higuchi W, Otsuka M. Self-setting hydroxyapatite cement: A novel skeletal drug-delivery system for antibiotics. *Journal of Pharmaceutical Sciences* 1992;81:529–531.
74. Ravelingien M, Mullens S, Luyten J, D'Hondt M, Boonen J, De Spiegeleer B, Coenye T, Vervaet C, Remon JP. Vancomycin release from poly(D,L-lactic acid) spray-coated hydroxyapatite fibers. *European Journal of Pharmaceutics and Biopharmaceutics* 2010 November;76(3):366–370.

75. Itokazu M, Matsunaga T, Kumazawa S, Oka M. Treatment of osteomyelitis by antibiotic impregnated porous hydroxyapatite block. *Clinical Materials* 1995;17(1994):173–179.
76. Stigter M, de Groot K, Layrolle P. Incorporation of tobramycin into biomimetic hydroxyapatite coating on titanium. *Biomaterials* 2002 October;23(20):4143–4153.
77. Stigter M, Bezemer J, de Groot K, Layrolle P. Incorporation of different antibiotics into carbonated hydroxyapatite coatings on titanium implants, release and antibiotic efficacy. *Journal of Controlled Release* 2004 September;99(1):127–137.
78. Shinto Y, Uchida A, Korkusuz F, Araki N, Ono K. Calcium hydroxyapatite ceramic used as a delivery system for antibiotics. *Journal of Bone and Joint Surgery Brittan* 1992;74-B(4):600–604.
79. Sudo A, Hasegawa M, Fukuda A, Uchida A. Treatment of infected hip arthroplasty with antibiotic-impregnated calcium hydroxyapatite. *Journal of Arthroplasty* 2008 January;23(1):145–150.
80. Krisanapiboon A, Buranapanikit B, Oungbho K. Biocompatibility of hydroxyapatite composite as a local drug delivery system. *Journal of Orthopaedic Surgery* 2006;14(3):315–318.
81. Ip M, Lui SL, Poon VKM, Lung I, Burd A. Antimicrobial activities of silver dressings: An *in vitro* comparison. *Journal of Medical Microbiology* 2006 January;55(Pt 1):59–63.
82. Joyce-Wöhrmann RM, Münstedt H. Determination of the silver ion release from polyurethanes enriched with silver. *Infection* 1999 January;27 (Suppl 1):S46–S48.
83. Ewald A, Hösel D, Patel S, Grover LM, Barralet JE, Gbureck U. Silver-doped calcium phosphate cements with antimicrobial activity. *Acta Biomaterialia* 2011 November;7(11):4064–4070.
84. Park H, Temenoff JS, Holland TA, Tabata Y, Mikos AG. Delivery of TGF-beta1 and chondrocytes via injectable, biodegradable hydrogels for cartilage tissue engineering applications. *Biomaterials* 2005 December;26(34):7095–7103.
85. Lu L, Yaszemski MJ, Mikos AG. TGF-beta1 release from biodegradable polymer microparticles: Its effects on marrow stromal osteoblast function. *Journal of Bone and Joint Surgery American* 2001 January;83-A(Suppl. 1; Pt 2):S82–S91.
86. Kim SE, Park JH, Cho YW, Chung H, Jeong SY, Lee EB, Kwon IC. Porous chitosan scaffold containing microspheres loaded with transforming growth factor-beta1: Implications for cartilage tissue engineering. *Journal of Controlled Release* 2003 September;91(3):365–374.
87. Kirby GTS, White LJ, Rahman CV, Cox HC, Qutachi O, Rose FRA. J, Hutmacher DW, Shakesheff KM, Woodruff MA. PLGA-based microparticles for the sustained release of BMP-2. *Polymers* 2011 March;3(1):571–586.
88. Park JS, Yang HN, Jeon SY, Woo DG, Na K, Park K-H. Osteogenic differentiation of human mesenchymal stem cells using RGD-modified BMP-2 coated microspheres. *Biomaterials* 2010;31:6239–6248.
89. Liu Y, de Groot K, Hunziker EB. BMP-2 liberated from biomimetic implant coatings induces and sustains direct ossification in an ectopic rat model. *Bone* 2005 May;36(5):745–757.
90. Schnettler R, Knöss PD, Heiss C, Stahl JP, Meyer C, Kilian O, Wenisch S, Alt V. Enhancement of bone formation in hydroxyapatite implants by rhBMP-2 coating. *Journal of Biomedical Materials Research Part B: Applied Biomaterials* 2009 July;90B(1):75–81.

91. Levine JP, Bradley J, Turk AE, Benedict JJ, Steiner G, Longaker MT, McCarthy JG. Bone morphogenetic protein promotes vascularization and osteoinduction in preformed hydroxyapatite in the rabbit. *Annals of Plastic Surgery* 1997;39:158–168.

92. Ripamonti U, Crooks J, Petit JC, Rueger DC. Periodontal tissue regeneration by combined applications of recombinant human osteogenic protein-1 and bone morphogenetic protein-2. A pilot study in Chacma baboons (*Papio ursinus*). *European Journal of Oral Sciences* 2001 August;109(4):241–248.

93. Suárez-González D, Barnhart K, Migneco F, Flanagan C, Hollister SJ, Murphy WL. Controllable mineral coatings on PCL scaffolds as carriers for growth factor release. *Biomaterials* 2012 January;33(2):713–721.

94. Lode A, Reinstorf A, Bernhardt A, Konig U, Pompe W, Gelinsky M. Calcium phosphate bone cements, functionalized with VEGF: Release kinetics and biological activity. *Journal of Biomedical Materials Research* 2007;81A:474–483.

95. Murphy WL, Peters MC, Kohn DH, Mooney DJ. Sustained release of vascular endothelial growth factor from mineralized poly(lactide-co-glycolide) scaffolds for tissue engineering. *Biomaterials* 2000 December;21(24):2521–2527.

96. Murphy WL, Hsiong S, Richardson TP, Simmons CA, Mooney DJ. Effects of a bone-like mineral film on phenotype of adult human mesenchymal stem cells in vitro. *Biomaterials* 2005 January;26(3):303–310.

97. Midy V, Rey C, Bres E, Dard M. Basic fibroblast growth factor adsorption and release properties of calcium phosphate. *Journal of Biomedical Materials Research* 1998;41:405–411.

98. Lee JS, Lu Y, Baer GS, Markel MD, Murphy WL. Controllable protein delivery from coated surgical sutures. *Journal of Materials Chemistry* 2010;20(40):8894–8903.

99. Blom EJ, Klein-Nulend J, Wolke JGC, van Waas MAJ, Driessens FCM, Burger EH. Transforming growth factor-b1 incorporation in a calcium phosphate bone cement: Material properties and release characteristics. *Journal of Biomedical Materials Research* 2002;59:265–272.

100. Hee SC, Park S-Y, Kim S, Sang KB, Duk SS, Ahn M-W. Effect of different bone substitutes on the concentration of growth factors in platelet-rich plasma. *Journal of Biomaterials Applications* 2008 May;22(6):545–557.

101. Arm DM, Tencer AF, Bain SD, Celino D. Effect of controlled release of platelet-derived growth factor from a porous hydroxyapatite implant on bone ingrowth. *Biomaterials* 1996 April;17(7):703–709.

102. Ramachandran R, Paul W, Sharma CP. Synthesis and characterization of PEGylated calcium phosphate nanoparticles for oral insulin delivery. *Journal of Biomedical Materials Research. Part B, Applied Biomaterials* 2009 January;88(1):41–48.

103. Paul W, Nesamony J, Sharma CP. Delivery of insulin from hydroxyapatite ceramic microspheres: Preliminary in vivo studies. *Journal of Biomedical Materials Research* 2002;61:600–662.

104. Yu X, Wei M. Preparation and evaluation of parathyroid hormone incorporated CaP coating via a biomimetic method. *Journal of Biomedical Materials Research. Part B, Applied Biomaterials* 2011 May;97(2):345–354.

105. Monkkonen J, Simila J, Rogers MJ. Effects of tiludronate and ibandronate on the secretion of proinflammatory cytokines and nitric oxide from macrophages in vitro. *Pharmacology Letters* 1998;62(8):L95–L102.

106. Thiébaud D, Sauty A, Burckhardt P, Leuenberger P, Sitzler L, Green JR, Kandra A, Zieschang J, Ibarra de Palacios P. An in vitro and in vivo study of cytokines in the acute-phase response associated with bisphosphonates. *Calcified Tissue International* 1997 November;61(5):386–392.

107. Elliott SN, Webb M, Davies NM, Macnaughton WK, Wallace JL. Alendronate induces gastric injury and delays ulcer healing in rodents. *Life Sciences* 1998;62(1):77–91.

108. Iafisco M, Palazzo B, Marchetti M, Margiotta N, Ostuni R, Natile G, Morpurgo M, Gandin V, Marzano C, Roveri N. Smart delivery of antitumoral platinum complexes from biomimetic hydroxyapatite nanocrystals. *Journal of Materials Chemistry* 2009;19(44):8385.

109. Aberg J, Brohede U, Mihranyan A, Strømme M, Engqvist H. Bisphosphonate incorporation in surgical implant coatings by fast loading and co-precipitation at low drug concentrations. *Journal of Material Science: Materials in Medicine* 2009 October;20(10):2053–2061.

110. Peter B, Pioletti DP, Laïb S, Bujoli B, Pilet P, Janvier P, Guicheux J, Zambelli P-Y, Bouler J-M, Gauthier O. Calcium phosphate drug delivery system: Influence of local zoledronate release on bone implant osteointegration. *Bone* 2005 January;36(1):52–60.

111. Gao Y, Zou S, Liu X, Bao C, Hu J. The effect of surface immobilized bisphosphonates on the fixation of hydroxyapatite-coated titanium implants in ovariectomized rats. *Biomaterials* 2009 March;30(9):1790–1796.

112. Sun S, Liu X, Ma B, Zhou Y, Sun H. Could local deliver of bisphosphonates be a new therapeutic choice for Gorham-Stout syndrome? *Medical Hypotheses* 2011 February;76(2):237–238.

113. Yu H, Qin A. Could local delivery of bisphosphonates be a new therapeutic choice for hemangiomas? *Medical Hypotheses* 2009 October;73(4):495–497.

114. Barroug A, Kuhn LT, Gerstenfeld LC, Glimcher MJ. Interactions of cisplatin with calcium phosphate nanoparticles: In vitro controlled adsorption and release. *Journal of Orthopaedic Research* 2004 July;22(4):703–708.

115. Otsuka M, Matsuda Y, Suwa Y, Fox JL, Higuchi WI. A novel skeletal drug delivery system using a self-setting calcium phosphate cement. 5: Drug release behavior from a heterogeneous drug-loaded cement containing an anticancer drug. *Journal of Pharmaceutical Sciences* 1994 November;83(11):1565–1568.

116. Otsuka M, Matsuda Y, Fox JL, Higuchi WI. A novel skeletal drug delivery system using self-setting calcium phosphate cement. 9: Effects of the mixing solution volume on anticancer drug release from homogeneous drug-loaded cement. *Journal of Pharmaceutical Sciences* 1995 June;84(6):733–736.

117. Lopez-Heredia MA, Kamphuis GJB, Thüne PC, Öner FC, Jansen JA, Walboomers XF. An injectable calcium phosphate cement for the local delivery of paclitaxel to bone. *Biomaterials* 2011 August;32(23):5411–5416.

118. Palazzo B, Iafisco M, Laforgia M, Margiotta N, Natile G, Bianchi CL, Walsh D, Mann S, Roveri N. Biomimetic hydroxyapatite–drug nanocrystals as potential bone substitutes with antitumor drug delivery properties. *Advanced Functional Materials* 2007 September;17(13):2180–2188.

119. Melville AJ, Rodríguez-Lorenzo LM, Forsythe JS. Effects of calcination temperature on the drug delivery behaviour of ibuprofen from hydroxyapatite powders. *Journal of Material Science: Materials in Medicine* 2008 March;19(3):1187–1195.

120. Choi S, Murphy WL. Sustained plasmid DNA release from dissolving mineral coatings. *Acta Biomaterialia* 2010 September;6(9):3426–3435.
121. Abrams L, Bajpai PK. Hydroxyapatite ceramics for continuous delivery of heparin. *Biomedical Sciences Instrumentation* 1994;30:169–174.
122. Bruder SP, Kraus KH, Goldberg VM, Kadiyala S. The effect of implants loaded with autologous mesenchymal stem cells on the healing of canine segmental bone defects. *Journal of Bone and Joint Surgery American* 1998 July;80(7):985–996.
123. Quarto R, Masrogiacomo M, Cancedda R, Kutepov S, Mukhachev V, Lavroukov A, Kon E, Maracci M. Repair of large bone defects with the use of autologous bone marrow stromal cells. *New England Journal of Medicine* 2001;344:385–386.
124. Okamoto M, Dohi Y, Ohgushi H, Shimaoka H, Ikeuchi M, Matsushima A, Yonemasu K, Hosoi H. Influence of the porosity of hydroxyapatite ceramics on in vitro and in vivo bone formation by cultured rat bone marrow stromal cells. *Journal of Material Science: Materials in Medicine* 2006;17:327–336.
125. Finkemeier CG. Bone-grafting and bone-graft substitutes. *Journal of Bone and Joint Surgery. American* 2002 March;84-A(3):454–464.
126. Hagio T, Tanase T, Akiyama J, Iwai K, Asai S. Formation and biological affinity evaluation of crystallographically aligned hydroxyapatite. *Journal of the Ceramic Society of Japan* 2008;116(1349):79–82.
127. Soballe K, Hansen ES, Brockstedt-Rasmussen H, Hjortdal VE, Juhl GI, Pedersen CM, Hvid I, Bunger C. Gap healing enhanced by hydroxyapatite coating in dogs. *Clinical Orthopaedics and Related Research* 1991;272:300–307.
128. Moilanen T, Stocks GW, Freeman MAR, Scott G, Goodier WD, Evans SJW. Hydroxyapatite coating of an acetabular prosthesis. *Journal of Bone and Joint Surgery Brittan* 1995;78-B:200–205.
129. Cao W, Hench LL. Bioactive materials. *Ceramics International* 1996;8842(95):493–507.
130. Benaqqa C, Chevalier J, Saädaoui M, Fantozzi G. Slow crack growth behaviour of hydroxyapatite ceramics. *Biomaterials* 2005 November;26(31):6106–6112.
131. Ruys AJ, Wei M, Sorrell CC, Dickson MR, Brandwood A, Milthorpe BK. Sintering effects on the strength of hydroxyapatite. *Biomaterials* 1995 March;16(5):409–415.
132. Bucholz RW, Carlton A, Holmes RE. Hydroxyapatite and tricalcium phosphate bone graft substitutes. *Orthopedic Clinics of North America* 1987;18(2):323–334.
133. Barroug A, Lemaitre J, Rouxhet P. Lysozyme on apatites: A model of protein adsorption controlled by electrostatic interactions. *Colloids and Surfaces* 1989;37:339–355.
134. Kawasaki T, Kobayashi W, Ikeda K, Takahashi S, Monma H. High-performance liquid chromatography using spherical aggregates of hydroxyapatite microcrystals as adsorbent. *European Journal of Biochemistry* 1986;157:291–295.
135. Kawasaki T, Niikura M, Kobayashi Y. Fundamental study of hydroxyapatite liquid chromatography high-performance. III: Direct experimental confirmation of the existence of two types of adsorbing surface on the hydroxyapatite crystal. *Journal of Chromatography* 1990;515:125–148.
136. Gorbunoff MJ. The interaction of proteins with hydroxyapatite. II: Role of acidic and basic groups. *Analytical Biochemistry* 1984;136:425–432.
137. Gorbunoff MJ. Protein chromatography on hydroxyapatite columns. *Methods in Enzymology* 1985;117:370–380.
138. Gorbunoff MJ, Timasheff SN. The interaction of proteins with hydroxyapatite. III: Mechanism. *Analytical Biochemistry* 1984 February;136(2):440–445.

139. Christoffersen J. Kinetics of dissolution of calcium hydroxyapatite. *Journal of Crystal Growth* 1980;49:29–44.
140. Koerten HK, van der Meulen J. Degradation of calcium phosphate ceramics. *Journal of Biomedical Materials Research* 1999 January;44(1):78–86.
141. Tang R, Hass M, Wu W, Gulde S, Nancollas GH. Constant composition dissolution of mixed phases. II: Selective dissolution of calcium phosphates. *Journal of Colloid and Interface Science* 2003 April;260(2):379–384.
142. Ducheyne P, Radin S, King L. The effect of calcium phosphate ceramic composition and structure on in vitro behavior: I. Dissolution. *Journal of Biomedical Materials Research* 1993;27:25–34.

6

Apatite-Coated Polymer Template for Implant and Drug Delivery

Zhe Zhang, Sam Zhang, and Lei Shang

CONTENTS

6.1 Introduction

The trend in biomaterial research mimics very closely to any other type of "conventional" material research: from bulk to thin film/coating, from single constituent to composite, and from macro- to micro- or even nanoscales. Hydroxyapatite (HA) demonstrates an excellent example of this trend that a material can be applied in different forms to solve various biomedical problems. This chapter is devoted to address a unique way that HA is used: as a coating material on polymeric scaffolds. Polymeric scaffolds, especially biodegradable ones, are extensively studied in recent years for guided tissue regeneration. Adhesion between the newly formed tissue and the promotion of cellular proliferation on the scaffold are important properties despite the general requirement of biocompatibility. It is well known that cellular proliferation and adhesion on biomaterials are both determined by cell type and biomaterial surface property. Studies are being conducted to improve those properties by material level modification and drug incorporation (e.g., growth factors). When it comes to musculoskeletal tissue regeneration, the material approach has a greater role to play because the major component of natural bone can be artificially synthesized with ease and high similarity. The nature already designed the musculoskeletal system in such a way that osteoblasts proliferate well on the existing tissue and the newly formed tissue integrates into the existing ones well. Therefore, the job left to researchers becomes relatively easier: provide an artificial interface between the scaffold and the tissue nearby so that in the eyes of the proliferating tissue the scaffold is of no difference to other host tissues. HA as the major mineral component of the natural bone serves this purpose well. The major drawback of HA when it is used alone is its poor mechanical strength; hence, it is often used as a coating material on substrates, especially metallic ones to improve the bone growth and adhesion, which is beneficial and crucial to bone replacement applications (e.g., hip and knee replacement surgery). In cases where repair is preferred rather than replacement, biodegradable polymer scaffolds are often used to provide the mechanical strength and allow new tissues to be formed. On the other hand, HA binds well to proteins, and HA itself is biodegradable as well. Those properties enable HA to be studied more than just as an artificial interface but also to deliver therapeutic agents locally for anti-inflammation or antibiotic purposes using polymer microspheres/nanospheres as drug carriers.

This chapter will introduce some of the commonly used biodegradable polymers as substrate first, followed by HA coating techniques on them. Properties of the HA-coated polymers will be illustrated in different practical applications as each application requires different properties.

6.2 Biodegradable Polymers

Biodegradable polymers, either synthetic or natural, are capable of being hydrolyzed into biocompatible small molecular byproducts through chemical or enzyme catalysts and do not induce an inflammatory reaction in the body; therefore, they are of great interest in the application of scaffold and fabrication of microspheres. Table 6.1 shows the list of popularly used biodegradable polymers in biomedical application.

6.2.1 Natural

Polysaccharides form a class of materials that have generally been underutilized in biomaterials. Apart from their biological activity, one of the more important properties of polysaccharides in general is their ability to form hydrogels. Hydrogel formation can occur by a number of mechanisms and is strongly influenced by the types of monosaccharide involved as well as the presence and nature of substituent groups. Polysaccharide gel formation is generally of two types: hydrogen bonded and ionic. Hydrogen-bonded gels are typical of molecules such as agarose (thermal gelation) and chitosan (pH-dependent gelation), whereas ionically bonded gels are characteristic of alginates and carrageenans. Natural hydrogel such as alginate and chitosan are also widely used to prepare microspheres by means of the ionic cross-linking method due to their good biodegradability, biocompatibility, and high mucoadhesive properties.[1–17]

TABLE 6.1

List of Biodegradable Polymers in Biomedical Application

	Classification	Polymer
Natural polymer	Protein-based polymers	Collagen, albumin, gelatin
	Polysaccharides	Agarose, alginate, carrageenan, hyaluronic acid, dextran, chitosan, cyclodextrins
Synthetic polymer	Polyester	Poly(lactic acid), poly(glycolic acid), poly(hydroxy butyrate), poly(ε- caprolactone), poly(α-malic acid), poly(dioxanones)
	Polyanhydrides	Poly(sebacic acid), poly(adipic acid), poly(terphthalic acid) and various copolymers
	Polyamides	Poly(imino carbonates), polyamino acids
	Phosphorous-based polymers	Polyphosphates, polyphosphonates, polyphosphazenes
	Others	Poly(cyano acrylates), polyurethanes, polyortho esters, polydihydropyrans, polyacetals

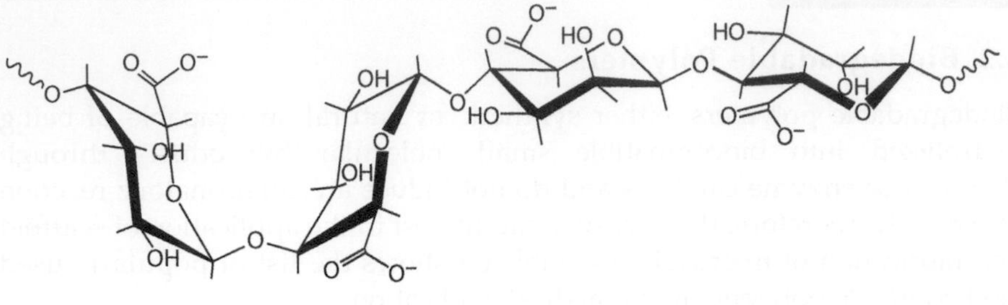

FIGURE 6.1
Chemical structure of alginate. (From Coviello, T. et al., *J. Control. Release*, 119(1), 5, 2007.)

6.2.1.1 Alginate

Commercially used alginate is extracted from three kinds of brown algae including *Laminaria hyperborea*, *Ascophyllum nodosum*, and *Macrocystis pyrifera*.[18] Sodium alginate is a water-soluble linear polysaccharide comprised of α-L-guluronic (G-block) and β-D-mannuronic (M-block) acid in alternate bond. Because of the particular structure of monomer units, the G-block and M-block regions exhibit completely different geometries. Two neighboring G-blocks create a diamond-shaped hole that is ready for conjugating with ions, as shown in Figure 6.1. Gelation of alginate is realized by a G-block cross-linked with divalent metal ions such as Ca^{2+}, Sr^{2+}, and Ba^{2+}, except Mg^{2+}.[20] The gelation process is mainly achieved by chelation with divalent ions. The Na ions from the G-block exchange with divalent Ca ions, and these G-blocks aggregate to form the egg-box structure.[21]

Some studies demonstrate that the alginate microsphere has large sized pore and shows the physical and chemical instability under high pH environment.[22] These limitations may lead to low drug encapsulation and quick release in drug delivery application.

6.2.1.2 Chitosan

Chitosan is the only cationic biodegradable polymer because of the amide groups in its chemical structure (Figure 6.2). It shows high drug encapsulation and excellently sustained drug release, as well as good bioadhesive properties. The most widely used chitosan is α-chitosan from the chitin of crab and shrimp shell.[23] It forms a gel with the popular ionic cross-linker pentasodium tripolyphosphate (TPP) in a mild aqueous environment. Glutaraldehyde is another agent chosen as a chemical cross-linker reacting with amide groups for gel formation.[24] In contrast to ionic gelation, chitosan microspheres prepared by chemical cross-linking have high density and low porosity.

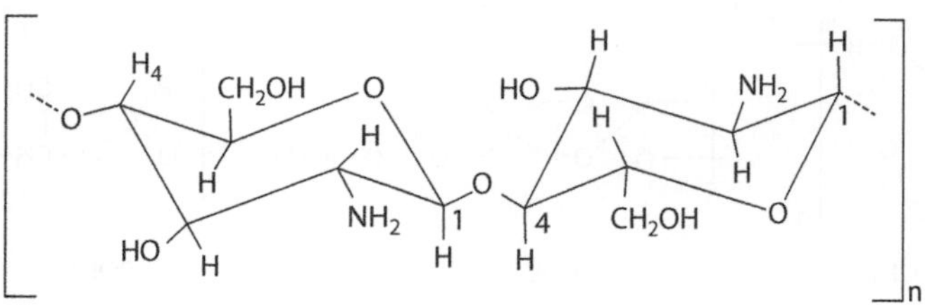

FIGURE 6.2
Chemical structure of chitosan. (From van der Merwe, S.M. et al., *Eur. J. Pharm. Biopharm.*, 58(2), 225, 2004.)

6.2.2 Synthetic

Most synthesized polymers used to prepare microspheres are hydrophobic. Thus, a scaffold or film is prepared by solvent evaporation; emulsion-evaporation method is commonly chosen for microsphere fabrication. This method also shows other advantages in the drug encapsulated microsphere process: it can maintain the chemical stability and bioactivity of the drugs, provide high encapsulation efficiency, and control the ideal size distribution of the microspheres.

Among the biodegradable synthesized polymers, the most commonly investigated in implant and drug delivery systems are the following polymers: poly(ortho ester) (POE), polyanhydride (PAH), poly(ε-caprolactone) (PCL), poly(lactic acid) (PLA), poly(glycolic acid) (PGA), and poly(lactic-co-glycolic acid) (PLGA).

6.2.2.1 Poly(Ortho Ester)

Since the 1970s, POE as a biodegradable polymer has evolved into four species: POE I, POE II, POE III, and POE IV.[26] The chemical structures of the species are shown in Figure 6.3. Compared to other polyesters that degrade homogeneously in aqueous solution throughout the polymer matrix, POE with high hydrophobicity and water impermeability performs degradation by surface erosion. The morphology of POE microspheres is shown in Figure 6.4, as reported by David.[27] Depending on the rate of surface erosion, the drug-loaded POE scaffold or microspheres maintain constant drug release rate without a significant burst release.[26] However, low pH environment could accelerate erosion rate due to the pH sensitivity of POE, which limits its application in drug delivery system.

6.2.2.2 Polyanhydride

PAH is a hydrophobic polymer that is linked with hydrolytically anhydride (Figure 6.5). The degradation of PAH is administrated by the polymer

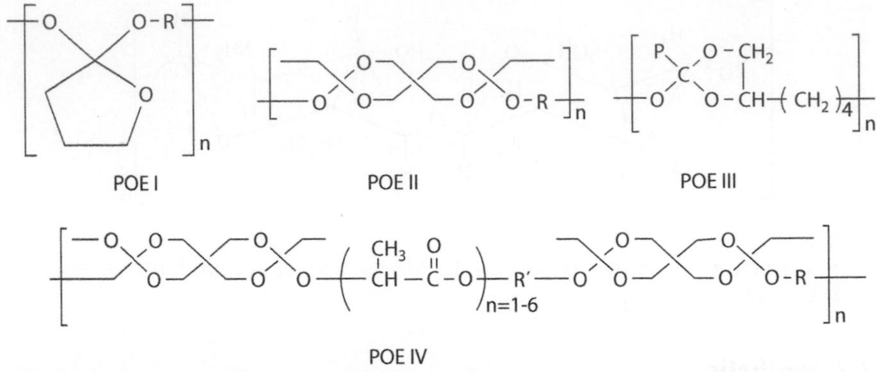

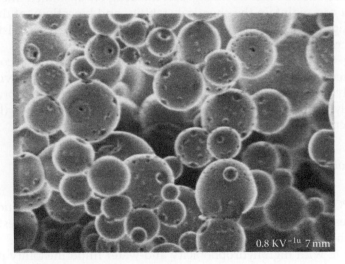

POE I POE II POE III

POE IV

FIGURE 6.3
Chemical structures of four POE species. (From Park, J. et al., *Molecules*, 10(1), 146, 2005.)

0.8 KV^{-1u} 7 mm

FIGURE 6.4
Micrograph of POE microspheres. (From Nguyen, D.N. et al., *Biomaterials*, 29(18), 2783, 2008.)

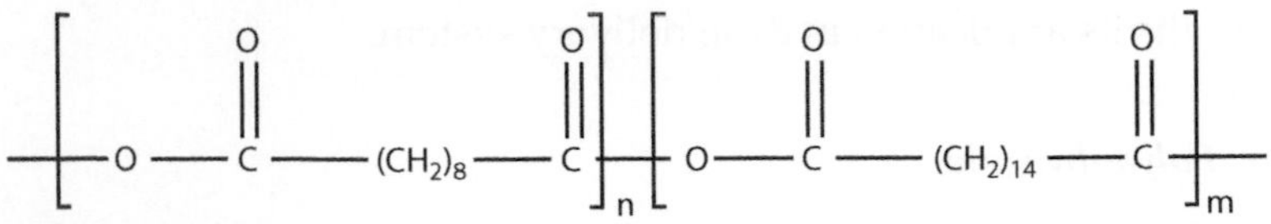

FIGURE 6.5
Chemical structure of PAH. (From Sokolsky-Papkov, M. et al., *Adv. Drug Deliver. Rev.*, 59(4–5), 187, 2007.)

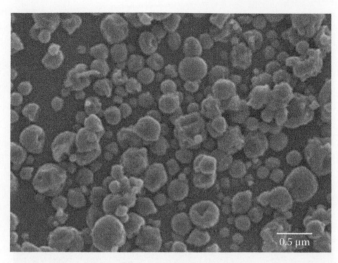

FIGURE 6.6
Micrograph of PAH microspheres. (From Lopac, S.K. et al., *J. Biomed. Mater. Res.: Part B Appl. Biomater.*, 91B(2), 938, 2009.)

monomer composition. Using the emulsion-evaporation method, PAH microspheres are harvested as shown in Figure 6.6.[30] The degradation of PAH depends on the water uptake rate and undergoes surface erosion in aqueous solution, which is the same mechanism as POE, and performs the zero-order drug release profile as the microsphere drug carriers. Since degraded monomer acids are nonmutagenic and noncytotoxic byproducts, PAH shows the minimal inflammatory reaction in vivo. One drawback of PAH is that it is difficult to make storage in anhydrous condition because of the hydrolytic instability of the anhydride bond.

6.2.2.3 Poly(ε-Caprolactone)

PCL is a semi-crystalline hydrophobic polymer, and its chemical structure is shown in Figure 6.7. Stephen et al. used PCL microspheres for taxol release, and the morphology of drug-loaded PCL microspheres is shown in Figure 6.8.[31] Due to its high crystallinity and hydrophobicity, the degradation of PCL is quite slow, which provides the drug for long-term delivery.

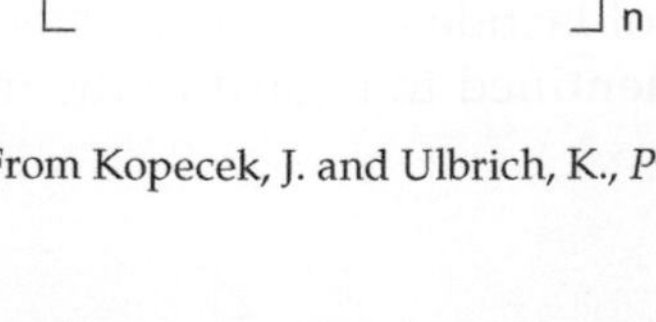

FIGURE 6.7
Chemical structure of PCL. (From Kopecek, J. and Ulbrich, K., *Prog. Polym. Sci.*, 9(1), 1, 1983.)

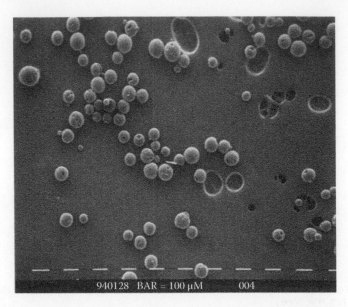

FIGURE 6.8
Micrograph of PCL microspheres. (From Dordunoo, S.K. et al., *Cancer Chemother. Pharmacol.*, 36(4), 279, 1995.)

A new design developed in blending PCL with other biodegradable polymers to provide opportunities to change the drug release rate.[32] There are two steps for biodegradation of PCL-blended polymers in vitro or in vivo. At first, the PCL-blended polymer maintains its morphology and mass. Degradation of the PCL-blended polymer starts with weight loss due to the continuous chain cleavage and small fragments' diffusion out of the polymer matrix. Its byproducts from biodegradation are non-acidic, which is good for biomedical application since non-acidic fragments do not induce local pH change.

6.2.2.4 Poly(Lactic-Co-Glycolic Acid)

PGA, PLA, and PLGA are all in the hydrophobic polyester family, which biodegraded by hydrolysis of the ester linkages. PLGA is a copolymer that is synthesized by means of random ring-opening polymerization of glycolic acid and lactic acid. The chemical structures of PGA, PLA, and PLGA are shown in Figure 6.9.

PLGAs have been used as Food and Drug Administration (FDA)-approved therapeutic devices due to their biodegradability and biocompatibility. Depending on the ratio of lactide to glycolide chosen, various PLGAs are harvested. PLGAs are identified in regard to the monomers' ratio: taking

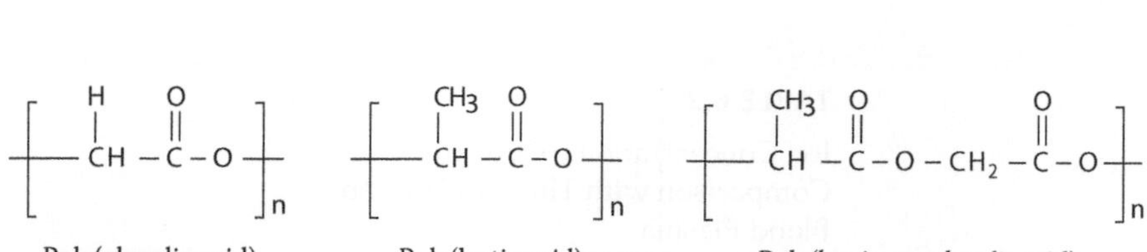

Poly(glycolic acid) Poly(lactic acid) Poly(lactic-co-glycolic acid)

FIGURE 6.9
Chemical structures of PGA, PLA, and copolymer PLGA. (From Park, J. et al., *Molecules*, 10(1), 146, 2005.)

PLGA (80:20) as an example, it is a copolymer comprised with 80% lactic acid and 20% glycolic acid. All PLGAs are amorphous and show a very low glass-transition temperature in the range of 40°C–60°C. In comparison with PGA and PLA, which show low solubility in some oil solvents, PLGA can be dissolved in most of common solvents such as chlorinated solvents, acetone, or ethyl acetate. PLGA as a member of polyester family also degrades as a result of the hydrolysis of ester linkages in the presence of water, which is the same as PLA and PGA. In general, the duration of the PLGA degradation is correlated to the ratio of glycolic acid and lactic acid. The PLGA with higher amount of glycolic acid performs the shorter duration needed for degradation. Therefore, widely used PLGA (50:50) exhibits faster degradation duration. Aside from the ratio of monomers, PLGA end-capped with carboxyl groups illustrates shorter degradation duration than that end-capped with ester groups.

Since the body can effectively deal with monomers of PLGA biodegradation, very minimal systemic toxicity was found to affect the application of PLGA in protein and DNA delivery.[34] One commercially marked drug delivery device by PLGA is Lupron Depot®, commonly used for the treatment of advanced prostate cancer.

6.3 Simulated Body Fluid

In 1991, Kububo proposed that the necessary requirement for artificial implants to adhere with bone cells is the formation of bone-like apatite or HA on their surface.[35] Like apatite formed in vivo, it also can be produced in simulated body fluid (SBF), a solution with ion concentrations nearly equal to those of human blood plasma. Table 6.2 shows the ion concentrations in SBF solution and human blood plasma as a reference. Since SBF or concentrated SBF can supply with abundant calcium and phosphate ions, apatite formed on the surface of the implant in SBF solution becomes a classical method for the apatite coating study.

TABLE 6.2

Ion Concentrations of SBF in
Comparison with Those in Human
Blood Plasma

Ion	Blood Plasma (mM)	SBF (mM)
Na^+	142.0	142.0
K^+	5.0	5.0
Mg^{2+}	1.5	1.5
Ca^{2+}	2.5	2.5
Cl^-	103.0	147.8
HCO_3^-	27.0	4.2
HPO_4^{2-}	1.0	1.0
SO_4^{2-}	0.5	0.5
pH	7.2–7.4	7.4

Source: Kokubo, T. and Takadama, H.,
Biomaterials, 27(15), 2907, 2006.

6.4 HA Deposition by Biomimetic Method

HA coating deposition by biomimetic method has been studied for quite
a few years. It can trigger the deposition of apatite coating on different
substrates, including polymer scaffolds, the surface of cement, ceramic, and
metal film or rods. In order to better understand the coating deposition
procedure in aqueous solution, Lu et al. summarized a theoretical analysis
of HA coating precipitation.[37]

6.4.1 Driving Force of Apatite Deposition

The thermodynamic driving force of apatite deposition is obtained based on
the free energy change in supersaturated solutions[38–39]:

$$\Delta G = -\frac{RT}{n}\ln(S) = -\frac{RT}{n}\ln\left(\frac{A_p}{K_{sp}}\right) \tag{6.1}$$

where
    ΔG is the Gibbs energy of apatite (CaP) ionic units in solution
    R is the universal gas constant (8.314 JK^{-1} mol^{-1})
    T is the absolute temperature
    n is the number of ion units in 1 mol of apatite
    S is the ratio of the activity product of ion units of precipitates (A_p) to the
        corresponding solubility product (K_{sp}), defined as supersaturation

Among all CaP products formed in the solution, HA is the most stable; other CaPs such as octacalcium phosphate or dicalcium phosphate dehydrate formed at the beginning tend to transform into HA. Therefore, only the equation of HA precipitation in aqueous solution is studied here:

$$10Ca^{2+} + 6PO_4^{3-} + 2OH^- = Ca_{10}(PO_4)_6(OH)_2 \tag{6.2}$$

The corresponding supersaturation (S) is calculated as follows:

$$S(HA) = \frac{a^5(Ca^{2+})a^3(PO_4^{3-})\, a(OH^-)}{K_{sp}(HA)} \tag{6.3}$$

where K_{sp}(HA) equals to 2.35×10^{-59}.[40]

The activity coefficient, γ_i, of each ion unit can be calculated from the modified Debye–Hückel equation:

$$\log \gamma_i = -Az_i^2 \left[\frac{I^{1/2}}{1+I^{1/2}} - 0.3I \right] \tag{6.4}$$

where

A is the Debye–Hückel constant varying with temperature, 0.5211 at 37°C

z_i is the charge of ions

I is the total strength of ions in the solution and is calculated as

$$I = \frac{1}{2} \sum_i c_i z_i^2 \tag{6.5}$$

where c_i is the mole number of each ion unit.

SBF was often chosen as a source of Ca and P for apatite formation by the biomimetic method; the equation for the ionic activity calculations would take into account magnesium and other ions in SBF that could affect HA precipitation.[41–43] In that case, classic theories of thermodynamics and kinetics would be not easy to apply since all chemical reactions in the solution must be considered.

6.4.2 Nucleation Rates

The CaP nanocrystal nucleation rate (J) can be calculated based on the classical equation of heterogeneous nucleation:

$$J = K \exp\left(-\frac{\Delta G}{kT}\right) = K \exp\left(-\frac{16\pi\vartheta^2\gamma^3 f(\theta)}{3k^3T^3(\ln S)^2}\right) \tag{6.6}$$

where

k is the Boltzmann constant

T is the absolute temperature

Depending on this equation, the nucleation rate (J) of apatite formation is proportional to the kinetic factor (K) and is determined by the activation energy of nucleation (ΔG). Here, ΔG is primarily affected by the interfacial energy γ, the supersaturation S, and the contact angle function $f(\theta)$. ν is defined as the CaP molecular volume and is determined by the crystal structure. The kinetic factor (K) is proportional to the probability (P) that ion units of CaP form a nucleus in the solution, i.e., $K = K'P$. Boistelle et al. proposed the equation to calculate the P value of HA precipitation as follows[44]:

$$P = \frac{9![Ca^{2+}]^5 \left[PO_4^{3-}\right]^3 [OH^-]}{5!3!\left([Ca^{2+}]+\left[PO_4^{3-}\right]+[OH^-]\right)^9} \tag{6.7}$$

Attained from their experiment, K' value could be calculated as $13.64 \times 10^{-24}\ cm^{-3}\ s^{-1}$. Based on the studies by Wu and Nancollas, the interfacial energy (γ) is also experimentally calculated as $10.4\ mJm^{-2}$ for HA.[45,46]

6.5 HA Coating on Polymer Scaffold

Recently, porous and 3-D polymer scaffolds are developed as a substrate for cell growth and proliferation in tissue engineering, and cells can secrete an extracellular matrix to induce new tissue formation.[47] Polymer scaffold is chosen as an artificial implant due to biocompatibility, biodegradability, high mechanical strength, and osteoconductivity. Biocompatibility and biodegradability are correlated with polymer nature and can be administrated by an appropriate choice of polymer. In order to improve mechanical strength and osteoconductivity, some researchers designed a layer of apatite coating on the surface of the polymer scaffold.

The polysaccharide chitosan with amino groups, as a natural biodegradable polymer, is widely applied as a porous 3-D scaffold.[48,49] Manjubala et al. used the freeze-dried method to fabricate macroporous chitosan scaffold (Figure 6.10a and b).[50,51] They designed a modified double diffusion chamber containing separate calcium and phosphate ions to induce apatite deposition on the chitosan scaffold. The CaP particles can be clearly observed on the top and cross section of the scaffold as shown in Figure 6.10c and d. Apatite layer formed on the surface of the chitosan scaffold results from the amino groups of chitosan. Positively charged amino groups can bind phosphate ions in the solution to supply the site for apatite nucleation. After the nuclei formation, apatite starts to grow by consuming calcium and phosphate ions in the solution.

Aside from the amino group, the carboxyl group is another effectively negative-charged chemical group that can induce apatite formation by means of absorption of calcium ions.[52] Alginate hydrogel by ionic or

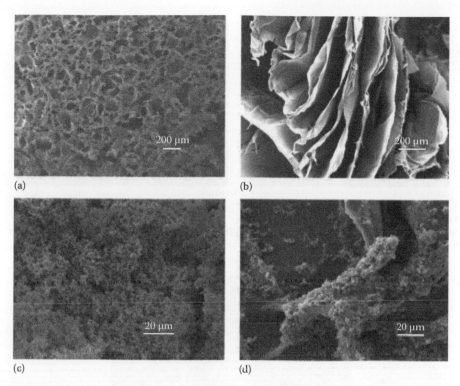

FIGURE 6.10
Micrographs of (a) and (c) top view of chitosan scaffold and mineralized scaffold, (b) and (d) cross section of chitosan scaffold and mineralized scaffold.

chemical cross-linking is used as the scaffold material for tissue engineering due to its biocompatibility and biodegradability.[21,53] Darilis et al. investigated the nucleation of HA on alginate scaffold for stem-cell-based bone tissue application,[54] and HA nanoparticles fully covering the surface of the alginate gel were observed, as shown in Figure 6.11. Since the alginate gel was prepared by using $CaCl_2$ as a cross-linker, phosphate ions in the solution were immediately absorbed on the surface of the gel; in addition, there are abundant carboxyl groups in the alginate structure that could absorb calcium ions in the solution simultaneously. Thus, HA particles exhibit rapid precipitation on alginate gel scaffold.

Recently, many researchers have shown interest in synthesized copolymer PLGA as a 3-D macroporous scaffold in tissue engineering since it is capable of controlling biodegradability by composition, in comparison to natural polymers. In order to increase cell bioadhesion and osteoconductivity, many efforts are made to deposit a layer of apatite on PLGA scaffold.[55-63] Murphy first reported HA coating on the PLGA scaffold by biomimetic method in SBF solution, as shown in Figure 6.12.[57] Later, they modified the NaOH treatment for PLGA scaffold before soaking to increase the surface

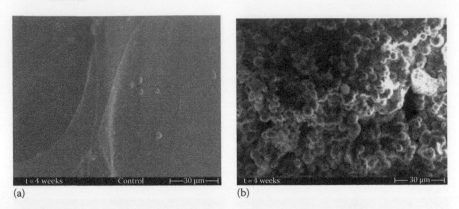

(a)   (b)

FIGURE 6.11
Micrographs for (a) control alginate gel and (b) HA coated on the surface of alginate gel after 4 weeks immersion in SBF.

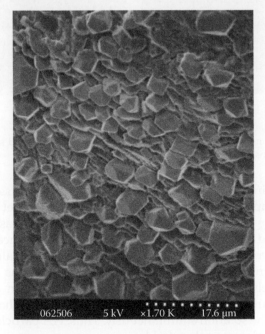

FIGURE 6.12
Micrograph of HA crystal deposition on PLGA scaffold after 16 days of incubation. (From Murphy, W.L. et al., *J. Biomed. Mater. Res.*, 50(1), 50, 2000.)

hydroxyl and carboxyl groups and then immersed the scaffold into SBF solution for HA crystal growth.[58] The presence of these increasing groups accelerates calcium ion binding to the polymer surface and improves mineral growth. The morphology of the mineral coated on modified PLGA is shown in Figure 6.13. Chou et al. carried out a two-step soaking method with

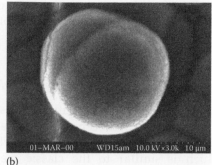

(a)   (b)

FIGURE 6.13
Micrographs of HA deposited on NaOH-treated PLGA scaffold after 16 days of incubation in SBF (a: magnification ×80; b: magnification ×3000). (From Murphy, W.L. and Mooney, D.J., *J. Am. Chem. Soc.*, 124(9), 1910, 2002.)

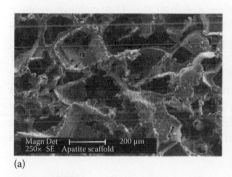

(a)   (b)

FIGURE 6.14
Micrographs of apatite-coated PLGA scaffold by sequentially immersing in (a) SBF1 and (b) SBF2. (From Chou, Y.F. et al., *J. Biomed. Mater. Res. B Appl. Biomater.*, 75(1), 81, 2005.)

supersaturated solution of 5× SBF (SBF1) and Mg^{2+} and HCO_3^- free 5× SBF (SBF2) for HA growth.[61] Unlike the apatite precipitated from original SBF, the coating obtained by two-step soaking method is comprised of plate-like structured HA nanocrystals (Figure 6.14).

6.6 HA Coating on Micro-/Nanospheres

Depending on the nucleation and growth theory of HA formation on scaffold, it can be applied to microparticle or nanoparticle as a substrate. The difference between their structures determines their biomedical application: HA coating on scaffold is applied to improve the bioactivity and bioadhesion, while HA-coated microparticles/nanoparticles are carriers for bioactive molecular drug delivery since HA can provide strongly chemical bonding with protein and DNA.

6.6.1 Supersaturated Solution

6.6.1.1 SBF or Concentrated Modified SBF

SBF or concentrated modified SBF solution, as a calcium and phosphate source, is usually chosen as the supersaturated solution for apatite formation. Murphy et al. prepared HA-coated PLGA microspheres using SBF solution, which is similar to the classic method of biomimetic HA coating on scaffolds.[64] After immersing PLGA microspheres in SBF solution for 7 days at 37°C, HA coating on the surface of microspheres was successfully produced. The morphology of uncoated and coated microspheres is shown in Figure 6.15. Mineral HA with a plate-like structure was clearly observed after microsphere soaking. It showed that the biomimetic method for HA coating formation is also suitable for particle as a substrate. The protein release rate from binding with HA and encapsulation in PLGA microspheres was also investigated (Figure 6.16). Both approaches showed sustained release; however, the total amount of bound protein release is significantly higher than the encapsulated protein after 30 days. Even more importantly, the bound protein release rate is nearly constant, as a linear kinetics showed in the release curve without a burst release stage. It demonstrated that the release of protein drug binding with HA coating did not result from surface diffusion and started with the dissolution of mineral HA. The dissolution

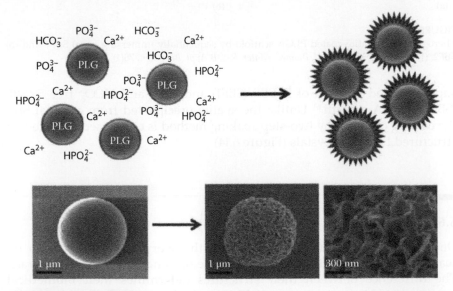

FIGURE 6.15
Schematic of mineral-coated PLGA microspheres and micrographs for uncoated microspheres.

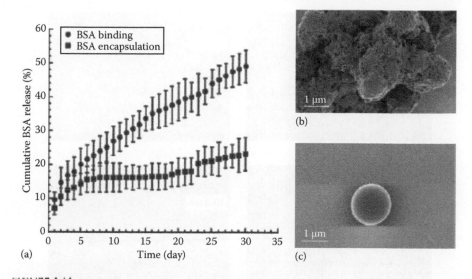

FIGURE 6.16
(a) Comparison of cumulative release of BSA bound to HA-coated PLGA microspheres and BSA encapsulated in PLGA microspheres. Micrographs of (b) HA-coated microspheres and (c) coated PLGA microsphere after the 30-day release period. (From Jongpaiboonkit, L. et al., *Adv. Mater.*, 21(19), 1960, 2009.)

rate of HA coating determined the bound protein release kinetics. This study opens a new view to control protein drug release in drug delivery systems.

They also studied the effect of the amount of the nucleating agent and surfactant in the supersaturated solution on the growth of HA coating.[65] The morphology of HA-coated microspheres with the addition of different concentrations of microspheres as nucleating agents is shown in Figure 6.17. 0.25% (w/v) of PLGA microspheres is the preferable amount added for HA coating growth without aggregation. During the growth of the mineral HA coating, the presence of surfactant (Tween 20) significantly inhibited the formation of the mineral and changed the surface morphology of the coating, as shown in Figure 6.18.

Kang et al. also chose PLGA microspheres, 150–300 µm in size, as substrates for HA growth.[66] While SBF with inorganic ion concentrations three times that of human blood plasma was modified as a source of Ca and P. After different immersion times, the surface morphology of uncoated PLGA and HA-coated PLGA microspheres is shown in Figure 6.19. HA coating is successfully formed on the surface after an immersion of 5 days; however, some obvious cracks are observed on the surface of the coating. Poor coverage of the coating is mainly caused by Mg^{2+}, HCO_3^-, and other ions in the SBF which are bonded with Ca^{2+} and PO_4^{3-} to affect the nucleation and growth of HA on the microspheres.

Apart from negatively charged PLGA microspheres, positively charged chitosan microspheres are also a good choice for HA growth because the $-NH_3$

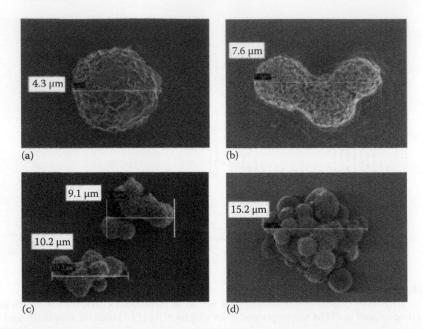

(a) (b)

(c) (d)

FIGURE 6.17
Micrographs of HA-coated microspheres after 7-day soaking in SBF solution with different amount of addition of microspheres: (a) 0.25% (w/v); (b) 0.50% (w/v); (c) 0.75% (w/v); (d) 1.00% (w/v). (From Jongpaiboonkit, L. et al., *ACS Appl. Mater. Interfaces*, 1(7), 1504, 2009.)

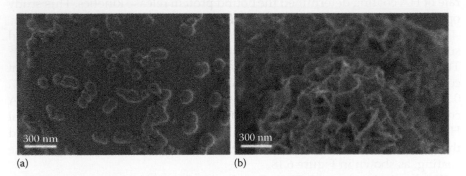

(a) (b)

FIGURE 6.18
Morphology of the surface of microspheres formed (a) in the presence of 0.1% (v/v) Tween 20 and (b) in the absence of Tween 20. (From Jongpaiboonkit, L. et al., *ACS Appl. Mater. Interfaces*, 1(7), 1504, 2009.)

group of chitosan can bond with phosphate ions to induce apatite nucleation in the solution. Jayasuriya et al. used chitosan as a template to prepare apatite-coated chitosan microspheres for the release of growth factor IGF-1.[67] The important point in their work is that they used 5 times the concentration of SBF for rapid biomineralization. Some researchers use an incubation period of 5–7 days in order to provide a complete HA coating precipitation

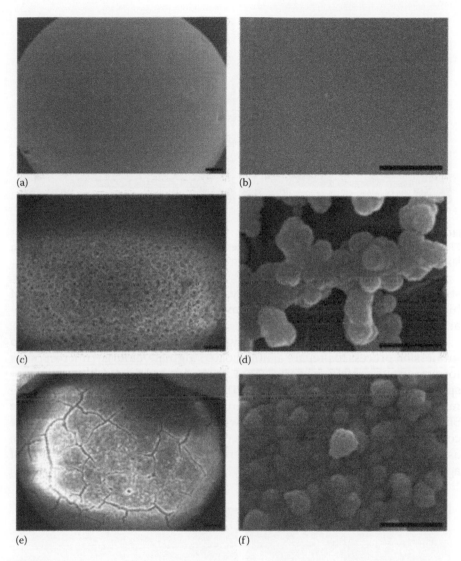

FIGURE 6.19
Micrographs of (a, b) PLGA microsphere and (c, d, e, and f) apatite-coated PLGA microsphere incubated in SBF for (c, d) 3 days and (e, f) 5 days. The scale bars indicate 10 μm in a, c, and e and 1 μm in b, d, and f. (From Kang, S.W. et al., *J. Biomed. Mater. Res. Part A*, 85A(3), 747, 2008.)

in 1.5× SBF or 3× SBF solution. It is quite a long immersion duration for the preparation of drug delivery devices. Jayasuriya et al. increased the concentration of SBF to five times to obtain a complete coverage of the HA coating in 24 h incubation. A thin layer comprising HA nanoparticles on the surface of chitosan is shown in Figure 6.20. However, the morphology of the HA coating is quite different from Murphy's work, which shows a plate-like

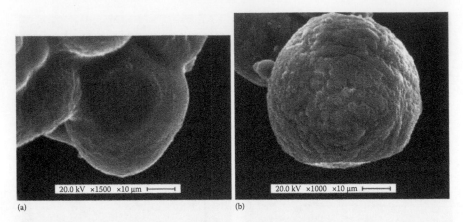

FIGURE 6.20
Morphology of chitosan microspheres before (a) and after (b) biomineralization. (From Jayasuriya, A.C. and Kibbe, S., *J. Mater. Sci.: Mater. Med.*, 21(2), 393, 2010.)

structure in SBF solution. Morphology difference is caused by the effect of high concentrations of Mg^{2+}, HCO_3^-, and other ions in 5× SBF on the nucleation and growth of HA.

Min Lee et al. modified Jayasuriya's work and designed a two-step immersion process to prepare apatite-coated alginate/chitosan microspheres for osteogenic protein delivery.[68] They first added alginate/chitosan microspheres into 5× concentrated SBF (SBF1) for a 6 h nucleation process; apatite-nuclei-coated microspheres were then poured into Mg^{2+} and HCO_3^- free SBF (SBF2) for 12 h of continuing growth. By a two-step method, an apatite coating with a plate-like structure is completely formed in less than 1 day, as shown in Figure 6.21b. The problem in this work is the high extent of aggregation that occurs (Figure 6.22a).

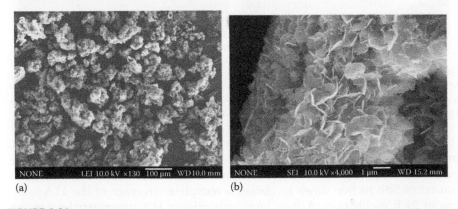

FIGURE 6.21
Micrographs of (a) alginate/chitosan microparticles after soaking in SBF1 and (b) after soaking in SBF2. (From Lee, M. et al., *Biomaterials*, 30(30), 6094, 2009.)

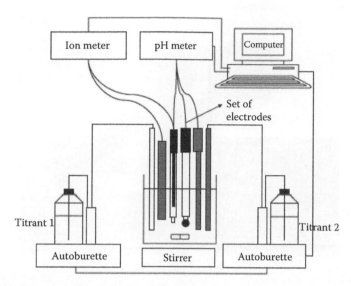

FIGURE 6.22
Diagram of computer-controlled constant composition method. (From Xu, Q. et al., *Biomaterials*, 28(16), 2687, 2007.)

6.6.1.2 Solution Containing Only Ca and P

A classic biomimetic method of HA coating formation is by immersing the substrate in SBF, but this usually takes up to several days to produce a complete coating. The other problems that need to be solved include high aggregation and poor coverage. The constant composition precipitation in the solution is a rapid method for HA particle formation, as first reported by Tomson and Nancollas,[69] Xu et al. successfully applied this method to prepare an HA coating fully covered on the surface of negative liposomes.[70] In this work, a stable supersaturated solution with respect to HA was prepared by mixing $CaCl_2$ and KH_2PO_4 (Ca/P ratio 1.67), at 37°C and pH 7.8, as neutral and basic pH tend to induce the formation of HA.[71] Then liposomes as nucleating agents were added to induce the nucleation and growth of HA nanocrystals. In order to monitor the whole process for HA precipitation, a computer-controlled system was designed as shown in Figure 6.22. An ion meter was used to measure calcium ion activity and a pH meter was used to monitor pH changes in the solution and to maintain a pH of 7.8 throughout the whole process by adding KOH. The titrants comprised of 25 mM $CaCl_2$ (Titrant 1), 15 mM KH_2PO_4, and 50 mM KOH (Titrant 2), which supplied Ca^{2+} and PO_4^{3-} to maintain the concentrations of Ca and P in the solution as constants. By using this system, an HA coating is successfully grown on liposomes in 6 h (Figure 6.23). However, regarding commercial development, this system is too complicated and is not available for mass production.

FIGURE 6.23
Bright field image and corresponding electron diffraction pattern of HA-coated liposome nanospheres. (From Xu, Q. et al., *Biomaterials*, 28(16), 2687, 2007.)

Anitha Ethirajan et al. developed a novel two-step process for HA coating formation, as shown schematically in Figure 6.24.[72] The addition of calcium ions for bonding with nanoparticles, then followed by a dropwise addition of phosphate ions, corresponds to the stoichiometric Ca/P ratio of 1.67 for HA formation. The concentration of Ca and P in the solution is very low: 0.5 mM Ca^{2+} and 0.3 mM PO_4^{3-}. The pH value in this work, maintained at a constant value of 10, leads to the generation of HA-coated nanoparticles. As a result, HA coating is formed after 24 h but is not fully covered, as shown in Figure 6.25. A pH of 10 throughout the whole experiment is not suitable for most biodegradable microspheres/nanospheres due to quick hydrolysis under alkaline condition.

In our lab, we created a novel two-step process under physiological temperature and pH for HA growth (as illustrated schematically in Figure 6.26). At first, after the addition of microspheres into the calcium solution, calcium ions in the solution are partly absorbed on the surface of microspheres due to bonding with carboxyl groups. This process would supply a Ca-rich surface for apatite nucleation. Residual calcium ions in the solution provide a supply for HA continuing growth. Secondly, when phosphate ions are added to the suspension, they would interact with the calcium ions at the microsphere surface to form the very first layer of HA, and thereafter, HA growth would take place. The whole growth is carried out at 37°C as the biomimetic process.

The PLGA microspheres we used as substrates were prepared by means of the oil in water (O/W) emulsion-evaporation method. The scanning electron microscope (SEM) micrographs show uniform distribution and smooth surface of the microspheres (Figure 6.27).

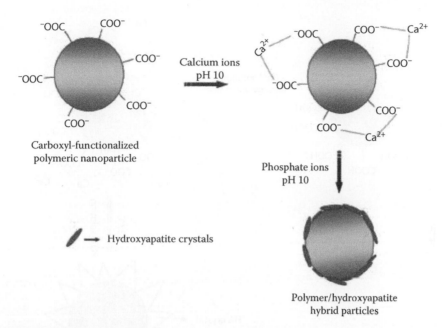

FIGURE 6.24
Schematic illustration of HA deposition on carboxyl-functionalized nanoparticle by consecutive addition of Ca^{2+} and PO_4^{3-}. (From Ethirajan, A. et al., *Chem. Mater.*, 21(11), 2218, 2009.)

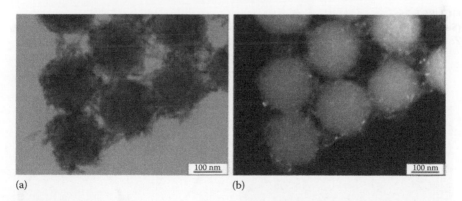

(a)   (b)

FIGURE 6.25
Micrographs of HA-coated nanoparticles, in bright field (a) and dark field (b) image modes. (From Ethirajan, A. et al., *Chem. Mater.*, 21(11), 2218, 2009.)

The average particle size of PLGA microspheres is measured as 6.67 ± 0.82 μm. The zeta potential of these microspheres is −27.0 ± 1.87 mV, which indicates the negatively charged surface caused by the carboxyl groups in the chemical structure of PLGA. It is an important point for choosing PLGA as a template for HA growth, since the rapid formation of HA crystals is

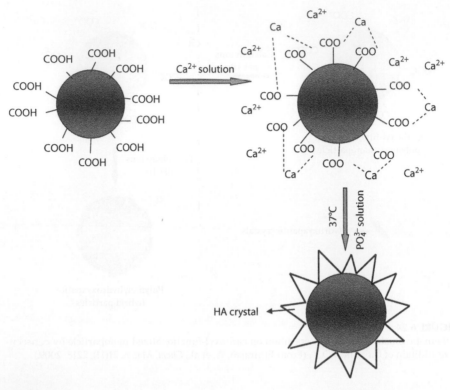

FIGURE 6.26
(**See color insert.**) Schematics of HA coating formation on PLGA microspheres. (From Zhang, Z. et al., *Nanosci. Nanotechnol. Lett.*, 3(4), 472, 2011.)

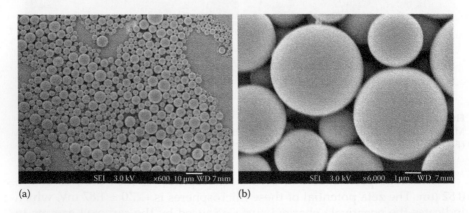

FIGURE 6.27
SEM micrographs of negatively charged PLGA microspheres (a) ×600 and (b) ×6,000 magnification. (From Zhang, Z. et al., *Nanosci. Nanotechnol. Lett.*, 3(4), 472, 2011.)

well pronounced for particles with a high amount of carboxyl groups.[72] Generally, calcium ions could be bonded with carboxyl groups on the surface of PLGA microspheres through electrostatic forces; followed by the addition of phosphate, HA crystals start to nucleate and grow on the microspheres.

The factors that affect HA formation are proposed as the amount of –COOH group, concentrations of Ca and P in the solution, and immersion time of the substrate. In our experimental design, we fixed the amount of the –COOH group on PLGA as a constant index. The study of HA growth focused on the ion concentration of the solution and immersion time of microspheres.

6.6.1.2.1 Effect of Concentrations of Ca and P in the Solution on the Formation of HA Coating

Two solutions containing Ca and P with different concentrations are shown in Table 6.3. We set the ratio of Ca and P as a constant of 1.67 with respect to HA for the two solutions. After directly mixing the Ca and P solution, Solution 2 shows a suspension with white particle precipitation, but Solution 1 is still clear, as shown in Figure 6.28. It indicates that HA particles are formed without a nucleating agent from Solution 2 of high concentration once Ca and P ions touch each other in the aqueous solution.

The formation of HA coating was investigated after the addition of PLGA microspheres in the two solutions for 16 h incubation. After soaking in Solution 1, a complete apatite coating is fully covered on the surface of microspheres (Figure 6.29a). From magnification images (Figure 6.29b and c), the surface of the coating on the microspheres shows a plate-like structure and the coated microspheres are still uniform without aggregation. In contrast to Solution 1, after soaking in Solution 2, the bare smooth surface of microspheres and large particles with irregular shapes can be observed as shown in Figure 6.29d and e.

Samples were also scanned by Fourier Transform Infrared (FTIR) analysis, as shown in Figure 6.30. FTIR spectra show phosphate absorption peaks of HA in both solutions at 1041 cm^{-1} (stretching vibration of PO_4^{3-}) and 601, 563 cm^{-1} (deformation vibration of PO_4^{3-}) in comparison to PLGA microspheres before immersion. It demonstrates that the coating with a plate-like structure in Solution 1 and large particles with irregular shapes in Solution 2 are both correlated with HA.

TABLE 6.3

Ca and P Solutions with Different Concentrations (Ca/P = 1.67)

Solution	Ca^{2+} (mM)	PO$_4^{3-}$ (mM)	Precipitation
1	1.25	0.75	No
2	2.5	1.5	Yes

Source: Zhang, Z. et al., *Nanosci. Nanotechnol. Lett.*, 3(4), 472, 2011.

FIGURE 6.28
(**See color insert.**) Photographs for Ca and P mixed solution with different concentrations:
(a) Solution1 and (b) Solution2.

Depending on SEM images and FTIR spectra, HA particles formed in Solution 2 could be quickly self-precipitated in the solution rather than on the surface of the microspheres due to the high concentration of Ca and P. Solution 1, with 1.25 mM of Ca^{2+} and 0.75 mM of PO_4^{3-}, has the preferable concentration for HA nanocrystal nucleation and growth on the microsphere substrate.

6.6.1.2.2 Effect of Immersion Time on the Formation of HA Coating

The effect of immersion time on HA coating growth was examined by FTIR based on samples at 1, 4, 8, 16, and 24 h, as shown in Figure 6.31. After the first hour of incubation, the peaks for stretching and deformation vibration of the phosphate groups are obtained, but are very weak. With lengthened immersion time, the phosphate groups show stronger peaks at 1041, 601, and 563 cm^{-1}. The peaks around 3500 cm^{-1} derived from the hydroxyl groups are very weak for all the samples, hinting that the HA coating is poorly crystallized.[74]

The morphology of PLGA microspheres before immersion as a control is shown in Figure 6.32a. Some HA crystals can be observed only after 1 h incubation (Figure 6.32b). Combined with the FTIR spectra, it is not difficult to see that a rapid nucleation and growth of HA takes place in the first hour. However, resulting from the short duration, HA crystals only grow on some parts of the microsphere. After 4 h and 8 h of immersion, HA crystals are formed more areas of the microsphere surface (Figures 6.32c and d), but still do

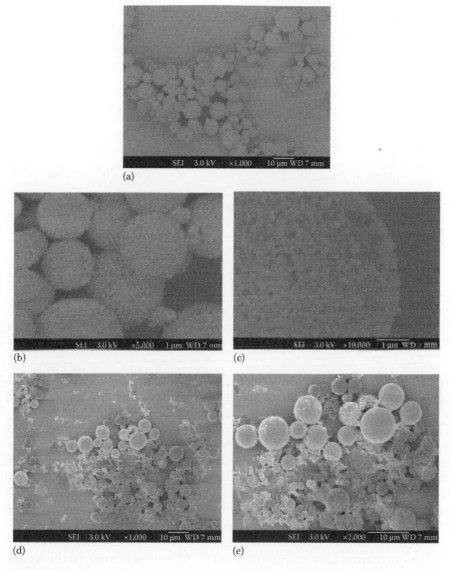

FIGURE 6.29
SEM micrographs of HA formation after PLGA microspheres immersed in Solution 1 (a through c) and Solution 2 (d and e). (From Zhang, Z. et al., *Nanosci. Nanotechnol. Lett.*, 3(4), 472, 2011.)

not cover completely. Until 16 h, complete coverage is seen as shown in Figure 6.32e. The coating is composed of HA crystals with plate-like structures on the surface of the microspheres. After soaking for 24 h, plate-like structures with sharp edges are clearly observed, as shown in Figure 6.32f.

Simultaneously, the size distribution of microspheres before and after coating is studied as shown in Table 6.4. After 16 h immersion, the thickness of

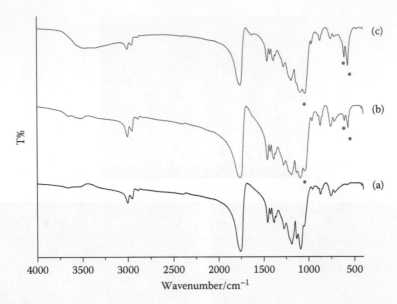

FIGURE 6.30

FTIR spectra for PLGA microspheres (a), and HA-coated PLGA microspheres immersed in Solution 1 (b) and Solution 2 (c). * represents phosphate groups. (From Zhang, Z. et al., *Nanosci. Nanotechnol. Lett.*, 3(4), 472, 2011.)

HA coating nearly reached 1.5 μm. However, there is a slight increase in 24 h incubation duration in comparison to 16 h. The reason is probably attributed to the consumption of Ca and P and decrease of ion concentration.

The phase analysis of the coating is carried out by x-ray diffraction (XRD) as shown in Figure 6.33. The strong characteristic peaks at 26° and 32° corresponding to HA are obtained for coated PLGA microspheres at different soaking times. The broadening and overlapping of peaks demonstrate that the coating is comprised of poorly crystallized HA, consistent with the FTIR result. Poor crystallization of HA coating contributing to higher dissolution rate may be exploited in controlling drug delivery.

Since HA is fully covered on the surface of PLGA microspheres after 16 h incubation, thermo gravimetric analysis (TGA) is used to analyze the weight of the HA coating after 16 h and 24 h, as shown in Figure 6.34. In the temperature range of 300°C–400°C, a significant weight loss is obtained, which corresponds to the thermal decomposition of PLGA. After heating at 500°C, there is no weight at all for PLGA microspheres, indicating that all microspheres were burned out. After 16 and 24 h incubations, 9.12% and 10.73% of HA remain on HA-coated microspheres, respectively. Depending on the data and spectra, there is a slight weight increase with incubation time from 16 to 24 h. However, the increase is very small because most of the calcium and phosphate ions in the solution are consumed and the kinetic balance is found between HA and Ca/P in the solution.

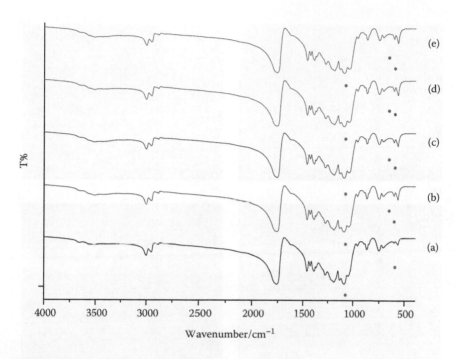

FIGURE 6.31
FTIR spectra for HA-coated microspheres with different immersion times: (a) 1 h, (b) 4 h, (c) 8 h, (d) 16 h, and (e) 24 h. * represents phosphate groups. (From Zhang, Z. et al., *Nanosci. Nanotechnol. Lett.*, 3(4), 472, 2011.)

6.6.2 Modified Chemical Groups of Microspheres

Anitha Ethirajan et al. also investigated the effect of the amount and category of surface chemical groups on HA growth.[72] They demonstrated that in the same incubation time, the amount of HA nanocrystals increased along with increases in the surface carboxyl groups of nanoparticles. Negatively charged surfactant (sodium dodecyl sulfate, SDS) present on the surface of nanoparticles can affect the bonding of calcium ions with carboxyl groups and inhibit HA growth. However, Xu et al. provided a different conclusion with respect to an SDS-modified microsphere template.[75] They prepared negatively charged SDS-modified PLGA microspheres with zeta potential of −79.1 mV in comparison to the commonly used poly(vinyl alcohol) (PVA) surfactant for PLGA preparation of only −10 mV. They illustrated that a complete HA coating could be formed on SDS-modified PLGA microspheres in 6 h immersion (shown in Figure 6.35), but there was almost no HA crystal observed on the PLGA microspheres with PVA surfactant. Thus, they concluded that SDS, as the polar and ionic head of −OSO_3, was a strong chelating group for calcium ions.

Since chitosan itself has a low capability to induce HA formation in normal SBF solution, Leonor I.B. et al. prepared silica-treated chitosan microspheres

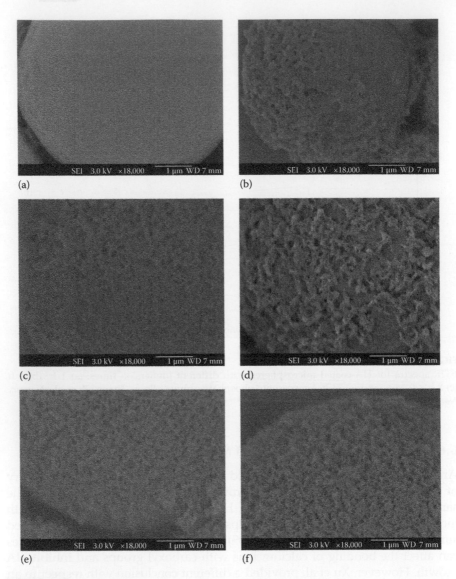

(a) (b) (c) (d) (e) (f)

FIGURE 6.32
SEM micrographs for uncoated microspheres (a) and HA-coated microspheres with different incubation times: (b) 1 h, (c) 4 h, (d) 8 h, (e) 16 h, and (f) 24 h. (From Zhang, Z. et al., *Nanosci. Nanotechnol. Lett.*, 3(4), 472, 2011.)

TABLE 6.4

Average Size of HA-Coated Microspheres with Different Soaking Times

Soaking time (h)	0	16	24
Size (µm)	6.67 ± 0.82	8.16 ± 1.34	8.41 ± 1.73

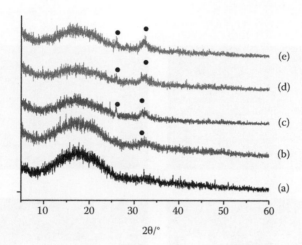

FIGURE 6.33
XRD spectra for uncoated PLGA microspheres (a) and HA-coated microspheres with different incubation times: (b) 1 h, (c) 4 h, (d) 8 h, and (e) 24 h (• represents HA characteristic peak). (From Zhang, Z. et al., *Nanosci. Nanotechnol. Lett.*, 3(4), 472, 2011.)

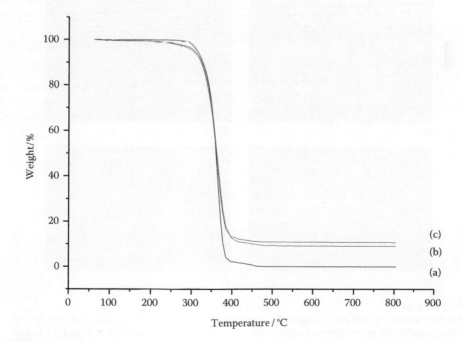

FIGURE 6.34
TGA spectra for (a) PLGA microspheres and (b) HA-coated PLGA microspheres after 16 h and (c) 24 h incubation.

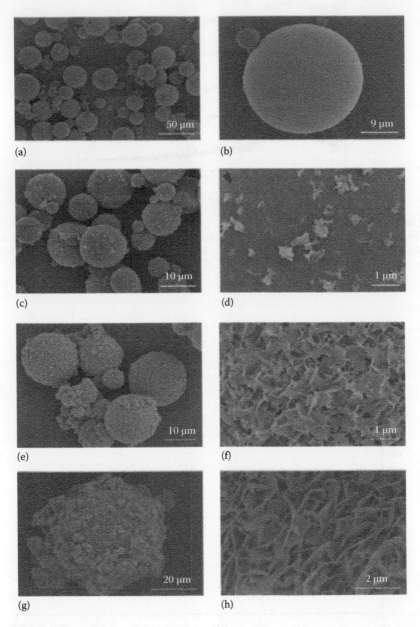

FIGURE 6.35
Micrographs of negatively charged PLGA microspheres with SDS surfactant (a and b) and HA-coated PLGA microspheres under different soaking times: (c and d) 1 h, (e and f) 3 h, and (g and h) 6 h.

as a template for HA growth.[76] From the image in Figure 6.36, HA particle formation was observed within 1 day; after 7 days of immersion, the HA coating fully covered on two kinds of silica-treated chitosan microspheres.

6.7 Protein Drug/HA Coprecipitation on the Microspheres

6.7.1 Coprecipitation of Drug and HA on the Metal or Scaffold

Using the biomimetic method, HA can be deposited on the templates from the solution at normal physiological temperature. This provides probability for coprecipitation of protein drug and HA because protein can maintain its bioactivity at normal physiological temperature. Many efforts were made to study the dissolution rate and crystal phase of drug/HA coprecipitation.[77–82]

Wen et al. investigated the effect of protein (bovine serum albumin, BSA) on the crystalline phase and morphology of coprecipitation coating on titanium alloy.[80] HA coating with the absence of protein shows straight plate-like units with sharp edges, while coprecipitation coating shows link-net-structured crystal plates with some organic-like substance (Figure 6.37).

Liu et al. focus on the dissolution rate of CaP/BSA coprecipitation in different bathing concentrations of BSA (Figure 6.38), which is directly correlated with the drug release rate.[82] They demonstrated that coprecipitation coating showed lower dissolution rate than CaP coating without BSA. However, there is no obvious difference in dissolution from coprecipitation with various bathing concentrations of BSA.

6.7.2 Drug-Co-Deposited HA Coating on the Microspheres

With the inspiration of drug/HA coprecipitation on metal, we adopted this coprecipitation method for microspheres to study the effect of drug on HA growth and drug release profile.

Bovine serum albumin (BSA), which is 66 kDa in molecular weight and 3.5 ± 0.3 nm in size, is chosen as a model protein drug. The size distribution of BSA is shown in Figure 6.39.

The nanosized BSA loading on the HA coating usually takes place on the surface of the coating, and the loading efficiency has its limitation. Due to protein bonding with HA, drug-co-deposited HA on the microspheres can be formed at the time of the HA nanocrystal deposition. The advantages of coprecipitation are that there is no need to reload the drug after HA formation, and the drug release duration may be lengthened.

The two-step biomimetic method is still suitable for drug-co-deposited HA preparation, due to its mild condition and maintenance of drug's bioactivity. This provides the possibility of studying drug-co-deposited HA coating on the polymer microspheres.

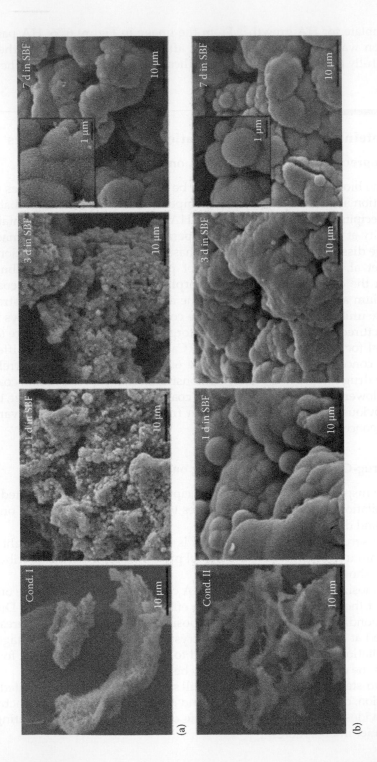

FIGURE 6.36
Micrographs of chitosan microparticles pretreated with a calcium silicate solution under condition I (a) or II (b) and subsequently soaked in SBF for 1, 3, and 7 days. (From Leonor, I.B. et al., *Acta Biomaterialia*, 4(5), 1349, 2008.)

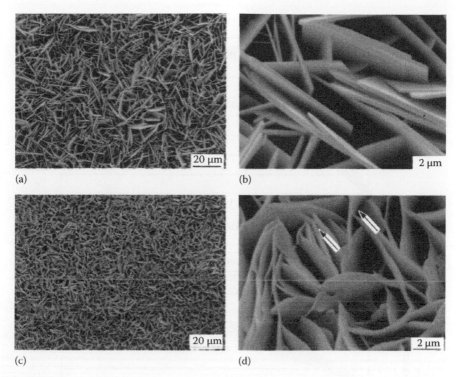

(a) (b)

(c) (d)

FIGURE 6.37
Micrographs of HA (a and b) and HA/BSA (bathing concentration 2 mg/mL) (c and d) coatings prepared on Ti alloy. (Wen, H.B. et al., *J. Biomed. Mater. Res.*, 46(2), 245, 1999.)

6.7.2.1 Effect of Drug on HA Formation

We propose the hypothesis that the drug has little or no effect on the nucleation and growth of HA, and loading efficiency is improved by increasing the initial drug concentration for drug-co-deposited HA. The different concentrations of the drug were designed to study drug loading and HA growth. After 24 h of incubation, the amount of drug incorporated in the coprecipitation coating was calculated as data shown in Table 6.5. The results are completely different from our hypothesis. The concentration of the initial drug increases from 100 to 500 µg/mL, and the amount of precipitated drug increases quite a little (from 43.5 to 55.0 µg/mL). When the initial drug increases up to 1000 µg/mL, there is no drug on the microspheres (0 µg/mL). This strange condition hints that BSA may strongly affect the growth of HA crystals, since the amount of the deposited drug is related to HA formation.

In order to confirm the drug effect on HA growth, FTIR measurements were used to analyze the coated microspheres without or with BSA drug coprecipitation as shown in Figure 6.40. For Figure 6.40a spectra as a control of HA-coated microspheres without BSA, the peaks at 1041, 601, and 563 cm^{-1} are

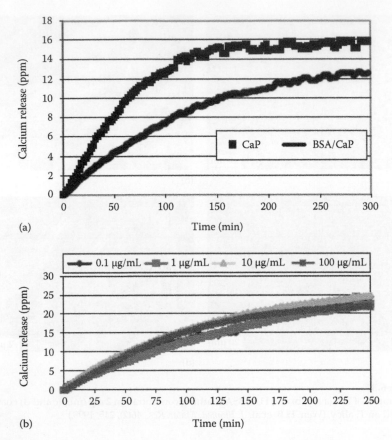

(a)

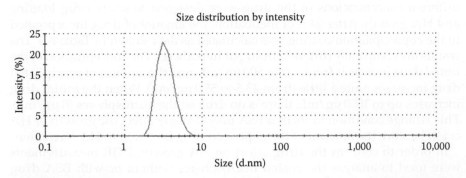

(b)

FIGURE 6.38
Release of Ca^{2+} ions (a) from CaP coatings prepared biomimetically in the absence or presence of BSA (bathing concentration of 1 mg/mL), (b) from CaP/BSA coprecipitated coatings with different bathing concentrations of BSA (0.1–100 µg/mL). (From Liu, et al., *Biomaterials*, 24(1), 65, 2003.)

FIGURE 6.39
Size distribution of BSA.

TABLE 6.5

Concentration of BSA before and after Coprecipitation

Sample	1	2	3	4
Concentration of drug in phosphate solution ($\mu g/mL$)	10	100	500	1000
Concentration of drug precipitated in the coating ($\mu g/mL$)	NA	43.5 ± 2.5	55.0 ± 6.2	0

associated with the phosphate groups. When 10, 100, and 500 $\mu g/mL$ BSA were added for coprecipitation, additional peaks at 1648 and 1541 cm^{-1} in FTIR spectra were observed, which can be ascribed to the amide I and amide II band of the protein, respectively (Figure 6.40b through d). The amide I band represents the stretching vibrations of C=O in the backbone of the protein, while amide II results from the combination of C–N stretching and N–H bending vibrations. These peaks demonstrate the incorporation of BSA drug into the HA coating. However, the condition differs after a 1000 $\mu g/mL$ addition for coprecipitation, as shown in Figure 6.40e. In the spectra, no characteristic peak that corresponded with either phosphate groups or amide

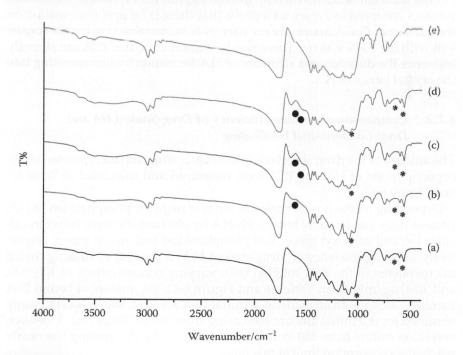

FIGURE 6.40

FTIR spectra for HA-coated microspheres (a) and coprecipitation of HA and BSA-coated microspheres with initial BSA concentration (b) 10 $\mu g/mL$, (c) 100 $\mu g/mL$, (d) 500 $\mu g/mL$, (e) 1000 $\mu g/mL$. (* represents phosphate groups; • represents amide groups.)

groups is observed. It illustrates that drug-co-deposited HA or even only HA is not formed after 24 h incubation, as the spectra is similar to uncoated bare PLGA microspheres.

The morphology of drug-co-deposited HA coating on the surface of PLGA microspheres was observed by FESEM. The fully covered HA coating without BSA, as a control, is shown in Figure 6.41a and b. The coating is comprised of many straight plate-like structured HA nanocrystals with sharp edges. However, some significant differences were found between the distinct concentrations of BSA drug for coprecipitation. At a concentration of 10 μg/mL of BSA, a completely uniform HA layer on the microspheres is observed, but the partial areas show the organic net-linked structure lacking a sharp edge (Figure 6.41c and d). When higher concentration of 100 μg/mL BSA was added, more net-linked structured coprecipitations appear, as shown in Figure 6.41e and f, and the coating is not as dense as the one obtained in the control. A marked difference appears when the concentration of initial BSA is increased to 500 μg/mL (Figure 6.41g and h). The surface of the microsphere is partially covered with apatite after 24 h of incubation. Moreover, all formed apatite shows organic net-linked structure, which is completely different from plate-like structured HA coating without the addition of BSA. When up to 1000 μg/mL BSA is added, the smooth surfaces observed in Figure 6.41i prove that there is no apatite formation on the surface of PLGA microspheres after 24 h of immersion, which is consistent with the FTIR spectra. These results demonstrate that BSA can strongly influence the duration and structure of HA formation by incorporating into the crystal lattice of HA.

6.7.2.2 Comparison of Loading Efficiency of Drug-Soaked HA and Drug-Co-Deposited HA Coating

The amount of the drug absorbed on the HA coating and incorporated in the coprecipitation of HA and BSA were measured and calculated in Table 6.6 and Figure 6.42.

Depending on the positive calcium ion and negative phosphate ion of HA, protein drug can strongly bond with HA by electrostatic force between calcium ion and carboxyl group, and phosphate ion and amino group, respectively. Loading efficiency of "drug-soaked HA" is studied by soaking coated microspheres in the BSA solution with varying concentrations of 100, 500, and 1000 μg/mL. From Table 6.6 and Figure 6.42, the amount of bound BSA increases along with an increase in initial BSA amount, which is very clearly obtained for the initial amount increasing from 100 to 500 μg/mL. However, condition differs from 500 to 1000 μg/mL, since the HA coating has nearly reached its absorption limit at this point.

The "drug-co-deposited HA" on the surface of PLGA microspheres provides a one-step method for HA growth and drug loading without an additional absorption process. However, the protein drug BSA, as a

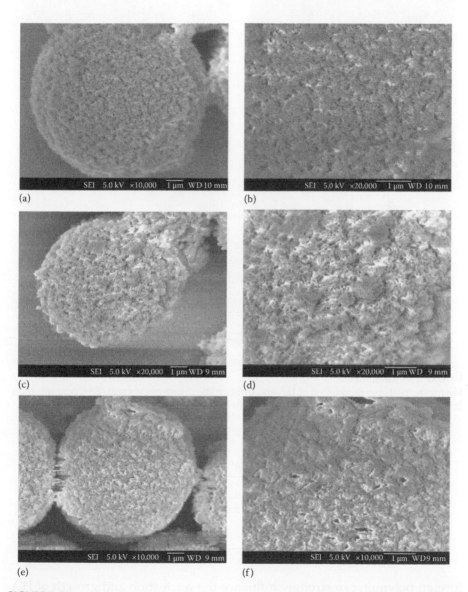

FIGURE 6.41
Micrographs of co-precipitation of HA and drug on the microspheres with BSA concentration (a) ([b] ×20,000 magnification) and coprecipitation of HA and drug on the microspheres with BSA concentration (c) 10 μg/mL ([d] ×20,000 magnification), (e) 100 μg/mL ([f] ×20,000 magnification).

(continued)

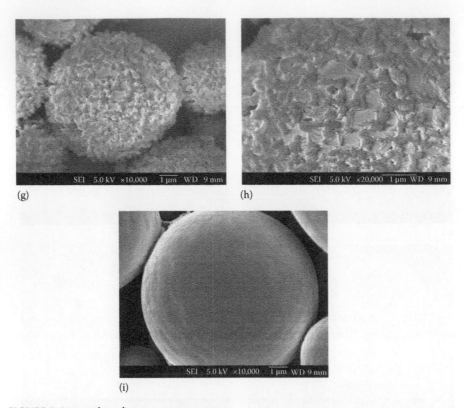

(g)   (h)

(i)

FIGURE 6.41 (continued)
Micrographs of co-precipitation of HA and drug on the microspheres with BSA concentration (g) 500 µg/mL ([h] ×20,000 magnification) and (i) 1,000 µg/mL.

TABLE 6.6

Amount of BSA from "Drug-Soaked HA" and "Drug-Co-Deposited HA"

Initial Loading (µg/mL)	100	500	1000
Absorbed on the surface of HA coating (µg/mL)	52.4±7.8	325.1±34.3	356.5±25.2
Deposited in coprecipitation of BSA and HA coating (µg/mL)	43.5±2.5	55.0±16.2	0

foreign polymer, can strongly influence the nucleation and growth of HA coatings. Thus, the loading profile of "drug-co-deposited HA" is different from that of "drug-soaked HA" because the amount of precipitated drug is related to HA formation. The amount of the precipitated drug in "drug-co-deposited HA" is also shown in Table 6.6 and Figure 6.42. Compared to the loading profile of "drug-soaked HA," there is little increase for the precipitated drug with the initial drug increasing from 100 to 500 µg/mL, and non-precipitated drug is measured against 1000 µg/mL of the

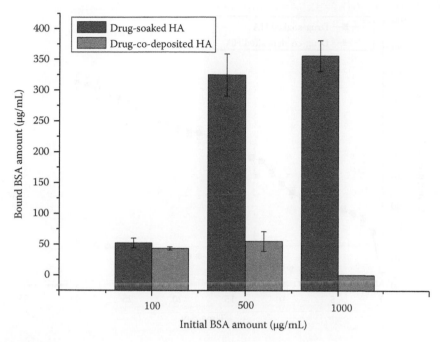

FIGURE 6.42
The amount of BSA from "drug-soaked HA" and "drug-co-deposited HA."

initial drug. These results illustrate that HA growth is inhibited by the presence of protein, and the amount of drug bound with HA decreases simultaneously.

6.7.2.3 Drug Release Profile of "Drug-Soaked HA" and "Drug-Co-Deposited HA"

In our work, BSA release profile from "drug-soaked HA" and "drug-co-deposited HA" with the same initial concentration (500 μg/mL) of BSA was investigated, as shown in Figure 6.43. The release kinetics of "drug-soaked HA" display two stages: a burst release after first day incubation and following a linear release until 30 days. After 1 day immersion, 20% of BSA releases from the HA coating. The rapid release rate is caused by diffusion of BSA out of the HA coating. Since BSA is absorbed on the surface of HA, in the beginning of release, partially instable BSA diffuses by exchanging with ions in the medium. While, bound BSA in the HA coating releases at a constant rate, which is correlated with the dissolution of HA. After 30 days of incubation, cumulative BSA release achieves 76%. Compared with "drug-soaked HA," BSA release from "drug-co-deposited HA" shows a linear kinetics without burst release stage. In addition, "drug-co-deposited HA" provides quite lower release rate than

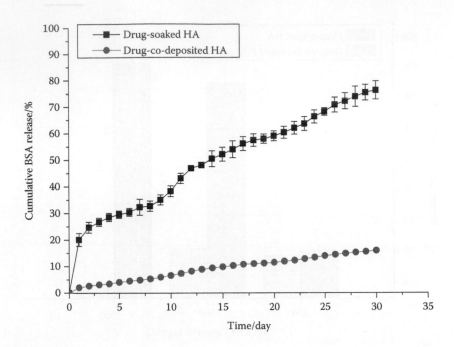

FIGURE 6.43
Comparison of cumulative release of BSA from "drug-soaked HA" and "drug-co-deposited HA."

"drug-soaked HA." After 30 days of incubation, cumulative BSA release from "drug-co-deposited HA" is obtained as 16%. It suggests that BSA is incorporated into the HA crystal lattice, which is consistent with the morphology shown in Figure 6.41. This novel drug delivery device offers the probability for long-term release profile.

We also studied the BSA release from "drug-co-deposited HA" with different initial concentrations of BSA (100 and 500 µg/mL). As shown in Figure 6.44, no difference in the BSA release rate is observed as a function of initial BSA concentration. It demonstrates that BSA release profile of "drug-co-deposited HA" is independent of the initial BSA concentration.

6.8 Future Research

HA coating on different substrates is well demonstrated in various studies. It is believed that the process-driven approach in current HA-scaffold coating research will be replaced by application-driven approaches: the focus is no longer on how to prepare HA coating on the substrates, rather the emphasis will be on what type of HA should be coated for a particular application.

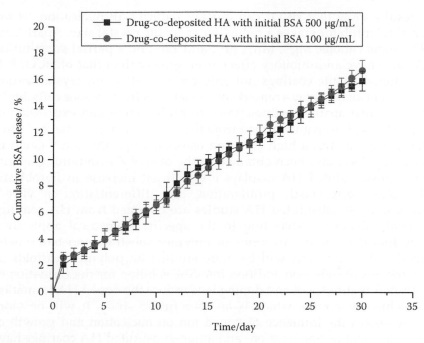

FIGURE 6.11

Cumulative release of BSA from "drug-co-deposited HA" with initial BSA concentration of 100 and 500 µg/mL.

This transition is critical because different application requires slightly different type of HA, hence modification of the HA coating itself is of more significance from the viewpoint of future application.

A classic example is the property difference between HA and doped HA coating by sol–gel method: fluorine-doped HA has a high dissolution resistance[83] and Mg-doped HA showed better bioactivity and adhesion in vitro[84] than pure HA. However, formation of HA is usually obtained at high temperature or serious condition, such as annealing, electrochemical deposition, liquid precursor plasma spraying, etc. It is not suitable for a polymer substrate due to polymer tends to chain cleavage under high temperature.

Biomimetic method for coating deposition was adopted involving mild conditions similar to those characteristic of biological environments and for the deposition of CaP coatings on many different objects. One of the main advantages of biomimetic deposition of coatings is the possibility of other ions being doped onto the CaP layer on a polymer substrate. Therefore, some researchers started the ion-doped HA coating study by biomimetic method. Carbonated HA, a popularly known anionic substituent, has a better physical attachment on the substrate. Barrere et al. studied apatite growth in the normal SBF solution and 3× HCO_3^- SBF solution.[85]

The results show that HCO_3^- can quickly induce CaP nucleation, as the HA crystal inhibitor makes a reduction of apatite crystal size. Apart from HCO_3^-, some cations, Mg^{2+}, Mn^{2+}, Sr^{2+}, and Zn^{2+}, by a partial substitution of Ca, also have an inhibitory effect, even greater than that of HCO_3^-.[86-89] These doped apatite coatings not only have an effect on crystallization structure but also display a remarkably beneficial effect on bone cells. Mg^{2+}- and Sr^{2+}-doped apatite coatings promote proliferation and expression of collagen. Also, the apatite formed from the relatively high content of Mn^{2+} in the phosphate has a high effect on osteocalcin production.[88] Good in vitro results have also been obtained in the case of Zn-containing apatite layers. Zn-substituted HA displays a significant increase in fibroblastic proliferation, osteoblastic proliferation, and differentiation in vitro.[89] Until now, most substituted HA studies are obtained from HA coatings on metal substrates. According to the specific biomedical application and technical feasibility in terms of polymer substrate and biomimetic method, substituted HA will be more studied on polymer scaffolds or microspheres. Besides ion addition into SBF solution for the formation of substituted apatite, Ca, P, and doping ions induce the doped HA formation by stoichiometric ratio, which is another future study. It will be more directly obtain the influence of doped ion on nucleation and growth of substituted apatite. Some cation- and anion-substituted HA coatings have lower crystallization and smaller crystal size than pure HA coatings as previously mentioned, which could affect the dissolution rate of apatite. It inspires a novel idea on controlling the drug release rate in drug delivery systems. After soaking the drug, ion-doped apatites may show different drug release rates related to the degradation of apatite. In the future, substituted apatite coatings may be chosen to control the protein drug release rate by different ion substitutions.

6.9 Summary

Hydroxyapatite is an important biocompatible ceramic. It has been successfully coated onto different biodegradable polymers. Processes to realize such coatings have been reviewed in this chapter. It is clear that the surface property of the substrate plays a critical role in the success of the coating. With the current understanding of the coating process, it is believed that future research should be focused on the modification of the property of HA coating itself. The findings from HA coating on metallic substrates should be applied and verified on polymeric substrates in the near future to allow HA coatings with various intrinsic properties to be coated on polymeric substrates for advanced applications, i.e., drug release rate control.

References

1. C. M. Lehr, J. A. Bouwstra, E. H. Schacht, and H. E. Junginger, In vitro evaluation of mucoadhesive properties of chitosan and some other natural polymers. *International Journal of Pharmaceutics*, 1992, 78(1–3): 43–48.
2. H. S. Ch'ng, H. Park, P, Kelly, and J. R. Robinson, Bioadhesive polymers as platforms for oral controlled drug delivery II: Synthesis and evaluation of some swelling, water-insoluble bioadhesive polymers. *Journal of Pharmaceutical Sciences*, 1985, 74(4): 399–405.
3. T. Espevik, M. Otterlei, G. S. Braek, L. Ryan, S. D. Wright, and A. Sundan, The involvement of CD14 in stimulation of cytokine production by uronic acid polymers. *European Journal of Immunology*, 1993, 23(1): 255–261.
4. D. R. Cole, M. Waterfall, M. McIntyre, and J. D. Baird, Microencapsulated islet grafts in the BB/E rat: A possible role for cytokines in graft failure. *Diabetologia*, 1992, 35(3): 231–237.
5. P. De Vos, B. De Haan, J. Pater, and R. Van Schilfgaarde, Association between capsule diameter, adequacy of encapsulation, and survival of microencapsulated rat islet allografts. *Transplantation*, 1996, 62(7): 893–899.
6. R. J. Mumper, A. S. Huffman, P. A. Puolakkainen, L. S. Bouchard, and W. R. Gombotz, Calcium-alginate beads for the oral delivery of transforming growth factor-β1 (TGF-β1): Stabilization of TGF-β1 by the addition of polyacrylic acid within acid-treated beads. *Journal of Controlled Release*, 1994, 30(3): 241–251.
7. W. R. Gombotz and S. F. Wee, Protein release from alginate matrices. *Advanced Drug Delivery Reviews*, 1998, 31(3): 267–285.
8. D. E. Chickering and E. Mathiowitz, Bioadhesive microspheres: I. A novel electrobalance-based method to study adhesive interactions between individual microspheres and intestinal mucosa. *Journal of Controlled Release*, 1995, 34(3): 251–261.
9. K. K. Kwok, M. J. Groves, and D. J. Burgess, Production of 5–15 µm diameter alginate-polylysine microcapsules by an air-atomization technique. *Pharmaceutical Research*, 1991, 8(3): 341–344.
10. R. Muzzarelli, V. Baldassarre, F. Conti, P. Ferrara, and G. Biagini, Biological activity of chitosan: Ultrastructural study. *Biomaterials*, 1988, 9(3): 247–252.
11. R. A. Muzzarelli, Human enzymatic activities related to the therapeutic administration of chitin derivatives. *Cellular and Molecular Life Sciences*, 1997, 53(2): 131–140.
12. O. Gåserød, Ian G. Jolliffe, F. C. Hampson, P. W. Dettmar, and G. S. Brak, The enhancement of the bioadhesive properties of calcium alginate gel beads by coating with chitosan. *International Journal of Pharmaceutics*, 1998, 175(2): 237–246.
13. C. Remuñan-López, A. Portero, M. Lemos, J. L. Vila-Jato, M. J. Nuñez, P. Riveiro, J. M. López, M. Piso, and M. J. Alonso, Chitosan microspheres for the specific delivery of amoxycillin to the gastric cavity. *S.T.P. Pharma Sciences*, 2000, 10(1): 69–76.
14. P. He, S. S. Davis, and L. Illum, In vitro evaluation of the mucoadhesive properties of chitosan microspheres. *International Journal of Pharmaceutics*, 1998, 166(1): 75–88.

15. J. Shimoda, H. Onishi, and Y. Machida, Bioadhesive characteristics of chitosan microspheres to the mucosa of rat small intestine. *Drug Development and Industrial Pharmacy*, 2001, 27(6): 567–576.

16. S. Miyazaki, A. Nakayama, M. Oda, M. Takada, and D. Attwood, Drug release from oral mucosal adhesive tablets of chitosan and sodium alginate. *International Journal of Pharmaceutics*, 1995, 118(2): 257–263.

17. C. Remuñán-López, A. Portero, J. L. Vila-Jato, and M. J. Alnoso, Design and evaluation of chitosan/ethylcellulose mucoadhesive bilayered devices for buccal drug delivery. *Journal of Controlled Release*, 1998, 55(2–3): 143–152.

18. O. Smidsrod and G. Skjak-Braek, Alginate as immobilization matrix for cells. *Trends in Biotechnology*, 1990, 8(3): 71–78.

19. T. Coviello, P. Matricardi, C. Marianecci, and F. Alhaique, Polysaccharide hydrogels for modified release formulations. *Journal of Controlled Release*, 2007, 119(1): 5–24.

20. D. A. Rees and E. J. Welsh, Secondary and tertiary structure of polysaccharides in solutions and gels. *Angewandte Chemie—International Edition*, 1977, 16(4): 214–224.

21. B. Dupuy, A. Arien, and A. Perrot Minnot, FT-IR of membranes made with alginate/polylysine complexes: Variations with the mannuronic or guluronic content of the polysaccharides. *Artificial Cells, Blood Substitutes, and Immobilization Biotechnology*, 1994, 22(1): 71–82.

22. M. L. Torre, P. Giunchedi, L. Maggi, R. Stefli, E. O. Machiste, and U. Conte, Formulation and characterization of calcium alginate beads containing ampicillin. *Pharmaceutical Development and Technology*, 1998, 3(2): 193–198.

23. R. S. Shepherd, S. Reader, and A. Falshaw, Chitosan functional properties. *Glycoconjugate Journal*, 1997, 14(4): 535–542.

24. S. R. Jameela and A. Jayakrishnan, Glutaraldehyde cross-linked chitosan microspheres as a long acting biodegradable drug delivery vehicle: Studies on the in vitro release of mitoxantrone and in vivo degradation of microspheres in rat muscle. *Biomaterials*, 1995, 16(10): 769–775.

25. S. M. van der Merwe, J. C. Verhoef, J. H. M. Verheijden, A. F. Kotzé, and H. E. Junginger, Trimethylated chitosan as polymeric absorption enhancer for improved peroral delivery of peptide drugs. *European Journal of Pharmaceutics and Biopharmaceutics*, 2004, 58(2): 225–235.

26. J. Heller, J. Barr, S. Y. Ng, H. R. Shen, K. Schwach-Abdellaoui, S. Emmahl, A. Rothen-Weinhold, and R. Gurny, Poly(ortho esters): Their development and some recent applications. *European Journal of Pharmaceutics and Biopharmaceutics*, 2000, 50(1): 121–128.

27. D. N. Nguyen, S. S. Raghavan, L. M. Tashima, E. C. Lin, S. J. Fredette, R. S. Langer, and C. Wang, Enhancement of poly(orthoester) microspheres for DNA vaccine delivery by blending with poly(ethylenimine). *Biomaterials*, 2008, 29(18): 2783–2793.

28. J. Park, M. Ye, and K. Park, Biodegradable polymers for microencapsulation of drugs. *Molecules*, 2005, 10(1): 146–161.

29. M. Sokolsky-Papkov, K. Agashi, A. Olaye, K. Shakesheff, and A. J. Domb, Polymer carriers for drug delivery in tissue engineering. *Advanced Drug Delivery Reviews*, 2007, 59(4–5): 187–206.

30. S. K. Lopac, M. P. Torres, J. H. Wilson-Welder, M. J. Wannemuehler, and B. Narasimhan, Effect of polymer chemistry and fabrication method on protein release and stability from polyanhydride microspheres. *Journal of Biomedical Materials Research: Part B Applied Biomaterials*, 2009, 91B(2): 938–947.

31. S. K. Dordunoo, J. K. Jackson, L. A. Arsenault, A. M. C. Oktaba, W. L. Hunter, and H. M. Burt, Taxol encapsulation in poly(ε-caprolactone) microspheres. *Cancer Chemotherapy and Pharmacology*, 1995, 36(4): 279–282.
32. V. R. Sinha, K. Bansal, R. Kaushik, R. Kumria, and A. Trehan, Poly-ε-caprolactone microspheres and nanospheres: An overview. *International Journal of Pharmaceutics*, 2004, 278(1): 1–23.
33. J. Kopecek and K. Ulbrich, Biodegradation of biomedical polymers. *Progress in Polymer Science*, 1983, 9(1): 1–58.
34. J. M. Anderson and M. S. Shive, Biodegradation and biocompatibility of PLA and PLGA microspheres. *Advanced Drug Delivery Reviews*, 1997, 28(1): 5–24.
35. T. Kokubo, Bioactive glass ceramics: Properties and applications. *Biomaterials*, 1991, 12(2): 155–163.
36. T. Kokubo and H. Takadama, How useful is SBF in predicting in vivo bone bioactivity? *Biomaterials*, 2006, 27(15): 2907–2915.
37. X. Lu and Y. Leng, Theoretical analysis of calcium phosphate precipitation in simulated body fluid. *Biomaterials*, 2005, 26(10): 1097–1108.
38. G. H. Nancollas, Kinetics of crystal growth from solution. *Journal of Crystal Growth*, 1968, 3–4: 335–339.
39. A. E. Nielsen, Electrolyte crystal growth mechanisms. *Journal of Crystal Growth*, 1984, 67(2): 289–310.
40. H. McDiwell, T. M. Gregory, and W. E. Brown, Solubility of $Ca_5(PO_4)_3OH$ in the system $Ca(OH)_2$-H_3PO_4-H_2O at 5, 15, 25, and 37 degree C. *Journal of Research National Bureau Standard Section A Physical Chemistry*, 1977, 81 A(2–3): 273–281.
41. F. Barrère, P. Layrolle, C. A. van Blitterswijk, and K. de Groot, Biomimetic calcium phosphate coatings on Ti6Al4V: A crystal growth study of octacalcium phosphate and inhibition by Mg^{2+} and HCO_3^-. *Bone*, 1999, 25(2, Supplement 1): 107S–111S.
42. F. Barrere, C. A. van Blitterswijk, K. de Groot, and P. Layrolle, Nucleation of biomimetic Ca-P coatings on Ti6Al4V from a SBF×5 solution: Influence of magnesium. *Biomaterials*, 2002, 23(10): 2211–2220.
43. F. Barrere, C. A. van Blitterswijk, K. de Groot, and P. Layrolle, Influence of ionic strength and carbonate on the Ca-P coating formation from SBF×5 solution. *Biomaterials*, 2002, 23(9): 1921–1930.
44. R. Boistelle and I. Lopez-Valero, Growth units and nucleation: The case of calcium phosphates. *Journal of Crystal Growth*, 1990, 102(3): 609–617.
45. W. Wu and G. H. Nancollas, The dissolution and growth of sparingly soluble inorganic salts: A kinetics and surface energy approach. *Pure and Applied Chemistry*, 1998, 70(10): 1867–1872.
46. W. Wu and G. H. Nancollas, Determination of interfacial tension from crystallization and dissolution data: A comparison with other methods. *Advances in Colloid and Interface Science*, 1999, 79(2–3): 229–279.
47. R. Langer and J. P. Vacanti, Tissue engineering. *Science*, 1993, 260(5110): 920–926.
48. S. V. Madihally and H. W. T. Matthew, Porous chitosan scaffolds for tissue engineering. *Biomaterials*, 1999, 20(12): 1133–1142.
49. H. Ehrlich, B. Krajewska, T. Hanke, R. Born, S. Hernemann, C. Knieb, and H. Worch, Chitosan membrane as a template for hydroxyapatite crystal growth in a model dual membrane diffusion system. *Journal of Membrane Science*, 2006, 273(1–2): 124–128.

50. I. Manjubala, I. Ponomarev, I. Wilke, and K. D. Jandt, Growth of osteoblast-like cells on biomimetic apatite-coated chitosan scaffolds. *Journal of Biomedical Materials Research—Part B Applied Biomaterials*, 2008, 84(1): 7–16.
51. I. Manjubala, S. Scheler, J. Bössert, and K. D. Jandt, Mineralisation of chitosan scaffolds with nano-apatite formation by double diffusion technique. *Acta Biomaterialia*, 2006, 2(1): 75–84.
52. M. Kawashita, M. Nakao, M. Minoda, H. M. Kim, T. Beppu, T. Miyamoto, T. Kokubo, and T. Nakamura, Apatite-forming ability of carboxyl group-containing polymer gels in a simulated body fluid. *Biomaterials*, 2003, 24(14): 2477–2484.
53. K. Hosoya, C. Ohtsuki, T. Kawai, M. Kamitakahara, S. Ogata, T. Miyazaki, and M. Tanihara, A novel covalently crosslinked gel of alginate and silane with the ability to form bone-like apatite. *Journal of Biomedical Materials Research Part A*, 2004, 71A(4): 596–601.
54. D. Suárez-González, K. Barnhart, E. Saito, R. Vanderby Jr., S. J. Hollister, and W. L. Murphy, Controlled nucleation of hydroxyapatite on alginate scaffolds for stem cell-based bone tissue engineering. *Journal of Biomedical Materials Research Part A*, 2010, 95A(1): 222–234.
55. A. C. Jayasuriya and C. Shah, Controlled release of insulin-like growth factor-1 and bone marrow stromal cell function of bone-like mineral layer-coated poly(lactic-co-glycolic acid) scaffolds. *Journal of Tissue Engineering and Regenerative Medicine*, 2008, 2(1): 43–49.
56. K. Shin, A. C. Jayasuriya, and D. H. Kohn, Effect of ionic activity products on the structure and composition of mineral self assembled on three-dimensional poly(lactide-co-glycolide) scaffolds. *Journal of Biomedical Materials Research Part A*, 2007, 83A(4): 1076–1086.
57. W. L. Murphy, D. H. Kohn, and D. J. Mooney, Growth of continuous bonelike mineral within porous poly(lactide-co-glycolide) scaffolds in vitro. *Journal of Biomedical Materials Research*, 2000, 50(1): 50–58.
58. W. L. Murphy and D. J. Mooney, Bioinspired growth of crystalline carbonate apatite on biodegradable polymer substrata. *Journal of the American Chemical Society*, 2002, 124(9): 1910–1917.
59. K. Shin, R. Guffey, A. C. Jayasuriya, L. Luong, M. Outslay, and D. H. Kohn, Formation and dissolution behavior of biomimetic mineral on 3-dimensional poly(lactide-co-glycolide) scaffolds. *Transactions of the 7th World Biomaterials Congress*, Sydney, Australia, 2004, p. 1841
60. R. Zhang and P. X. Ma, Biomimetic polymer/apatite composite scaffolds for mineralized tissue engineering. *Macromolecular Bioscience*, 2004, 4(2): 100–111.
61. Y. F. Chou, J. C. Y. Dunn, and B. M. Wu, In vitro response of MC3T3-E1 preosteoblasts within three-dimensional apatite-coated PLGA scaffolds. *Journal of Biomedical Materials Research—Part B Applied Biomaterials*, 2005, 75(1): 81–90.
62. A. C. Jayasuriya, C. Shah, V. Goel, and N. A. Ebraheim, Characterization of biomimetic mineral coated 3D plga scaffolds. *ASME International Mechanical Engineering Congress and Exposition*, Seattle, WA, 2007, pp. 105–110.
63. A. C. Jayasuriya, C. Shah, N. A. Ebraheim, and A. H. Jayatissa, Acceleration of biomimetic mineralization to apply in bone regeneration. *Biomedical Materials*, 2008, 3(1): 015003.

64. L. Jongpaiboonkit, T. Franklin-Ford, and W. L. Murphy, Mineral-coated polymer microspheres for controlled protein binding and release. *Advanced Materials*, 2009, 21(19): 1960–1963.

65. L. Jongpaiboonkit, T. Franklin-Ford, and W. L. Murphy, Growth of hydroxyapatite coatings on biodegradable polymer microspheres. *ACS Applied Materials and Interfaces*, 2009, 1(7): 1504–1511.

66. S. W. Kang, H. S. Yang, S. W. Seo, D. K. Han, and B. S. Kim, Apatite-coated poly(lactic-co-glycolic acid) microspheres as an injectable scaffold for bone tissue engineering. *Journal of Biomedical Materials Research Part A*, 2008, 85A(3): 747–756.

67. A. C. Jayasuriya and S. Kibbe, Rapid biomineralization of chitosan microparticles to apply in bone regeneration. *Journal of Materials Science: Materials in Medicine*, 2010, 21(2): 393–398.

68. M. Lee, W. Li, R. K. Siu, J. Whang, X. Zhang, and C. Soo, Biomimetic apatite-coated alginate/chitosan microparticles as osteogenic protein carriers. *Biomaterials*, 2009, 30(30): 6094–6101.

69. M. B. Tomson and G. H. Nancollas, Mineralization kinetics: A constant composition approach. *Science*, 1978, 200(4345): 1059–1060.

70. Q. Xu, Y. Tanaka, and J. T. Czernuszka, Encapsulation and release of a hydrophobic drug from hydroxyapatite coated liposomes. *Biomaterials*, 2007, 28(16): 2687–2694.

71. A. W. Pederson, J. W. Ruberti, and P. B. Messersmith, Thermal assembly of a biomimetic mineral/collagen composite. *Biomaterials*, 2003, 24(26): 4881–4890.

72. A. Ethirajan, U. Ziener, and K. Landfester, Surface-functionalized polymeric nanoparticles as templates for biomimetic mineralization of hydroxyapatite. *Chemistry of Materials*, 2009, 21(11): 2218–2225.

73. Z. Zhang, S. Zhang, S. S. Venkatraman, L. Shang, Growth of hydroxyapatite coating on polymer microspheres. *Nanoscience and Nanotechnology Letters*, 2011, 3(4): 472–476.

74. C. Du, P. Klasens, R. E. Haan, J. Bezemer, F. Z. Cui, K. de Groot, and P. Layrolle, Biomimetic calcium phosphate coatings on Polyactive® 1000/70/30. *Journal of Biomedical Materials Research*, 2002, 59(3): 535–546.

75. Q. Xu and J. T. Czernuszka, Controlled release of amoxicillin from hydroxyapatite-coated poly(lactic-co-glycolic acid) microspheres. *Journal of Controlled Release*, 2008, 127(2): 146–153.

76. I. B. Leonor, E. T. Baran, M. Kawashita, R. L. Reis, T. Kokubo, and T. Nakamura, Growth of a bonelike apatite on chitosan microparticles after a calcium silicate treatment. *Acta Biomaterialia*, 2008, 4(5): 1349–1359.

77. L. N. Luong, S. L. Hong, R. J. Patel, M. E. Outslay, and D. H. Kohn, Spatial control of protein within biomimetically nucleated mineral. *Biomaterials*, 2006, 27(7): 1175–1186.

78. Y. Liu, P. Layrolle, J. de Bruijn, C. van Blitterswijk, and K. de Groot, Biomimetic coprecipitation of calcium phosphate and bovine serum albumin on titanium alloy. *Journal of Biomedical Materials Research*, 2001, 57(3): 327–335.

79. I. B. Leonor, C. M. Alves, H. S. Azevedo, and R. L. Reis, Effects of protein incorporation on calcium phosphate coating. *Materials Science and Engineering: C*, 2009, 29(3): 913–918.

80. H. B. Wen, J. R. de Wijn, C. A. van Blitterswijk, and K. de Groot, Incorporation of bovine serum albumin in calcium phosphate coating on titanium. *Journal of Biomedical Materials Research*, 1999, 46(2): 245–252.

81. H. S. Azevedo, I. B. Leonor, C. M. Alves, and R. L. Reis, Incorporation of proteins and enzymes at different stages of the preparation of calcium phosphate coatings on a degradable substrate by a biomimetic methodology. *Materials Science and Engineering: C*, 2005, 25(2): 169–179.

82. Y. Liu, E. B. Hunziker, N. X. Randall, K. de Groot, and P. Layrolle, Proteins incorporated into biomimetically prepared calcium phosphate coatings modulate their mechanical strength and dissolution rate. *Biomaterials*, 2003, 24(1): 65–70.

83. H. W. Kim, H. E. Kim, and J. C. Knowles, Fluor-hydroxyapatite sol-gel coating on titanium substrate for hard tissue implants. *Biomaterials*, 2004, 25(17): 3351–3358.

84. G. C. Qi, S. Zhang, K. A. Khor, C. M. Liu, X. T. Zeng, W. J. Weng, and M. Qianshi, In vitro effect of magnesium inclusion in sol-gel derived apatite. *Thin Solid Films*, 2008, 516(16): 5176–5180.

85. F. Barrere, C.A. van Blitterswijk, K. de Groot, and P. Layrolle, Influence of ionic strength and carbonate on the Ca-P coating formation from SBF ×5 solution. *Biomaterials*, 2002, 23(9): 1921–1930.

86. F. Barrere, C.A. van Blitterswijk, K. de Groot, and P. Layrolle, Nucleation of biomimetic Ca-P coating on Ti6Al4V from a SBF ×5 solution: Influence of magnesium. *Biomaterials*, 2002, 23(10): 2211–2220.

87. A. L. Oliveira, R. L. Reis, and P. Li, Strontium-substituted apatite coating grown on Ti6Al4V substrate through biomimetic synthesis. *Journal of Biomedical Materials Research Part B: Applied Biomaterials*, 2007, 83B(1): 258–265.

88. B. Bracci, P. Torricelli, S. Panzavolta, E. Boanini, R. Giardino, and A. Bigi, Effect of Mg^{2+}, Sr^{2+}, and Mn^{2+} on the chemico-physical and in vitro biological properties of calcium phosphate biomimetic coatings. *Journal of Inorganic Biochemistry*, 2009, 103(12): 1666–1674.

89. X. P. Wang, A. Ito, Y. Sogo, X. Li, and A. Oyane, Zinc-containing apatite layers on external fixation rods promoting cell activity. *Acta Biomaterialia*, 2010, 6(3): 962–968.

7

Biofunctionalization of NiTi Shape Memory Alloy Promoting Osseointegration by Chemical Treatment

Yanli Cai, Xianjin Yang, Zhenduo Cui, Minfang Chen, Kai Hu, and Changyi Li

CONTENTS

7.1 Introduction

Nickel titanium (NiTi, also known as nitinol) is a metal alloy of nickel and titanium, in which the two elements are present in roughly equal atomic percentages. It undergoes a phase transformation in its crystal structure when cooled from the stronger, high-temperature form (austenite) to the weaker, low-temperature form (martensite). This thermoelastic martensitic phase transformation in the material is responsible for its extraordinary properties. These properties include shape memory effect and superelasticity (also called pseudoelasticity). Shape memory refers to the ability of NiTi to undergo deformation at one temperature and then recover its original, unde-formed shape upon heating above its "transformation temperature." The superelasticity occurs at a narrow temperature range just above its trans-formation temperature; in this case, no heating is necessary to cause the undeformed shape to recover, and the material exhibits enormous elasticity.

The combination of shape memory effect, superelasticity, and good bio-compatibility makes NiTi alloy especially suitable for biomedical applica-tions such as catheter shafts, orthodontic guidewires, valve frames, stents, filters, and endoscopic guide tubes. The application of the shape memory device not only simplifies the operation techniques but also firmly secures the fracture area [1]. The superelasticity of NiTi alloy allows for the design of shape memory devices with a constant stress over a wide range of strains/ shapes, which can accelerate healing without applying undue forces during the operation process. However, there are still some concerns. One is that the potential release of nickel ions from NiTi implant may cause harmful allergic, toxic, or carcinogenic reactions [2]. Although equiatomic NiTi is an intermetallic compound with well-defined bulk and surface properties, nickel ions can still be dissolved from bulk material due to corrosion. The corrosion behavior of NiTi alloy depends on the surface topography and surface chemistry. In general, surfaces tend to be highly corrosion resistant with a good topographical finish (polished or electropolished) and homo-geneous composition (free from second phase like Ni-rich particles, coarse carbides, and nitrides) [3]. Also, it is suggested to further improve the cor-rosion resistance by chemical treatments, passivation, etc., to build up a protective TiO_2 layer. Moreover, the good biocompatibility of NiTi alloy is a direct consequence of lower nickel ion release rate due to the protective TiO_2 layer. Another concern is the bioinert characteristic of NiTi alloy. After it is implanted into the living body, it is usually encapsulated by fibrous tissues that isolate the implant from the surrounding bone [4,5]. This results in small gaps between the natural bone and the implant and leads to movement at the implant–bone interface. Finally, it causes implantation failure, and revision surgery is needed to replace the loosened implant [6].

To make NiTi alloy more applicable as a hard tissue implant, one feasible solution is to coat bioactive ceramics on its surface, which can bond to and integrate with living bone spontaneously in the body. Since synthetic hydroxyapatite (HA) ceramic has chemical similarity to bone and teeth min-eral, it has been used as coatings on metallic prostheses to provide bioactive fixation since the mid-1980s [7,8]. Many techniques have been developed for the preparation of HA coatings onto metallic implant surfaces. The meth-ods applied on NiTi alloy include plasma spraying [9,10], radio-frequency magnetron-sputtered deposition [11], sol–gel deposition [12], electrodeposi-tion [13], biomimetic growth [14–16], and so on. Each method has its own advantages and disadvantages with regard to the coating properties, such

TABLE 7.1

Comparison of Different Methods for Hydroxyapatite Coating Formation
on NiTi Alloy

Method	Advantages	Disadvantages	References
Plasma spraying	High deposition rates; various coating thicknesses; low cost	High temperature induces decomposition; rapid cooling produces amorphous coatings	[9,10]
Radio-frequency magnetron-sputtered deposition	Thin, dense, and pore-free coating on flat substrates	Higher Ca/P ratio of coating than synthetic HA; expensive; time consuming	[11]
Sol–gel deposition	Low processing temperature; can coat complex shapes; thin coating; high purity and homogeneity; low cost	Some processes may require controlled atmosphere and high sintering temperature	[12]
Electrodeposition	Rapid deposition rate; can coat complex substrates	Nonuniform thickness; impurity; poor biological fixation to metal substrates	[13]
Biomimetic growth	Low processing temperature; form bone-like apatite, can coat complex shapes; can incorporate growth factors	Time consuming; require replenishment or control of pH value of simulated body fluid	[14–16]

as coating chemistry, phase composition, crystallinity, mechanical properties, and biological properties. A comparison of different methods applied on NiTi alloy is summarized in Table 7.1.

Although a HA-coated metallic implant combines both good biocompatibility of HA and excellent mechanical properties of a metal substrate, significant differences still exist between synthetic HA and natural bone. For example, the bioactivity, in vivo resorption, and degradation of synthetic HA and HA coatings are inferior to those of bone mineral. Thus, it is attractive to use a bone-like structure or composition on a metal substrate.

Chemical treatment technology is a low-temperature method to modify metals and prepare coatings from chemical routes, which is independent of substrate shape and harmless to substrate properties. Various types of bioactive metals with different functions can be developed by the simple chemical treatment, for example, Ti, NiTi, Ti6Al4V, Ti-Zr-Nb-Ta, and so on [17–21]. Basically, the metal is treated with an acid or an alkali solution to create a micro- or nanoscale structure. Then, after soaking in a simulated body fluid (SBF) with ion concentrations nearly equal to those in human blood plasma [22], a bone-like apatite layer forms on the treated metal surface and thus facilitates the bonding to bone in the body [23,24]. In this chapter, NiTi alloy was first treated with an acid and alkali solution and followed by a precalcification process. The ability to induce the formation of bone-like apatite coating was examined by soaking the NiTi alloy in the SBF. To further resemble natural bone, apatite/collagen composite coating was prepared on the NiTi alloy. In vitro and in vivo studies were performed to investigate both the biological and histological performances of bioactive apatite-coated NiTi alloys.

7.2 Chemical Treatment of NiTi Alloy

As mentioned earlier, NiTi alloys have been used as orthopedic devices due to their shape memory effect, superelasticity, and biocompatibility. However, two factors restrict their applications: the bioinert surface is difficult to integrate with bone, and the released nickel ions may cause toxic and allergic responses [25,26]. As such, surface modification is necessary to make it bioactive and lower the nickel release rate. One of the most effective methods is chemical treatment. In this work, nitric acid (HNO_3) and sodium hydroxide (NaOH) solutions were employed in sequence to create a micro- or even nanostructured network on NiTi alloy followed by immersing in saturated disodium hydrogen phosphate (Na_2HPO_4) and calcium hydroxide $Ca(OH)_2$ solutions to induce the formation of bone-like apatite on NiTi alloy.

7.2.1 Acid Treatment

Aqueous acid solution (HNO_3) was adopted to increase surface roughness and form a dense TiO_2 layer on NiTi alloy. In the aqueous HNO_3 solution, Ni is active, while Ti is inert, thus NiTi alloy can be highly passivated when n is 6 according to the n/8 principle. When NiTi alloy is immersed in the aqueous HNO_3 solution, Ni atoms are dissolved due to the chemical reaction (1). At the same time, the surrounding Ti atoms are oxidized to form a layer of TiO_2 (2). The process is illustrated in Figure 7.1.

$$3Ni + 8HNO_3 = 3Ni(NO_3)_2 + 2NO\uparrow + H_2O \qquad (7.1)$$

$$3Ti + 4HNO_3 + H_2O = 3TiO_2 \cdot H_2O + 4NO\uparrow \qquad (7.2)$$

The acid treatment creates a porous structure on NiTi surface, as shown in Figure 7.2. Numerous pores with the diameters of 1–2 μm are present on NiTi alloy after it is immersed in the aqueous HNO_3 solution at 60°C for 20 h. The porous surface increases the surface roughness and provides numerous nucleation sites for apatite as well as increases adherence between the coating and the substrate. Figure 7.3 shows x-ray diffraction (XRD) patterns

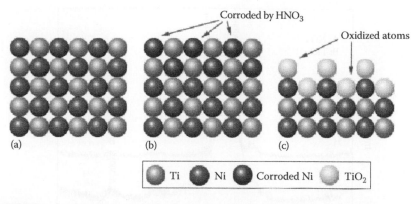

FIGURE 7.1
Schematic illustration of the formation of a protective TiO_2 layer on NiTi alloy when etched with the aqueous HNO_3 solution. (a) Atom arrangement of untreated NiTi alloy; (b) atom arrangement of corroding NiTi alloy; (c) atom arrangement of corroded NiTi alloy.

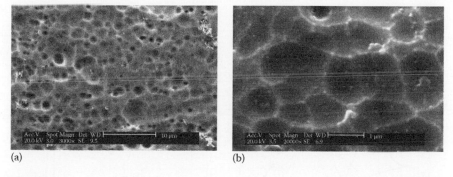

FIGURE 7.2
SEM images of NiTi samples treated by aqueous HNO_3 solution for 20 h (b is the high magnification of a).

of NiTi samples before and after the acid treatment. After the acid treatment, the main peaks of NiTi alloy at 39.80° and 60.84° shift toward a higher degree with increased width and weakened intensity. Also, the presence of typical peaks of TiO_2 (39.98°, 61.28°, 36.76°, and 37.40°) indicates the formation of a thicker TiO_2 layer on NiTi surface. The chemical states of NiTi alloy after the acid treatment were examined by X-ray photoelectron spectroscopy (XPS). Figure 7.4a shows the narrow scan of Ti 2p. The peaks at the binding energy of 455.1 and 458.8 eV are attributed to Ti^{2+} and Ti^{4+}, of which the area percentage is 25.7% and 74.2%, respectively. The higher percentage of Ti^{4+} suggests the selective oxidation of Ti atoms in NiTi alloy. The narrow scan of Ni 2p is shown in Figure 7.4b. Only Ni^0 at 852.9 eV is detected, meaning that the remaining Ni atoms in NiTi alloy are not oxidized by the aqueous HNO_3 solution.

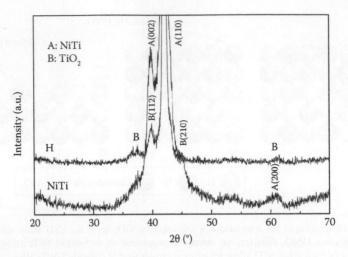

FIGURE 7.3
XRD patterns of NiTi alloy before (NiTi) and after the acid treatment (H).

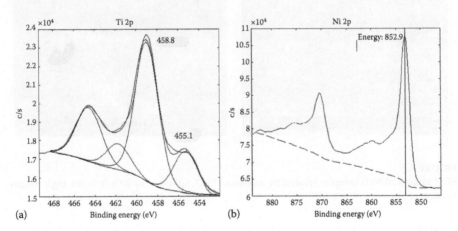

FIGURE 7.4
XPS spectra of NiTi alloy after the acid treatment: (a) Ti 2p and (b) Ni 2p.

7.2.2 Alkali Treatment

After the acid treatment, aqueous alkali solution (NaOH) was used to further activate NiTi alloy. Scanning electron microscope (SEM) observation reveals the effect of NaOH concentration on the surface morphology of NiTi alloy (Figure 7.5). Needle- or rod-like substances appear on the acid treated surface after it is immersed in 0.5–5 M aqueous NaOH solution at 60°C for 24 h. Moreover, the higher the NaOH concentration, the bigger and the more rods form on the surface. In Figure 7.5a, when the NaOH concentration is 0.5 M,

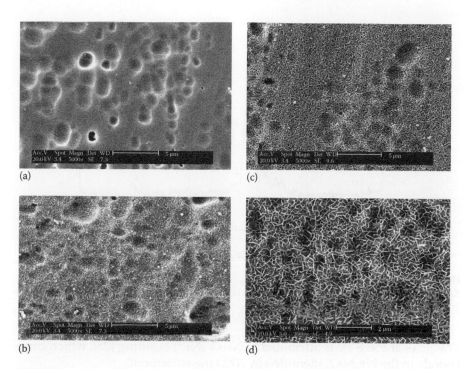

FIGURE 7.5
SEM images of NiTi samples treated in aqueous NaOH solution at different concentrations: (a) 0.5 M, (b) 1.0 M, (c) 2.0 M, and (d) 5.0 M, respectively.

it is almost the same as that after the acid treatment (Figure 7.2). After the treatment in the 1.0 M NaOH solution, a layer can be seen with the needle size of 100 nm (Figure 7.5b). When the concentration increases to 2 M, it becomes a network structure (Figure 7.5c). A nanoporous network in the order of 100 nm pore size covers the whole surface after the treatment in the 5 M NaOH solution (Figure 7.5d).

XRD patterns of NiTi samples after the acid–alkali treatment are shown in Figure 7.6. Compared with that after the acid treatment (Figure 7.3, H), two new phases appear after the treatment in the 1 M aqueous NaOH solution, i.e., $NaTi_2O_4$ and $Na_2Ti_3O_7$. When the concentration of NaOH solution increases to 5 M, Na_5NiO_4 forms together with $Na_2Ti_3O_7$. However, there is no TiO_2 peak detected, indicating that the previously formed TiO_2 layer in the acid treatment reacts with NaOH to produce sodium titanate. XPS analysis was performed to further confirm the reactions during the alkali treatment [27]. The survey scan (Figure 7.7a) shows all the elements on the NiTi surface after the NaOH treatment, including Ni, Ti, Na, O, and C (due to contamination in air). In Figure 7.7b, the narrow scan of Ti 2p, the peak at 453.5 eV is ascribed to Ti–Ti bonding [28], and the peak at 458.9 eV can be assigned to Ti–O (Ti^{4+}) bonding. Figure 7.7c shows the narrow san of Ni 2p.

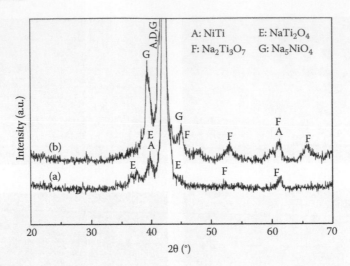

FIGURE 7.6
XRD patterns of NiTi samples treated in aqueous NaOH solution at different concentrations: (a) 1.0 M and (b) 5.0 M.

The peak at 856.8 eV is attributed to Ni–O (Ni^{3+}) bonding [28], which corresponds to the Na$_5$NiO$_4$ identified by XRD measurement.

In the initial stage of the alkali treatment, the following reactions happen:

$$TiO_2 + OH^- + nH_2O = HTiO_3^- \cdot nH_2O \tag{7.3}$$

$$Ti + 3OH^- = Ti(OH)_3^+ + 3e \tag{7.4}$$

$$Ti(OH)_3^+ + e = TiO_2 \cdot H_2O + 1/2H_2 + 4e \tag{7.5}$$

$$Ti(OH)_3^+ + OH^- = Ti(OH)_4 \tag{7.6}$$

$$Ti(OH)_4 + OH^- = HTiO_3^- \cdot 2H_2O \tag{7.7}$$

Because of high activation energy for homogeneous nucleation of apatite in human body fluid, it is necessary to introduce a type of functional group as an effective site for faster apatite nucleation on the material surface in the living body. Kokubo et al. [21,29] found that Ti–OH groups were capable of inducing apatite nucleation. After the alkali treatment, the sodium titanate transforms into hydrated titania by ion exchange, and many Ti–OH groups form on the surface of NiTi alloy. The Ti–OH groups play a role as effective nucleation sites by absorbing Ca^{2+} and PO$_4^{3-}$ ions in the SBF, and the ionic

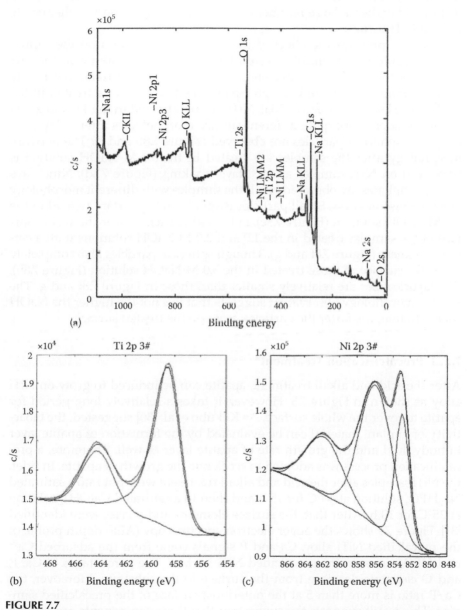

FIGURE 7.7
XPS profiles of NiTi alloy after the acid and alkali treatments: (a) XPS survey scan, (b) Ti 2p, and (c) Ni 2p. (Reprinted from *Surf. Coat. Technol.*, 173, Chen, M.F., Yang, X.J., Liu, Y., Zhu, S.L., Cui, Z.D., Man, H.C., Study on the formation of an apatite layer on NiTi shape memory alloy using a chemical treatment method, 229, Copyright 2003, with permission from Elsevier.)

activity product of apatite in the surrounding fluid is increased due to the increase in OH⁻ ion concentration in the SBF. As a consequence of what was mentioned earlier, a large number of apatite nuclei form on the chemically treated NiTi surface.

Figure 7.8 presents the effect of the NaOH concentration on the formation of apatite on chemically treated NiTi alloy. After soaking in the SBF for 3 days, the needle-like substances formed in the acid treatment totally disappear; however, there is no precipitation on NiTi surface treated in the 0.5 M NaOH solution (Figure 7.8a). NiTi samples treated in the 1.0 and 2.0 M NaOH solutions give quite different surface morphologies after 3 days, i.e., a layer of spherical particles are observed (Figure 7.8b and c). The network morphology after the samples are treated in the 5.0 M NaOH solution is preserved on NiTi samples after 3 days' soaking (Figure 7.8d). Nine days later, precipitates are observed on all the samples with different morphology and coating thickness. Small particles disperse on the surface treated in the 0.5 M NaOH solution (Figure 7.8e). Spherical particles cover the whole surface of the samples treated in the 1.0 and 2.0 M NaOH solutions with a certain thickness (Figure 7.8f and g). Though spherical particles also completely cover the surface of those treated in the 5.0 M NaOH solution (Figure 7.8h), the particle sizes are relatively smaller than those in Figure 7.8f and g. The surface morphology observation suggests that it is not the higher the NaOH concentration, the faster the particles grow on the treated surface.

7.2.3 Precalcification Treatment*

After the acid and alkali treatment, apatite can be induced to grow on NiTi alloy as shown in Figure 7.8. However, it takes a relatively long period for apatite to cover the whole surface. As Kokubo et al. [30] suggested, the bioactivity of implant material can be evaluated by the formation of apatite layer in body fluid and the growth rate of apatite layer as well. Therefore, a precalcification process was adopted to accelerate the growth of apatite. In brief, the NiTi samples after the acid and alkali treatment were put into a saturated Na_2HPO_4 solution at 40°C for 15 h and then in a saturated $Ca(OH)_2$ solution at 25°C for 10 h. After that, the surface elements and phases were identified [31]. Figure 7.9 shows the auger electron spectroscopy (AES) depth profile of the precalcified NiTi alloy. Ca and P signals come from the adsorbed Ca^{2+} and HPO_4^{2-} ions on the precalcified NiTi surface. The percentages of Ca, P, and O elements decrease from the surface to the substrate. Moreover, the Ca/P ratio is more than 5 at the outermost surface of the precalcified samples. This result supports the suggestion that these components are present as ions on the surface, since there is no stable calcium phosphate with Ca/P

---

* *Source: Scrip. Mater.*, 54, Yang, X.J., Hu, R.X., Zhu, S.L., Li, C.Y., Chen, M.F., Zhang, L.Y., and Cui, Z.D., Accelerating the formation of a calcium phosphate layer on NiTi alloy by chemical treatments, 1457, Copyright 2006, with permission from Elsevier.

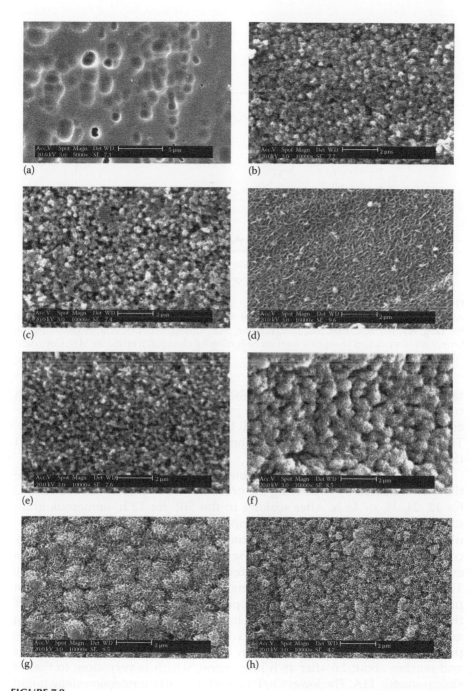

FIGURE 7.8

SEM images of NiTi alloys after soaking in the SBF for 3 days (a–d) and 9 days (e–h). The samples were treated with 0.5 M (a and e), 1.0 M (b and f), 2.0 M (c and g), 5.0 M (d and h) aqueous NaOH solution before soaking in the SBF.

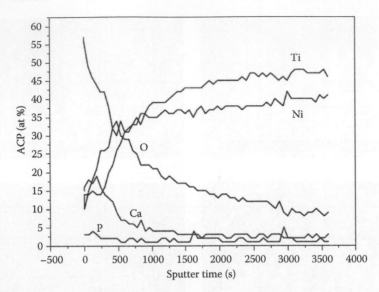

FIGURE 7.9
AES elements depth profile of the precalcified NiTi alloy. (Reprinted from *Scripta Materialia*, 54, Yang, X.J., Hu, R.X., Zhu, S.L., Li, C.Y., Chen, M.F., Zhang, L.Y., Cui, Z.D., Accelerating the formation of a calcium phosphate layer on NiTi alloy by chemical treatments, 1457, Copyright 2006, with permission from Elsevier.)

ratio higher than 2 [32]. Figure 7.10 shows XRD patterns of the acid–alkali treated samples before and after the precalcification. No significant difference is found between the acid–alkali treated and the precalcified samples. The main phases on the surfaces of both samples are sodium titanate and titanium oxides (TiO_2 and TiO).

The effect of precalcification treatment on the growth of bone-like apatite was evaluated by immersing the precalcified samples in the SBF at 37°C for 12, 24, 36, or 48 h. After soaking in the SBF for 12 h, numerous granules appear in a network on the sample surface (Figure 7.11a). More granules form with increasing soaking time (Figure 7.11b and c). A relatively complete coating forms after 36 h (Figure 7.11c), and the sample surface is entirely covered by compact and uniform particles after 48 h (Figure 7.11d). Thus, the formation of a perfect coating is accelerated by the precalcification process. Figure 7.12 shows energy dispersive spectrometer (EDS) spectra of the precalcified NiTi alloy immersed in the SBF for various periods. The Ca and P peaks are detected during the whole immersion period, of which the percentages increase with immersion time. Also, the Ca/P ratio increases and reaches about 1.6 after 48 h immersion, which is very close to that of stoichiometric HA. The lower Ca/P ratio at the early immersion stage might be due to the presence of other calcium phosphates in the deposited layer. Figure 7.13 shows XRD patterns of the precalcified samples immersed in the SBF for various periods. The main phase on the precalcified surface is poorly

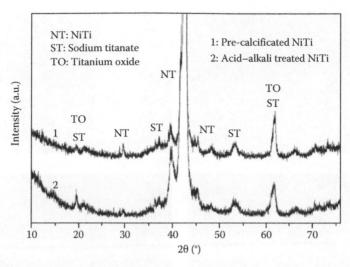

FIGURE 7.10

XRD patterns of the acid–alkali treated samples before and after precalcification. (Reprinted from *Scripta Materialia*, 54, Yang, X.J., Hu, R.X., Zhu, S.L., Li, C.Y., Chen, M.F., Zhang, L.Y., Cui, Z.D., Accelerating the formation of a calcium phosphate layer on NiTi alloy by chemical treatments, 1457, Copyright 2006, with permission from Elsevier.)

crystallized HA, which is in good agreement with the EDS results. However, strong peaks of HA can only be detected after 24 h immersion.

Wen et al. [18] also used a precalcification step prior to immersion, and they found that calcium phosphate precipitation can be dramatically speeded up on the acid–alkali treated titanium surface. Meanwhile, Jonasova et al. [32] reported that only an inhomogeneous and nonuniform apatite layer is formed on the NaOH treated titanium even after 20 days immersion in the SBF. These findings reflect the fact that the applied treatment parameters before immersion in the SBF significantly influence the apatite formation. In our work, the unchanged surface phases after the precalcification indicate that the surface features induced by the acid and/or alkali treatment remain, such as the microporous surface, nanosized porous network. They still contribute to the high nucleation efficiency of apatite due to increased surface area, pore size, and pore volume. The abundant adsorption of Ca^{2+} and HPO_4^{2-} ions onto the microporous surface causes local supersaturation of Ca^{2+} and HPO_4^{2-} ions and thus increases the driving force for apatite nucleation [6,32]. On the other hand, the concentrated locations of Ca and P ions might act as the preseeded nuclei for the heterogeneous nucleation of apatite. Once in apatite nuclei form, they spontaneously grow by consuming Ca and P ions from the SBF to form a dense and uniform apatite layer [21]. The rapid formation of apatite layer on the precalcified NiTi alloy implies shortened time to establish the expected physical–chemical bond between the implant and the bone, which is beneficial to the faster osseointegration.

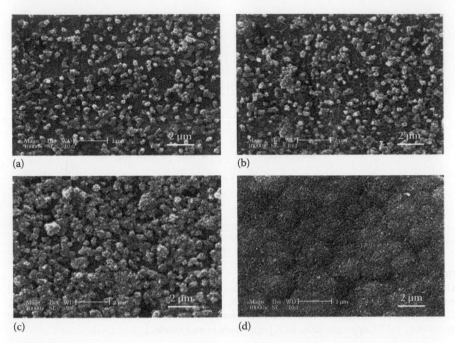

(a) (b) (c) (d)

FIGURE 7.11

SEM micrographs of the precalcified samples immersed in the SBF for 12 h (a), 24 h (b), 36 h (c), and 48 h (d). (Reprinted from *Scripta Materialia,* 54, Yang, X.J., Hu, R.X., Zhu, S.L., Li, C.Y., Chen, M.F., Zhang, L.Y., Cui, Z.D., Accelerating the formation of a calcium phosphate layer on NiTi alloy by chemical treatments, 1457, Copyright 2006, with permission from Elsevier.)

7.3 Bioactive Apatite Coating on Chemically Treated NiTi Alloy

Chemical treatments are crucial for the biomimetic growth of apatite on NiTi alloy. The acid immersion is advantageous to increase the surface roughness of NiTi alloy by producing many acid etched pits or grooves. The alkali treatment at the second step might have two concurrent effects on the acid-etched surface. One is the formation of a microporous surface layer, and the other is the formation of more titanium oxides in this microporous layer including TiO_2 and TiO. Each step of the treatment is important for acquiring the microporous titanium oxide surface, which cannot be produced by either acid or alkali treatment individually. The precalcification does not change the surface features of the acid–alkali treated surface but results in the abundant adsorption of Ca^{2+} and HPO_4^{2-} ions onto the microporous surface. The local supersaturation of Ca^{2+} and HPO_4^{2-} ions contributes to the faster nucleation and growth of apatite.

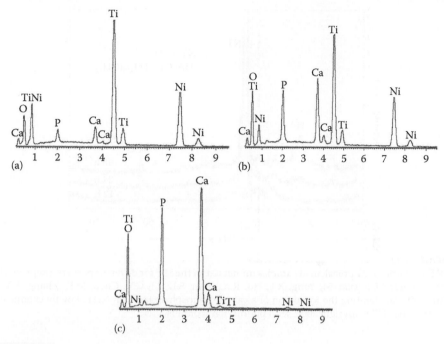

FIGURE 7.12

EDS spectra of the precalcified samples immersed in the SBF for 12 h (a), 24 h (b), and 48 h (c). (Reprinted from *Scripta Materialia*, 54, Yang, X.J., Hu, R.X., Zhu, S.L., Li, C.Y., Chen, M.F., Zhang, L.Y., Cui, Z.D., Accelerating the formation of a calcium phosphate layer on NiTi alloy by chemical treatments, 1457, Copyright 2006, with permission from Elsevier.)

It is important to understand the performances of both natural and synthetic biomaterials for the development of new and more biocompatible implant materials. The evaluation involves both in vitro and in vivo assessments. in vitro assessment uses cells and/or tissues derived from animals or humans and maintained in tissue culturing conditions. In vivo test uses animal models to evaluate histological responses to biomaterials in living animals. By obtaining the clinical data during the trial period and histological data from the examination of postmortem tissues, we are able to assess the tissue responses to the implanted materials. In this section, the formation of apatite layer and the growth mechanism of apatite on chemically treated NiTi alloy are discussed, and also the in vitro and in vivo biological and histological performances are investigated.

7.3.1 Biomimetic Growth of Apatite on Chemically Treated NiTi Alloy

A series of optimal parameters of chemical treatments are selected: 32.5% HNO_3 solution, 60°C, 20 h; 1.2 M NaOH solution, 103°C–105°C, 5 h;

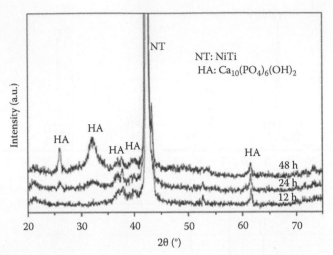

FIGURE 7.13
XRD patterns of the precalcified samples immersed in the SBF for different periods. (Reprinted from *Scripta Materialia*, 54, Yang, X.J., Hu, R.X., Zhu, S.L., Li, C.Y., Chen, M.F., Zhang, L.Y., Cui, Z.D., Accelerating the formation of a calcium phosphate layer on NiTi alloy by chemical treatments, 1457, Copyright 2006, with permission from Elsevier.)

saturated Na_2HPO_4 solution, 40°C, 15 h, saturated $Ca(OH)_2$ solution, 25°C, 10 h. Figure 7.14 shows surface morphologies of the chemically treated NiTi samples after soaking in the SBF for different periods. Before soaking (Figure 7.14, 0 h), the sample surface preserves the microporous morphology. After soaking for 9 h (Figure 7.14, 9 h), some particles with the size of 300–500 nm disperse on the surface. Later, these particles continuously grow to be spherical and become bigger in size (Figure 7.14, 12 and 18 h) and almost cover the whole surface of the sample after soaking in the SBF for 24 h (Figure 7.14, 24 h). These spherical particles connect with each other to form a complete coating (Figure 7.14, 72 h) when the soaking time is up to 3 days. The coating becomes thicker with some light cracks after 5 days (Figure 7.14, 120 h), and the particles grow to about 2 µm in diameter and more cracks appear in the coating after 14 days (Figure 7.14, 296 h). EDS was used to identify the elements in the coating. Ca, P, and O are detected, of which the intensities increase with increasing soaking time. It means that the precipitated particles are calcium phosphates and the coating becomes thicker and thicker. The atomic percentages of Ca and P as well as the Ca/P ratio are listed in Table 7.2. In the initial period, the Ca/P ratio is low, only around 1.0. It increases with the soaking time and reaches 1.60 after 120 h, which is closer to that of HA. The cross-sectional views of the coated NiTi alloys are shown in Figure 7.15. The coating thickness is about 5 µm after soaking in the SBF for 3 days and up to about 15 µm after 14 days.

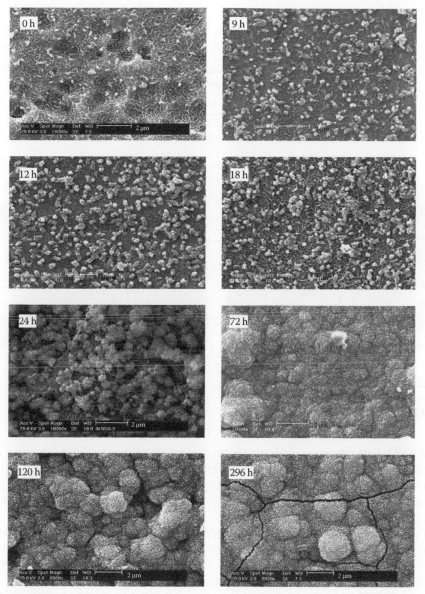

FIGURE 7.14
SEM images of chemically treated NiTi samples after soaking in the SBF for different periods.

The phase evolution on the NiTi alloy during the immersion is shown in Figure 7.16. XRD pattern of pure HA powder is included as reference. From Figure 7.16, it can be seen that the (002) peak of the sample soaked in the SBF for 5 days is sharper than that for 3 days, while other diffraction peaks of HA, such as (112), (310), (213), and (004), are still broad. The broadness of

TABLE 7.2

Ca/P Ratio of NiTi Alloys after Soaking in the SBF for Different Periods

	Soaking Time (h)			
	12	24	72	120
Ca (at%)	1.86	5.43	11.95	16.20
P (at%)	1.73	4.63	8.62	10.15
Ca/P	1.07	1.15	1.39	1.60

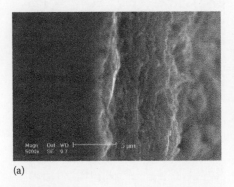 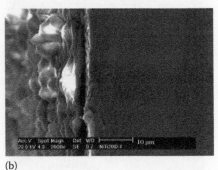

(a) (b)

FIGURE 7.15
Cross-sectional views of the chemically treated NiTi alloy after soaking in the SBF for 3 days (a) and 14 days (b).

the XRD diffraction peak depends on the crystal size and/or perfection or strain: the broader the peak, the smaller or less perfect or more strained the crystal. The broad peaks indicate the poorly crystallized HA with small size, while the sharp (002) peak shows the preferential orientation of HA crystals grown in the latter period.

Fourier transform infrared spectroscopy (FTIR) was used to identify the chemical groups in the coatings after soaking in the SBF for 72 h (HNBPS72) and 120 h (HNBPS120). The sample before soaking (HNBP) and pure HA powder were also examined for comparison, as shown in Figure 7.17. There is no chemical group on HNBP. However, similar peaks appear on both the samples after soaking in the SBF for 72 and 120 h. Compared with those of pure HA powder, the peaks at 1009, 599, and 557 cm^{-1} are attributed to PO_4^{3-} groups. It is noted that the peak of OH at 632 cm^{-1} in the pure HA does not exist in HNBPS72 but appears in HNBPS120. Moreover, the typical peaks of CO_3^{2-} are also detected at 1483, 1421, and 872 cm^{-1}, meaning the formation of carbonated apatite that is also called bone-like apatite.

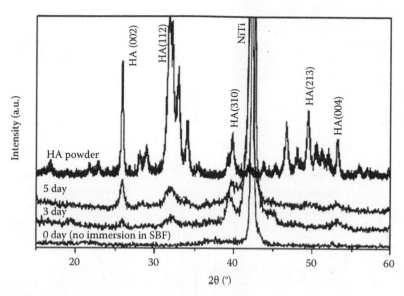

FIGURE 7.16
XRD patterns of the chemically treated NiTi after soaking in the SBF for 0, 3, and 5 days; XRD pattern of HA powder is included for reference. (Reprinted from *Surf. Coat. Technol.*, 173, Chen, M.F., Yang, X.J., Liu, Y., Zhu, S.L., Cui, Z.D., Man, H.C., Study on the formation of an apatite layer on NiTi shape memory alloy using a chemical treatment method, 229, Copyright 2003, with permission from Elsevier.)

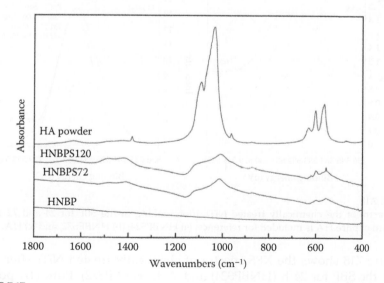

FIGURE 7.17
FTIR patterns of the chemically treated NiTi before and after soaking in the SBF for 72 and 120 h; pure HA powder is taken as reference.

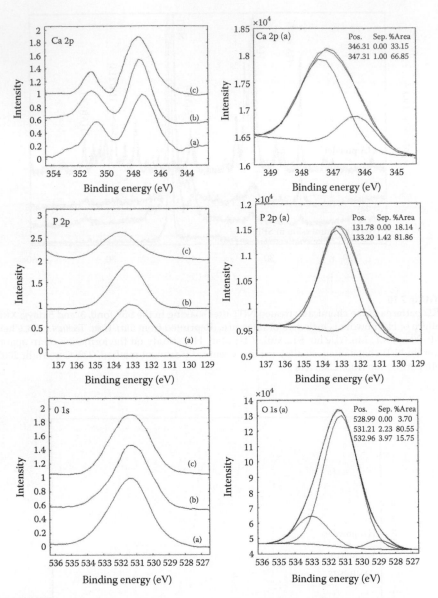

FIGURE 7.18
XPS spectra of the chemically treated NiTi after soaking in the SBF for 24 and 72 h; XPS spectrum of pure HA is included for reference: (a) HNBPS24, (b) HNBPS72, and (c) HA.

Figure 7.18 shows the XPS spectra of chemically treated NiTi after soaking in the SBF for 24 h (HNBPS24) and 72 h (HNBPS72). Pure HA powder was also measured as a reference. Ca 2p spectrum of HNBPS72 is almost the same as that of pure HA, but that of HNBPS24 shifts a little to lower binding energy. The fitting analysis of HNBPS24 shows that 66.85% of the peak is attributed to that at 347.1 eV (typical binding energy of Ca 2p in HA) [33,34], and the peak at 346.3 eV might belong to other calcium phosphate. Comparing P 2p spectra, the P 2p of HNBPS 24 also shifts a little to lower binding energy, of which 81.9% is due to PO_4^{3-} (the binding energy is 133.1 eV) [34]. Again, in the spectra of O 1s, HNBPS24 shifts a little to lower binding energy compared with that in pure HA and HNBPS72. The peak at 531.1 and 532.9 eV is attributed to O–H and O–P bonds, respectively. And the peak at 529.0 eV belongs to that in metallic oxide, indicating the existence of non-apatite phase on the chemically treated NiTi, which might be $Ca_2P_2O_7 \cdot 4H_2O$ as measured by XRD.

SBF is a kind of saturated solution containing Ca^{2+} and PO_4^{3-} ions. Calcium phosphates can be precipitated homogeneously at certain conditions, but homogeneous nucleation generally occurs with much more difficulty: much higher energy and supersaturation. However, when the chemically treated NiTi alloy is immersed in the SBF, the bioactive surface provides a number of nucleation sites for heterogeneous nucleation of calcium phosphates, which can promote the precipitation of calcium phosphates. Assume that the following reaction takes place in the SBF:

$$10Ca^{2+} + 6PO_4^{3-} + 2OH^- = Ca_{10}(PO_4)_6(OH)_2 \tag{7.8}$$

The thermodynamic driving forces for hydroxyapatite precipitation are calculated based on the classical equation of free energy change in super-saturated solutions [35]:

$$\Delta G_v = -RT \ln\left(A_p / K_{sp}\right) \tag{7.9}$$

$$\Delta G_v = -RT \ln\left(\frac{a_{Ca^{2+}}^{10} \times a_{OH^-}^2 \times a_{PO_4^{3-}}^6}{K_{sp,HA}}\right) \tag{7.10}$$

where
  R is the gas constant
  T is the absolute temperature
  A_p is the ionic activity product
  $K_{sp,HA}$ is the solubility product
  A_p/K_{sp} is the supersaturation
  a is the activity

When the chemically treated NiTi alloy is immersed in the SBF, the treated surface provides a number of nucleation sites that are beneficial for the heterogeneous nucleation of apatite. The Gibbs free energy of the heterogeneous nucleation is expressed as [35]

$$\Delta G^* = \frac{16\pi\sigma^3}{3(\Delta G_v)^2} f(\theta) \tag{7.11}$$

$$= \frac{16\pi\sigma^3}{3R^2T^2(\ln A_p/K_{sp})^2} f(\theta) \tag{7.12}$$

where
　ΔG^* is the free energy for the formation of an embryo of critical size
　σ is the interface energy between the nucleus and the solution
　$f(\theta)$ is a function of the contact angle between the nucleus and the substrate

As reported by Kokubo et al. [21,36], an amorphous sodium titanate hydrogel layer formed on the surface of titanium or titanium alloys by the alkali treatment. When exposed to the SBF, the alkali treated surface releases Na+ ions into the SBF via exchanges for H_3O^+ ions in the SBF. Thus, Ti–OH groups form on the surface and then induce the formation of apatite in the SBF. In our study, a microporous network is built on the NiTi surface after the alkali treatment, which contains most of the sodium titanate and a little amount of sodium nickelate. When the treated NiTi is immersed in the SBF, similar reactions take place. Since the release of Na+ ions into the SBF would cause the increase of pH, the changes in pH values in the SBF are monitored as shown in Figure 7.19. The increase in pH after 6 h indicates the beginning of the ion exchange between Na+ and H_3O^+ ion. The pH becomes stable until the apatite layer covers the whole surface of the substrate. The growth of the apatite layer on the treated NiTi can be explained by Equation 7.12. Firstly, the ion exchange between Na+ and H_3O^+ ions makes microscale concave surfaces with negative charges, providing numerous nucleation sites for the heterogeneous nucleation of apatite. Also, in the case of forming the crystal nucleus with same critical size and contact angle, it is the easiest to nucleate on the concave surface, followed by smooth and convex surface. Thus apatite nucleus prefers to be formed in the small concave sites. Furthermore, the negatively charged Ti–OH groups absorb Ca^{2+} and PO_4^{3-} ions in the SBF, which decreases the contact angle between the nucleus and the substrate, i.e., decreases $f(\theta)$. Consequently, ΔG^* is decreased, and it is easier to form apatite crystals. Secondly, the precalcification results in abundant adsorption of Ca^{2+} and HPO_4^{2-} ions onto the microporous surface and consequent local supersaturation of Ca^{2+} and HPO_4^{2-} ions on the sample surface. The increase in the local supersaturation finally decreases ΔG_v. Once the free energy barrier is overcome, the apatite formation is favored.

　According to the earlier EDS and XRD analysis, $Ca_2P_2O_7 \cdot 4H_2O$ firstly forms on the treated surface, which might be due to the increase in local

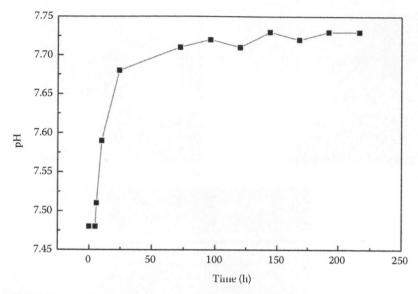

FIGURE 7.19
Measured pH values during the soaking of the chemically treated NiTi in the SBF.

supersaturation. Generally, when the concentrations of Ca^{2+} and PO_4^{3-} ions in the solution are low, and the pH is higher than 4.3, the solution is only supersaturated to HA. However, when both the concentration and the pH are high, the SBF is supersaturated to several kinds of calcium phosphates. In this case, the formation of $Ca_2P_2O_7 \cdot 4H_2O$ is preferred. $Ca_2P_2O_7 \cdot 4H_2O$ continues to grow, and the concentrations of Ca^{2+} and PO_4^{3-} decrease. After about 20 h, the supersaturation of the SBF is low enough to form HA. At the same time, the as-grown calcium phosphates act as the nucleation sites for HA formation. Thus, HA crystals nucleate and grow rapidly.

7.3.2 In Vitro Biological Properties*

Biocompatibility of metallic implant plays an important role in the long-term survival of prosthetic implant fixation [37], and cell attachment and spreading on material surface is a major indicator in implant technology [38]. Although cell culture studies cannot directly mimic the cellular response and the environment in vivo, the toxic effects of different materials can be well quantified in vitro [39]. The effects of bioactive surface treatment on

---

* *Source: Mater. Sci. Eng. C*, 28, Cui, Z.D., Chen, M.F., Zhang, L.Y., Hu, R.X., Zhu, S.L., and Yang, X.J., Improving the biocompatibility of NiTi alloy by chemical treatments: An in vitro evaluation in 3T3 human fibroblast cell, 1117, Copyright 2008, with permission from Elsevier.

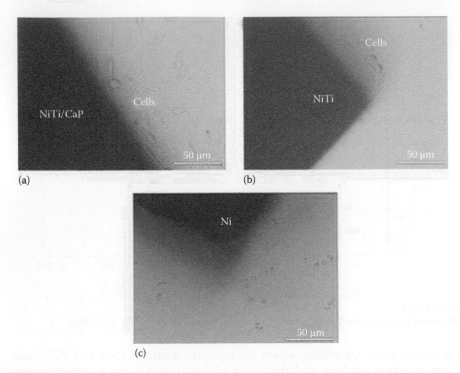

FIGURE 7.20
Light micrographs of fibroblasts cultured for 3 days on (a) NiTi/CaP, (b) MP NiTi, and (c) Ni. Black areas in the images are the borders and shadows of the test samples. (Reprinted from *Mat. Sci. Eng. C*, 28, Cui, Z.D., Chen, M.F., Zhang, L.Y., Hu, R.X., Zhu, S.L., Yang, X.J., Improving the biocompatibility of NiTi alloy by chemical treatments: An in vitro evaluation in 3T3 human fibroblast cell, 1117, Copyright 2008, with permission from Elsevier.)

the biocompatibility of NiTi alloy were investigated by in vitro cell culture tests. In this test, 3T3 human embryonic fibroblast cells were used because they are substrate-dependent, nonspecific cell lines [26]. The in vitro cell responses on the apatite-coated NiTi (NiTi/CaP), mechanically polished NiTi (MP NiTi), and pure Ni samples were studied to compare their biological properties [40].

Figure 7.20 shows the light micrographs of fibroblasts cultured for 3 days near the samples of NiTi/CaP, MP NiTi, and Ni, respectively. Black areas at the corners of the images are the borders and shadows of the test samples. As shown in Figure 7.20a, the fibroblasts attach to the NiTi/CaP sample with the shape of a shuttle and/or elongated polygon, which are the normal individual form of the 3T3 fibroblast cells. Cell morphologies near the MP NiTi are similar to those near the NiTi/CaP, but the number of attached cells to the MP NiTi is slightly lower than that on the NiTi/CaP. Significant inhibition of cell attachment and growth is observed near the Ni sample. Most fibroblasts cultured on the Ni present a retracted and completely deteriorated morphology

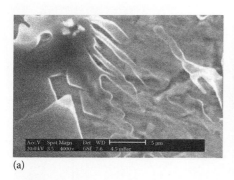
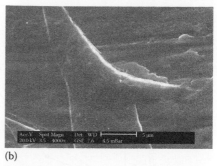

(a)   (b)

FIGURE 7.21
SEM micrographs of fibroblasts cultured for 3 days on (a) NiTi/CaP and (b) MP NiTi. (Reprinted from *Mat. Sci. Eng. C*, 28, Cui, Z.D., Chen, M.F., Zhang, L.Y., Hu, R.X., Zhu, S.L., Yang, X.J., Improving the biocompatibility of NiTi alloy by chemical treatments: An in vitro evaluation in 3T3 human fibroblast cell, 1117, Copyright 2008, with permission from Elsevier.)

(Figure 7.20c) without any surface evaginations. The lower attachment and spreading on the Ni will finally lead to cell death. Figure 7.21 shows the SEM micrographs of the fibroblasts cultured for 3 days on the samples. Cells are well flattened on the NiTi/CaP. The lamellipodia presents and extends radially from the central area and strongly adheres to the substrate and draws the cellular body by cytoplasmic contraction (Figure 7.21a). Cells cultured on the MP NiTi are also well flattened, but the lamellipodia is still in the early stage without extending to adjacent areas (Figure 7.21b).

After 6 day culture, the cell morphologies near the NiTi/CaP and the MP NiTi samples are as shown in Figure 7.22. The fibroblasts attach very well to both the NiTi/CaP and the MP NiTi, but cell density near the NiTi/CaP is

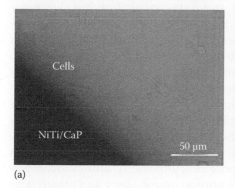
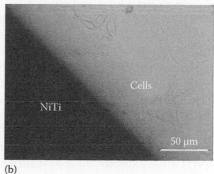

(a)   (b)

FIGURE 7.22
Light micrographs of fibroblasts cultured for 6 days on (a) NiTi/CaP and (b) MP NiTi. Black areas in the images are the borders and shadows of the test samples. (Reprinted from *Mat. Sci. Eng. C*, 28, Cui, Z.D., Chen, M.F., Zhang, L.Y., Hu, R.X., Zhu, S.L., Yang, X.J., Improving the biocompatibility of NiTi alloy by chemical treatments: An in vitro evaluation in 3T3 human fibroblast cell, 1117, Copyright 2008, with permission from Elsevier.)

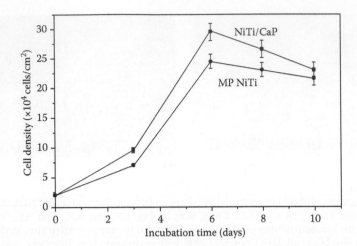

FIGURE 7.23

Cell numbers of fibroblasts cultured on NiTi/CaP and MP NiTi samples for various periods. (Reprinted from *Mat. Sci. Eng. C*, 28, Cui, Z.D., Chen, M.F., Zhang, L.Y., Hu, R.X., Zhu, S.L., Yang, X.J., Improving the biocompatibility of NiTi alloy by chemical treatments: An in vitro evaluation in 3T3 human fibroblast cell, 1117, Copyright 2008, with permission from Elsevier.)

higher than that near the MP NiTi. Also, the fibroblasts on the NiTi/CaP are in more close contact with each other but grow slightly less close to the MP NiTi. Unhomogeneous cell distribution is observed near the MP NiTi sample, and the even worse case is that the cell-free region is found (Figure 7.22b). However, no evident membrane lesions and cytolysis are observed.

Cell proliferation on different surfaces is compared in terms of the cell density after culturing for different days (Figure 7.23). The mean cell number per area increases significantly after 3 and 6 day incubation but decreases after 8 and 10 day incubation for both the NiTi/CaP and the MP NiTi samples. The decrease in cell density is ascribed to the culture medium, which is not replaced over the specific incubation periods. However, there is no statistical difference in the optical density of cells between the NiTi/CaP and the MP NiTi at all the designated incubation points.

The aforementioned results are in agreement with previously published work. It is reported that the biocompatibility of NiTi alloy is promising and comparable to that of pure titanium [26,39,41], and NiTi is well tolerated by the 3T3 human embryonic fibroblasts. Moreover, the surface modification of NiTi alloy exhibits no major difference on cytotoxicity and cytocompatibility. The 3T3 human fibroblasts show normal morphology and attach onto both the NiTi/CaP and the MP NiTi surfaces over the incubation periods. The biocompatibility of NiTi alloy is attributed to the thermodynamically and mechanically stable oxide layer composed mainly of TiO_2 on its surface [38], and nickel ions are not easily released out of NiTi because of a strong interatomic bonding between nickel and titanium atoms [42].

Surface treatment also has a critical importance for the biocompatibility of NiTi alloy [43]. Toxic and immunologic reactions of NiTi are generally due to inadequate surface treatment and a breakdown of the oxide layer under certain experimental conditions [42]. These may be responsible for the cell-free regions observed on the MP NiTi incubated for 6 days (Figure 7.22b). In addition, Bogdanski et al. [41] reported that the toxic effects to cells could be a result of increasing nickel content and/or Ni-rich intermetallic phases in the NiTi alloy. They also observed cell-free areas on the high nickel sections of the functionally graded Ni-NiTi-Ti samples. NiTi containing more than 60% Ni and Ni_3Ti alloys may not be passivated as fast as titanium or NiTi with a layer of TiO_2, in which local corrosion is enhanced and more nickel ions are released in the cell culture medium. As a result, they show a low biocompatibility, because locally concentrated nickel ions due to the corrosion of NiTi induce adverse tissue reactions or alter cellular functions such as cell proliferation [44].

The cell culture results reveal that the NiTi/CaP samples have an excellent biocompatibility. When implants are implanted into the human environment, the implant surface makes the first contact with the living tissue [43], and the cell behaviors on implant surface depend upon the cell–implant interactions [38]. The nature of material surface directly influences the interactions between the living tissue and the implant material. In this work, chemical treatments are used to modify the surface of NiTi alloy to induce the deposition of a bone-like apatite layer on the alloy surface. The good biocompatibility of this apatite layer is confirmed by the results of in vitro cell attachment and proliferation tests that the NiTi/CaP samples are well tolerated by fibroblasts. The fibroblasts attach and spread very well on the NiTi/CaP with normally elongated shape, and an increased adherence of fibroblasts is observed throughout on the NiTi/CaP samples compared with the MP NiTi samples. The increased cell adherence can be ascribed to the evolution in the surface chemistry of the treated samples because calcium phosphates have an intrinsic good biocompatibility [6,20,45]. Moreover, the changes in surface morphology associated with the deposition of apatite layer such as the increased surface area also contribute to the increased cell adherence [24,46,47]. On the other hand, the apatite-coated NiTi alloy has beneficial properties including sealing against nickel release and absorbing bioactive molecules such as growth factors [44]. Consequently, the NiTi/CaP samples exhibit a better cytocompatibility as compared with the MP NiTi samples.

Redey et al. [48] reported that several surface requirements are necessary for a biomaterial to have osteoconductive properties. The material surface needs to be biocompatible and, if possible, similar with the chemical composition of the bone. Cell attachment and proliferation are important indices for osteoconductive biomaterials. An osteoconductive biomaterial allows osteoprogenitor cells to attach, proliferate, and differentiate at its surface and then induces the formation of an interface of mineralized collagenous bone matrix between bone tissue and bulk implant. The critical parameter for achieving osteoconduction is the initial number of well-attached osteoblastic cells to the bone substitute. It has been well known that calcium phosphates such as HA are generally used to coat metallic implants to improve the osteoconductivity of the implants, because they have a similar chemical composition with the mineral component of the bone [20,49]. As shown in Figure 7.21, the fibroblasts cultured on the NiTi/CaP show a normally elongated form with radially extended lamellipodia. This is an indication of good adhesion according to Maitz et al. [45]. The cell density near the NiTi/CaP samples is higher than that near the MP NiTi samples (shown in Figures 7.20 and 7.22). These results reveal the osteoconductivity of the apatite-coated NiTi alloy.

In other words, pure nickel is strongly toxic to 3T3 human embryonic fibroblasts, and the mechanically polished and the apatite-coated NiTi alloy exhibit good cytocompatibility with the cells. Moreover, the apatite-coated NiTi has a better biocompatibility as compared with the mechanically polished NiTi and shows a good osteoconductivity.

7.3.3 In Vivo Histological Performances*

In vivo histological performances of apatite-coated NiTi alloy (48 h immersion in the SBF after the chemical treatments) were evaluated by rabbit models, with untreated NiTi implants as controls. China white rabbits, weighing 2.5–3.0 kg, were prepared by intramuscularly injecting ketamine (15 mg/kg) and diazepam (0.5 mg/kg). Under sterile surgical conditions, the femur was exposed through a medical incision. Three slightly oversized holes (2.1 mm in diameter) separated by about 10 mm were made in the latered corex bone using a low-speed dental burr under saline irrigation. The implants (cylindrical shape, 2 mm in diameter, and 6 mm in length) were randomly inserted in the holes with light pressure. The long axis of the cylinder was perpendicular to the long axis of the femur. Three apatite-coated cylinders were implanted in one leg of the rabbit and three uncoated ones were implanted in the other leg as a paired control. The rabbits were sacrificed with the physical method, at 6, 13, 26, 53 weeks as well as 2 years postoperation, respectively, followed by the histological evaluations [50,51].

X-rays of the rabbit femurs were taken to confirm the stability of the implants at 6 weeks postimplantation. All implants remained seated at the original site, indicating a good interference fit. There was no abnormal behavior, skin irritations, infections, or tumors in any of the test animals. Histology

---

* *Source: Mat. Sci. Eng. C*, 27, Li, C.Y., Yang, X.J., Zhang, L.Y., Chen, M.F., and Cui, Z.D., In vivo histological evaluation of bioactive NiTi alloy after two years implantation, 122, Copyright 2007, with permission from Elsevier.

  Mat. Sci. Eng. C, 24, Chen, M.F., Yang, X.J., Hu, R.X., Cui, Z.D., Man, H.C., Bioactive NiTi shape memory alloy used as bone bonding implants, 497, Copyright 2004, with permission from Elsevier.

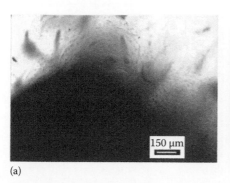
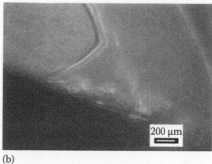

(a)   (b)

FIGURE 7.24
(See color insert.) Nondecalcified histological cross section of apatite-coated NiTi implant at 6 weeks after implantation. (a) Using a light microscope, showing direct contact of bone and coated implant (original magnification ×100). (b) Using a fluorescent light microscope, showing active transformation of bone (osseous tissue is red and zones signed by Achromycin are yellow, original magnification × 75). (Reprinted from *Mat. Sci. Eng. C*, 24, Chen, M.F., Yang, X J., Hu, R.X., Cui, Z.D., Man, H.C., Bioactive NiTi shape memory alloy used as bone bonding implants, 497, Copyright 2004, with permission from Elsevier.)

shows that the osteoblasts directly contact with the apatite coating on the NiTi rod, without any intervention between the femurs and the firm implants at 6 weeks after implantation (Figure 7.24a). The zones labeled by Achromycin are clearly seen in the microscope under fluorescent light, indicating active transformation of bone and osseointegration in the front of the implants (Figure 7.24b). At 13 weeks after implantation, more close direct contacts were found in the apatite-coated samples (Figure 7.25a) and a new bone layer, about 120 μm, had formed. Furthermore, a green bone layer between the marrow and the coated implant was also seen by fluorescent microscope (Figure 7.25b). In contrast, a fibrous layer (red connective tissue) of variable thickness appeared between the new bone and the uncoated NiTi implants under fluorescent light (Figure 7.26a). Under SEM, the gaps between the implant and the fibrous layer were seen to be about 15–20 μm in width (Figure 7.26b). Some island osteohistology, about 10% of surface area, were found on the surface of cleaved specimen with the uncoated NiTi implants. Some bulged cells, organic matrix, and a little collagen fibers were observed on the implants (Figure 7.27a), indicating certain proliferation of osteoblast. Osteoblast proliferation was very active on the surface of the apatite-coated implants. About 90% of the surface areas on the coated implants were directly covered with new bone. Some osteoblasts were present, and a large number of collagen fibers (about 0.1 μm in size) were around the cells (Figure 7.27b), which suggests that new bone was growing luxuriantly on the apatite-coated surface.

For metallic implants, it is important to examine whether the implant materials and metal ions or worn particles released from the surface have negative effects on bone formation. The in vivo results have shown that untreated NiTi has less bone contact [52], a slower osteogenesis process,

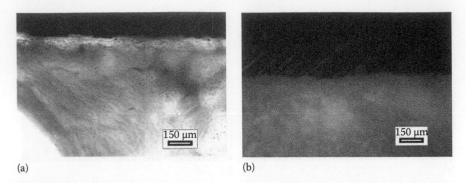

(a)   (b)

FIGURE 7.25
(See color insert.) Nondecalcified histological cross section of apatite-coated NiTi implant at 13 weeks after implantation. (a) Using a light microscope, showing a new bone layer contacting with coated implant (original magnification ×100). (b) Using a fluorescent light microscope, showing a green bone layer between the marrow and the coated implant (original magnification × 75). (Reprinted from *Mat. Sci. Eng. C*, 24, Chen, M.F., Yang, X.J., Hu, R.X., Cui, Z.D., Man, H.C., Bioactive NiTi shape memory alloy used as bone bonding implants, 497, Copyright 2004, with permission from Elsevier.)

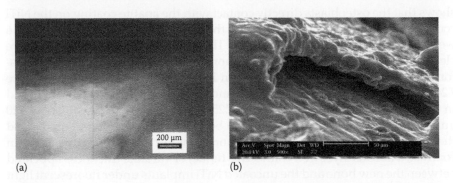

(a)   (b)

FIGURE 7.26
The gaps between new bone layer and uncoated implant. (a) Using light microscopy (original magnification ×100). (b) Using SEM. (Reprinted from *Mat. Sci. Eng. C*, 24, Chen, M.F., Yang, X.J., Hu, R.X., Cui, Z.D., Man, H.C., Bioactive NiTi shape memory alloy used as bone bonding implants, 497, Copyright 2004, with permission from Elsevier.)

and lower activity of osteonectin synthesis than pure titanium or other titanium alloys [53]. Though the NiTi implant with a mechanically ground surface has no negative effects on the total new bone formation [54], attention must be directed to the fact that a fibrous layer forms between the implants and the bone (Figure 7.26a). An incontestable conclusion is that the interface between the uncoated implant and the bone is analogous to a fibrous capsule and bone. The fibrous encapsulation around the uncoated implants induces weak bone–implant interface. Therefore, gaps at the interface are produced when the specimen is cleaved into two parts for SEM examination.

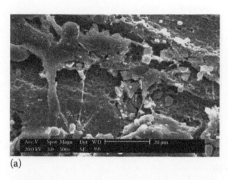
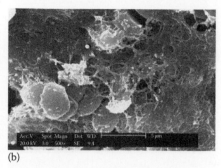

(a) (b)

FIGURE 7.27
Surface morphologies of implants at 13 weeks after implantation. (a) A little bulged cell and collagen fibrous networks on uncoated NiTi implant. (b) A large number of bone cells and collagen fibrous networks on apatite-coated NiTi implant. (Reprinted from *Mat. Sci. Eng. C*, 24, Chen, M.F., Yang, X.J., Hu, R.X., Cui, Z.D., Man, H.C., Bioactive NiTi shape memory alloy used as bone bonding implants, 497, Copyright 2004, with permission from Elsevier.)

On the other hand, there are no gaps observed between the apatite-coated implants and the bone, indicating that the coated implants are integrated with the bone tissue after periodical implantation. This hypothesis is supported by previous in vivo studies [55–58]. Direct contact between the coated NiTi and the bone after 6 and 13 weeks of implantation is observed in our study. It is thought that some chemical interaction may occur between the apatite coating and the bony tissue. This chemical bonding provides a substantial union that can transmit all types of forces [58], so the bonding strength at the interface is high.

In contrast to the uncoated implant, faster osteoblast proliferation occurs on the surface of the coated implant at 13 weeks of implantation. It shows early bone remodeling on the apatite-coated implant. Apatite coating demonstrates its superiority to help the implant get stable quickly at the early period after implantation. This is due to the fact that apatite is the main inorganic constituent of hard tissues, and the free Ca and P ions can be used in metabolism, so apatite coating shows good osteoconduction. New osteoblasts with more projection directly adhere on the apatite coating, and their projection penetrates into the apatite crystals. At the same time, an important role of the osteoblast is to secrete osseous collagen fibers. The more active osteoblasts are, the more osseous collagen fibers are, and the faster new bone forms, as demonstrated in Figure 7.27. According to the size and distributed formation of osseous collagen fibers, it is judged that the new bone layer shown in Figure 7.27b is woven bone that first forms on apatite-coated implants. After that, the parallel fibroid bone forms (see Figure 7.26a). Since there are big cavities in the woven bone, its strength is lower. The fragmentation occurs in the woven bone when the specimen is split into two parts along the long axis of the implant.

At 26 weeks after implantation, there is no obvious gap or interface between the apatite-coated implant and the host bone, and a mature bone–implant

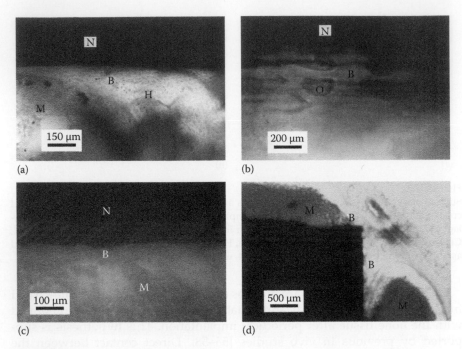

FIGURE 7.28
(See color insert.) Nondecalcified histological cross section of apatite-coated NiTi implant at 26 weeks after implantation. (a) Using a light microscope, showing a mature bone layer. (b) Using a fluorescent light microscope, showing pink osteoid on apatite-coated NiTi implant. (c) Formation of new bone layer in medullary cavity. (d) A special specimen. N, NiTi; B, Bone; M, Marrow; O, Osteoid; H, Haversian.

union is observed (Figure 7.28a). At the same time, pink osteoid still presents and undergoes mineralization (Figure 7.28b). This suggests that the new bone formation and remodeling take place simultaneously on the apatite-coated implant. Moreover, a layer of green bone with a thickness of about 100 μm forms at the interface of the apatite-coated NiTi implant and the medullary cavity (Figure 7.28c), suggesting the transformation from stem cells in the medullary cavity into new bone directed by the apatite layer on NiTi implant after 26 weeks. Compared with the apatite-coated NiTi implant, the uncoated NiTi implant shows weaker ability in inducing the new bone formation. Although a new bone layer forms between the bone and the uncoated NiTi implant (Figure 7.29a), it is not continuous (Figure 7.29c). Also, the pink osteoid further indicates weaker osteoinduction ability of the uncoated NiTi implant than the apatite-coated one (Figure 7.29b). Figure 7.28d shows a special case that the apatite-coated NiTi implant passes through the cortical bone and the medullary cavity at one side to reach the cortical bone at the other side. Thus, both the apatite-coated surface and the uncoated surface are exposed in the marrow. After 26 weeks, a new bone layer forms and covers the coated surface about 1.5 mm along the coating but only 100 μm on the

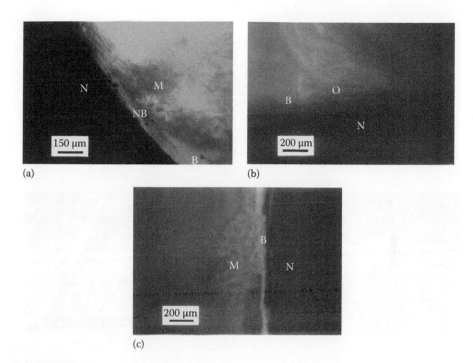

(a) (b)

(c)

FIGURE 7.29
(See color insert.) Nondecalcified histological cross section of uncoated NiTi implant at 26 weeks after implantation. (a) Using a light microscope, showing a new bone layer between bone and NiTi implant. (b) Using a fluorescent light microscope, showing new bone and mineralizing osteoid. (c) Noncontinuous new bone in medullary cavity. N, NiTi; B, bone; M, marrow; O, osteoid.

uncoated surface. As such, the osteoinduction ability of apatite coating on the NiTi implant is clearly illustrated.

SEM was also used to observe the surface and interface morphologies after the implantation for 26 weeks. New lamellar bone forms on both the apatite-coated and the uncoated NiTi implants (Figure 7.30a and b). Collagen fibers bundle together and arrange in the lamellar shape. Osteoblasts are embedded in the matrix and mineralized to new bone. Although similar surface morphology is observed on the apatite-coated NiTi and the uncoated NiTi, they exhibit distinct interfacial views. The apatite-coated NiTi implant integrates with the host bone closely without any gaps (Figure 7.30c); however, there is a clear gap between the uncoated NiTi implant and the host bone (Figure 7.30d). Though the gap might be caused by the specimen preparation, it still suggests the poor bonding between the metallic implant and the host bone without effective mechanical locking or chemical bonding.

After nearing 1 year's implantation (53 weeks), both the apatite-coated and the uncoated NiTi implants are surrounded by new bone, as shown in Figure 7.31a and b. Cortical bone grows along the long axis of both the implants, but much more new bone forms around the apatite-coated one.

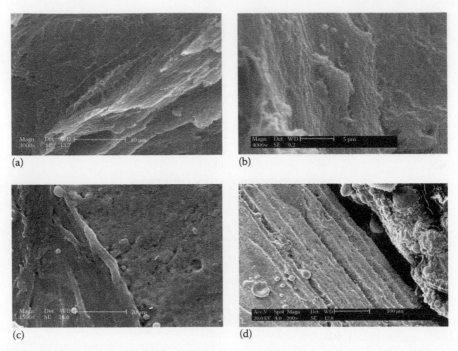

FIGURE 7.30
SEM surface and interface morphologies at 26 weeks after implantation: lamellar bone on the surfaces of apatite-coated NiTi implant (a) and uncoated NiTi implants (b); the interface between apatite-coated NiTi implant and host bone (c), and the gap between uncoated NiTi implant and host bone (d).

Under the fluorescent light microscope, there is no fibrous layer (red connective tissue) on the uncoated NiTi implant as observed after 13 weeks' implantation (Figure 7.26a), but a new layer of green bone contacts with the implant (Figure 7.31e). However, the interface between the uncoated implant and the bone is quite smooth, which might be easily fractured. However, the apatite-coated NiTi implant interlocks with the bone (Figure 7.31d), suggesting stronger bonding. It is worth noting that osteolysis happens at the interface of the uncoated NiTi implant and the bone (Figure 7.31b). Under higher magnification (Figure 7.31c), the bone structure becomes unclear without obvious texture. However, there is no such osteolysis happening around the apatite-coated NiTi implant.

It is well known that the hard tissue implant should have higher strength, bigger Young's modulus, and stronger antifatigue property, because the metallic implants cannot regulate by themselves. But the huge differences between the bone and the implant would cause mismatch and stress concentration. New bone formed around the implant would also change due to the applied forces on the host bone, and consequently, the internal bone structure would also change. In general, bone growth is promoted when tension force

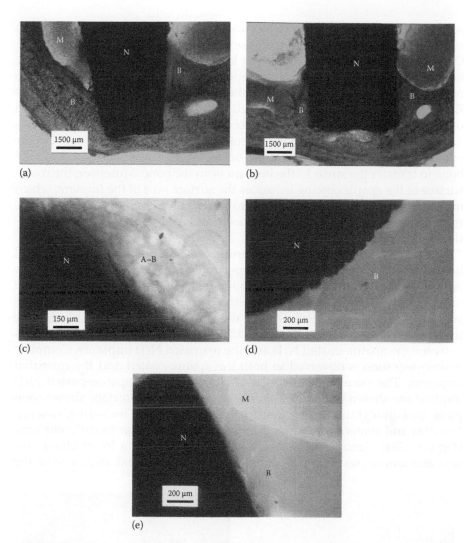

FIGURE 7.31
(See color insert.) Nondecalcified histological cross section of apatite-coated NiTi and uncoated NiTi implants at 53 weeks after implantation. Using a light microscope, showing a mature bone layer encompasses apatite-coated NiTi implant (a), and absorbable bone at one end of uncoated NiTi implant (b and c); using a fluorescent light microscope, showing rough interface between host bone and apatite-coated NiTi implant (d), and smooth interface between host bone and uncoated-NiTi (e). N, NiTi; B, bone; M, marrow; A-B, absorbable bone.

is applied, while bone resorption happens under compression force. When the uncoated NiTi implant is inserted into the femur of the rabbit, a vertical force generated by the normal load on the leg transmits onto the long axis of the implant. Since the implant is in cylindrical shape, the stress concentrates at the end of the implant where it contacts with the cortical bone. The stress accumulated during 1 year finally exceeds the limit that natural bone can sustain; thus, osteolysis and bone resorption take place (Figure 7.31b and c). Consequently, the implant would be loosened or even detached from the host bone, which will cause implantation failure. It is good that there is no osteolysis or bone resorption around the apatite-coated NiTi implant (Figure 7.31a), and the implant integrates well with the new bone, showing good osseointegration (Figure 7.31d). It is obvious that the apatite coating plays an important role in reducing the osteolysis and the bone resorption. On the one hand, the apatite coating acts as a buffer layer between the metal and the bone to transfer the stress in the implant or in the bone. Moreover, the rough surface of the apatite coating increases the surface area of the implant, which may transfer the longitudinal stress to the surrounding bones. As a result, the stress concentration is largely reduced, and the osteolysis or bone resorption does not happen.

After 2 years' implantation, the animals still appeared to be in good health conditions throughout the experimental period. After the animals were sacrificed and the implantation area was exposed, it was found that all the implants were steady, and no signs of inflammation or adverse tissue reaction were observed around the implants. Figure 7.32 reveals the undecalcified histological images. There are no significant differences between the apatite-coated NiTi and the uncoated NiTi implants. Complete osseointegration is observed in both the apatite-coated and the uncoated implants. The decalcified histological images of the apatite-coated NiTi implant are shown in Figure 7.33. The implant–bone interface shows complete osseointegration, and mature bone structures including osseous lamellas and cement line formed at bone reconstruction in different time (Figure 7.33a). Among the osseous lamellas, there are a lot of blood vessels and osteocytes. Figure 7.33b shows the histological structure of the

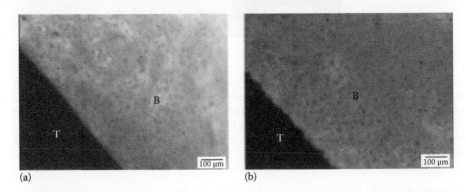

(a)   (b)

FIGURE 7.32
Optical-microscopic images of undecalcified sections (T, NiTi alloy; B, bone). (a) apatite-coated NiTi; (b) uncoated NiTi. (Reprinted from *Mat. Sci. Eng. C*, 27, Li, C.Y., Yang, X.J., Zhang, L.Y., Chen, M.F., Cui, Z.D., In vivo histological evaluation of bioactive NiTi alloy after two years implantation, 122, Copyright 2007, with permission from Elsevier.)

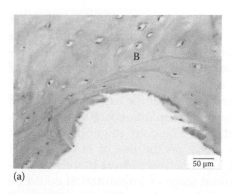
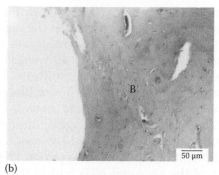

(a)   (b)

FIGURE 7.33
Optical-microscopic images of decalcified sections (B: bone). (a) apatite-coated NiTi; (b) uncoated NiTi. (Reprinted from *Mat. Sci. Eng. C*, 27, Li, C.Y., Yang, X.J., Zhang, L.Y., Chen, M.F., Cui, Z.D., In vivo histological evaluation of bioactive NiTi alloy after two years implantation, 122, Copyright 2007, with permission from Elsevier.)

uncoated NiTi implant. Compared with Figure 7.33a, the osseous lamella structure is less mature. There are some cell concentrated zones, blood vessel, and mature marrow in the bone. SEM was performed to examine the implant–bone interface. Figure 7.34a shows the implant–bone interface of the apatite-coated NiTi. The implant–bone interface is completely osseointegrated including compact bone and osseous lamella structures. Fibrous tissue is absent. Compared with the apatite-coated NiTi, the implant–bone interface of the uncoated NiTi is also completely osseointegrated, but the osseous lamella structures are obscure. This is ascribed to the relatively slow osseointegration at the uncoated NiTi–bone interface during the early

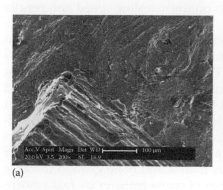
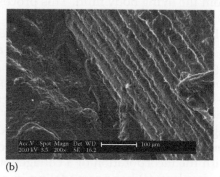

(a)   (b)

FIGURE 7.34
SEM micrographs of the implant–bone interface. (a) apatite-coated NiTi; (b) uncoated NiTi. (Reprinted from *Mat. Sci. Eng. C*, 27, Li, C.Y., Yang, X.J., Zhang, L.Y., Chen, M.F., Cui, Z.D., In vivo histological evaluation of bioactive NiTi alloy after two years implantation, 122, Copyright 2007, with permission from Elsevier.)

implantation stage. As shown in Figure 7.26, there are gaps and fibrous layer at the interface between the bone and the uncoated NiTi implant. On the contrary, osteoblasts proliferate actively on the surface of the apatite-coated NiTi implant after 13 weeks' implantation, and a large amount of new bone directly contact with the host bone. The results described earlier indicate the relatively lower biocompatibility of the uncoated NiTi and the significantly enhanced bioactivity of the apatite-coated NiTi, especially within the early stage of implantation. These are in good agreement with the results reported by another study on the use of HA-coated Ti implants at initial healing period [59]. The relatively lower biocompatibility of the uncoated NiTi within the initial stage of implantation could be ascribed to several factors including nickel ion release from the alloy, cell adhesion ability, and osteogenesis mechanism. Wever et al. [60] demonstrated that nickel release from NiTi alloy in the SBF at 37°C measured by atomic absorption spectrophotometry reduced in time during the measuring period with a high release rate to a level that could not be detected anymore after 10 days immersion. The released nickel ions with a high concentration might induce negative effects on the biocompatibility of the uncoated NiTi alloy during the initial implantation period. Apatite coating on the implant increases cell adhesion [9] and thus accelerates the new bone formation especially during the initial stage of osseointegration [61]. Therefore, the apatite-coated NiTi shows better biocompatibility at the early implantation stage. The function of two-direction osteogenesis of apatite coatings was verified by Weinlaender [62]. Bone matrix deposits not only from the implant bed to the surface of implant but also in the reverse direction simultaneously. In the case of the uncoated titanium implant, however, the bone matrix deposits only from the implant bed to the surface of the implant. The osseointegration of titanium containing implants depends on the surface titanium oxide that absorbs calcium phosphate from the living body fluid. This process occurs at a slow speed and results in prolonged osseointegration [63].

In summary, HA or apatite coating on NiTi alloy plays an important role in new bone formation. HA coating achieves a strong bonding with living bone in a relatively short period. The formation of the bone–implant bonding is schematically shown in Figure 7.35. Briefly, it can be described as follows: (1) initial dissolution of HA coating causes a rise of calcium and phosphate ion concentration in a local environment around the coating; (2) crystal precipitation on HA coating and ion exchange with surrounding tissues; (3) formation of a carbonated calcium phosphate layer with the incorporation of a collagenous matrix and bone growth towards the implant; (4) bone remodeling in area of stress transfer by fast resorption of both carbonated HA in bone mineral and HA coating; (5) further bone ingrowth and remodeling at the bone–implant interface and biological fixation through the bidirectional growth of a bonding layer [64]. Although both the apatite-coated and the uncoated NiTi implants exhibit

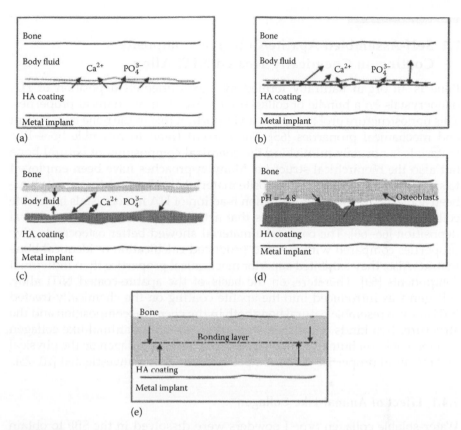

FIGURE 7.35
Schematic diagram of the formation of bone–implant bonding. (a) Partial dissolution of HA coating causing an increase of Ca^{2+} and PO_4^{3-} ion concentrations in the local area around the coating. (b) Precipitation of crystals on HA coating and ion change with surrounding tissues. (c) Formation of a carbonated calcium phosphate layer with the incorporation of a collageneous matrix and bone growth toward the implant. (d) Bone remodeling-osteoblasts resorb normal bone, creating a local pH of −4.8, leading to faster resorption of both carbonated HA in bone and the HA coating. (e) Bidirectional growth and formation of a bonding layer between bone and HA coating through further bone remodeling. (From Sun, L., Berndt, C.C., Gross, K.A., Kucuk, A.: Material fundamentals and clinical performance of plasma-sprayed hydroxyapatite coatings: A review. *J. Biomed. Mat. Res.* 2001. 58. 570. Copyright Wiley-VCH Verlag GmbH & Co. KGaA. Reproduced with permission.)

good biocompatibility after 2 years' implantation in this study, the apatite-coated NiTi implants present better histological performances than the uncoated NiTi implants at the early implantation stage. Both the in vitro cell culture and the in vivo animal models indicate that the chemical treatment is an effective method to biofunctionalize NiTi alloy, and the apatite-coated NiTi alloy is promising used as hard tissue implant in the orthopedic field.

7.4 Self-Assembled Apatite/Collagen Composite Coating on Chemically Treated NiTi Alloy

Bone is an organ naturally formed by *c*-axis oriented deposition of HA nanocrystals on a bundle of collagen with excellent mechanical properties. This nanostructure probably plays an important role in the bone metabolism and mechanical properties [65]. The current trend to resemble bone-like materials is not only mimicking the chemical composition of natural bone but also the hierarchical structure. Many approaches have been employed to synthesize HA/collagen composite materials, and promising results have been achieved by the co-precipitation reaction of HA nanocrystals in soluble collagen macromolecular matrices that act as template and orient mineral deposition [66–68]. The composite material showed better osteoconductive properties compared with HA and produced calcification of identical bone matrix. Also, they exhibited superior mechanical properties than individual components [69]. Therefore, on the basis of the apatite-coated NiTi alloy, collagen was introduced into the apatite coating on the chemically treated NiTi alloy to resemble natural bone both in the chemical composition and the structure. Two kinds of collagen were used, including animal-like collagen and recombinant human-like collagen. The effects of collagen on the physical and chemical properties of the composite coating were investigated [70–73].

7.4.1 Effect of Animal-Like Collagen*

Water-soluble collagen type I powders were dissolved in the SBF to obtain 1.0 g/L collagen simulated body fluid (CSBF). The chemically treated NiTi alloy was immersed in the SBF and CSBF for 3 and 7 days. Figure 7.36 shows the surface morphologies after immersion. After soaking in the SBF for 3 days, some calcium phosphate particles begin to nucleate and grow, but they do not cover the whole surface of the sample. Whereas after soaking in the CSBF for 3 days, the calcium phosphate particles cover the whole surface of the sample uniformly with some colloidal and elliptic particulates that might be ascribed to the addition of collagen. This is demonstrated by the EDS analysis of the elliptic particulates in Figure 7.37. The main elements of

---

* *Source*: With kind permission from Springer Science+Business Media: *Front. Mater. Sci. China*, Comparison of physical characteristics and cell culture test of hydroxyapatite/collagen composite coating on NiTi SMA: Electrochemical deposition and chemically biomimetic growth, 1, 2007, 229, Hu, K., Yang, X., Cai, Y., Cui, Z., and Wei, Q.

    Scrip. Mater., 54, Cai, Y., Liang, C., Zhu, S., Cui, Z., and Yang, X., Formation of bonelike apatite-collagen composite coating on the surface of NiTi shape memory alloy, 89, Copyright 2006, with permission from Elsevier.

    Surf. Coat. Technol., 201, Hu, K., Yang, X.J., Cai, Y.L., Cui, Z.D., and Wei, Q., Preparation of bone-like composite coating using a modified simulated body fluid with high Ca and P concentrations, 1902, Copyright 2006, with permission from Elsevier.

FIGURE 7.36

SEM micrographs of the coatings deposited in the CSBF (a), SBF (b) for 3 and 7 days (a_1, b_1 for 3 days and a_2, b_2 for 7 days). (Reprinted from *Scripta Materialia*, 54, Cai, Y., Liang, C., Zhu, S., Cui, Z., Yang, X., Formation of bonelike apatite-collagen composite coating on the surface of NiTi shape memory alloy, 89, Copyright 2006, with permission from Elsevier.)

elliptic particulates are C and O, of which the weight per cents is more than 50% of the total weight. The ratio of Ca/P is 1.94, which is much higher than that of normal calcium phosphate compounds. It could be concluded that the elliptic particulates might be a hybrid of collagen, calcium, and phosphate ions. When the soaking time is up to 7 days, spherical calcium phosphate particles uniformly spread on the surface (with some small cracks) of the sample soaked in the SBF, whereas a compact calcium phosphate coating with more colloidal and elliptic particulates forms after the biomimetic growth in the CSBF. As the isoelectric point of the collagen used in the experiment is pH 5.9–6.4, which is lower than pH 7.4 of the CSBF, most collagen in the CSBF is negative. The terminal groups ($-COO^-$) might absorb Ca^{2+} easily; consequently, the growth of calcium phosphate particles is promoted, especially in the initial growth period.

Figure 7.38a and b are the high magnification images of Figure 7.36a_2 and b_2 respectively, which reveal the nanoscale character of the coatings. In the former (Figure 7.38a), the calcium phosphate particles grow in the manner of

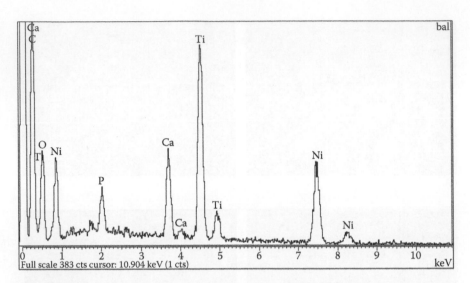

FIGURE 7.37

EDS profile of elliptic particulates on the surface of sample. (Reprinted from *Scripta Materialia*, 54, Cai, Y., Liang, C., Zhu, S., Cui, Z., Yang, X., Formation of bonelike apatite-collagen composite coating on the surface of NiTi shape memory alloy, 89, Copyright 2006, with permission from Elsevier.)

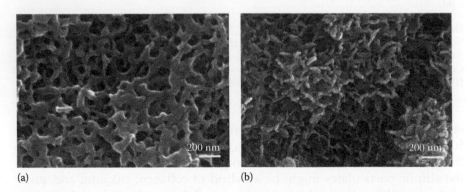

(a)   (b)

FIGURE 7.38

SEM images of the coatings deposited in the CSBF (a), SBF (b) for 7 days. (Reprinted from *Scripta Materialia*, 54, Cai, Y., Liang, C., Zhu, S., Cui, Z., Yang, X., Formation of bonelike apatite-collagen composite coating on the surface of NiTi shape memory alloy, 89, Copyright 2006, with permission from Elsevier.)

lamella that connected with each other to form a network, the structure of which is similar to that of bone [74]. Whereas in the latter, calcium phosphate particles grow into clusters formed by small club-shaped and needle-like substances. The different micromorphologies of coatings after soaking in the CSBF and SBF suggest that collagen might affect the nucleation mode of calcium phosphate particles (and in this way influence the micromorphology).

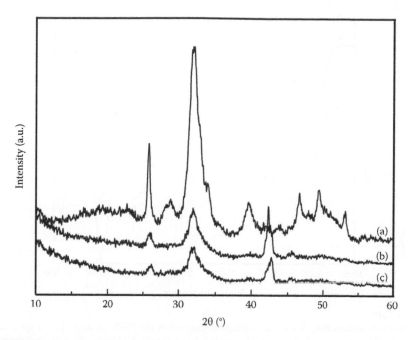

FIGURE 7.39
XRD patterns of human thighbone (a), coatings deposited in the CSBF (b) and SBF (c) for 7 days. (Reprinted from *Scripta Materialia*, 54, Cai, Y., Liang, C., Zhu, S., Cui, Z., Yang, X., Formation of bonelike apatite-collagen composite coating on the surface of NiTi shape memory alloy, 89, Copyright 2006, with permission from Elsevier.)

Figure 7.39 shows the XRD patterns of human thighbone and coatings after the biomimetic growth in the CSBF as well as in the SBF for 7 days. The main phase of human thighbone (a) is apatite, of which the main peaks are (002) around 26° and the integrated peaks of (211), (112), and (300) between 31° and 33°. The intensities, shapes, and positions of the peaks of (b) and (c) are almost the same, and the main peaks are the same as human thighbone. Broad peaks in spectra (b) and (c) indicate poorly crystallized HA with small particle sizes. From XRD patterns shown in Figure 7.39, it can be speculated that bone-like apatite is the main phases of the coatings after the biomimetic growth in the CSBF and SBF.

Figure 7.40a exhibits an IR spectrum of human thighbone that shows the typical peaks of phosphate at the positions of 1095, 1020, 602, and 561 cm^{-1}. These bands are assigned to molecular vibrations of phosphate, of which 1095 and 1020 cm^{-1} are PO_4^{3-} $v3$ bands, and 602 and 561 cm^{-1} are PO_4^{3-} $v4$ bands [75]. It can be observed that the typical amide bands of proteins are present in spectrum (a): C=O stretching vibration at 1651 cm^{-1} is ascribed to amide I band (1600–1700 cm^{-1}), C–N stretching vibration and N–H bending vibration at 1547 cm^{-1} to amide II band (1500–1550 cm^{-1}), and 1242 cm^{-1} for C–N stretching vibration and N–H bending vibration to

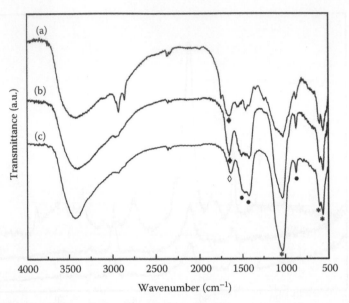

FIGURE 7.40
IR spectra of human thighbone (a), coatings deposited in the CSBF (b) and SBF (c) for 7 days:
*, PO_4^{3-}; •, CO_3^{2-}; ♦, amide I; ◊, water. (Reprinted from *Scripta Materialia*, 54, Cai, Y., Liang, C.,
Zhu, S., Cui, Z., Yang, X., Formation of bonelike apatite-collagen composite coating on the surface of NiTi shape memory alloy, 89, Copyright 2006, with permission from Elsevier.)

amide III band (1200–1300 cm⁻¹) [76]. Moreover, weak bands at 1456, 1418,
and 873 cm⁻¹ are ascribed to CO_3^{2-}, of which 1456 and 1418 cm⁻¹ are carbonate $v3$ bands (1400–1580 and 1350–1550 cm⁻¹), and 872 cm⁻¹ is the carbonate
$v2$ bands (850–890 cm⁻¹). IR spectra of coatings deposited in the CSBF and
SBF are shown in Figure 7.40b and c, respectively. Compared with the peaks
of human thighbone (Figure 7.40a through c), both show the main peaks
of phosphate at the positions of 1030, 604, 563 and 1034, 600, 563 cm⁻¹ and
carbonate bands at 1508, 1420, 872 and 1473, 1420, 872 cm⁻¹, from which it
could be concluded that the coatings deposited in the CSBF and SBF have
the same inorganic components as bone. Furthermore, there is a band at 1650
cm⁻¹ in spectrum (b), which is similar to that of bone and should be ascribed
to amide I band of collagen; whereas the band at 1637 cm⁻¹ in spectrum (c)
should be assigned to absorbed and occluded water in the coating [77]. The
lack of amide II band and amide III band in spectrum (b) might be related to
the amount of collagen added to the CSBF or the fact that the vibrations of
C–N and N–H are more liable to be obstructed by combination with ions in
solution than those of C=O. The IR spectra indicate that the coatings deposited in the CSBF not only have the inorganic component of bone but also
the organic component, i.e., a bone-like apatite–collagen composite coating
has been formed on the surface of NiTi alloy. This coating may be useful in
enhancing the bioactivity of NiTi shape memory alloys.

To shorten the immersion periods while maintaining the proper concentrations and proportions of apatite and collagen, the SBF was modified by increasing the Ca and P ion concentrations and denoted by MSBF. The ion concentrations of the MSBF were identical to Kokubo's SBF [17], but the [Ca^{2+}] and [HPO$_4{}^{2-}$] concentrations were increased fivefold (see Table 7.3). The pH of the MSBF was controlled at 6.13 with 1 M HCl. The water-soluble type I collagen powder was dissolved at a concentration of 0.556 g/L in the MSBF. In the process of biomimetic growth, the pH of the MSBF was adjusted by tris-hydroxymethyl aminomethane [(CH$_2$OH)$_3$CNH$_3$] to rise slowly to 7.15 within 24 h. The chemically treated NiTi samples were kept in the MSBF at 50°C, and the MSBF was changed everyday.

Figure 7.41 shows the surface morphologies of NiTi alloy after the biomimetic growth in the MSBF for 3 days. A uniform coating forms on the surface of the substrate (Figure 7.41a). It could be seen that the coating is composed of many small spherical particles, forming a loose arrangement. The average diameter of the single particle is about 1 μm. There are colloid-like connections among the small particles (Figure 7.41b). The stability of the MSBF containing

TABLE 7.3

Ionic Concentration of SBF and MSBF Solutions, in Comparison with Those of Human Blood Plasma (HBP) (mM)

	Na$^+$	K$^+$	Ca^{2+}	Mg^{2+}	HCO$_3^-$	Cl$^-$	HPO$_4^{2-}$	SO$_4^{2-}$
HBP	142.0	5.0	2.5	1.5	27.0	103.0	1.0	0.5
SBF	142.0	5.0	2.5	1.5	4.2	148.0	1.0	0.5
MSBF	142.0	5.0	12.5	1.5	4.2	159.8	5.0	0.5

Source: Hu, K. et al., *Surf. Coat. Technol.,* 201, 1902, 2006.

(a)   (b)

FIGURE 7.41
SEM micrographs of composite coating grown biomimetically for 3 days in the MSBF, (a) and (b) are at low and high magnifications, respectively. (Reprinted from *Surf. Coat. Technol.,*201, Hu, K., Yang, X.J., Cai, Y.L., Cui, Z.D., Wei, Q., Preparation of bone-like composite coating using a modified simulated body fluid with high Ca and P concentrations, 1902, Copyright 2006, with permission from Elsevier.)

high calcium and phosphate concentration is due to the acidic condition (pH = 6.13) adjusted by HCl. It is known that the relationship between CaP precipitation in solution and CaP coating growth is competitive [22]. If the calcifying solutions reach their supersaturation point rapidly, it will lead to CaP precipitate into the solution and hold CaP coating back from growing on the substrate. The gradual controlled release of trishydroxymethyl aminomethane into the MSBF within 24 h delays the CaP precipitation in solution and ensures that a CaP coating forms simultaneously. On the other hand, the isoelectric point of collagen used in the experiment is pH 5.9–6.4. Accompanying the rising pH, collagen molecules assemble themselves into collagen fibrils and become negatively charged ions whose terminal groups are mostly –COO⁻ that might absorb Ca^{2+}. Therefore, a calcium phosphate–collagen composite coating forms on the substrate.

The IR spectra of the coating and a mechanical mixture of pure HA and collagen are shown in Figures 7.42 and 7.43. Figure 7.42a and b exhibits the typical peaks of collagen. In the spectrum of mechanically mixed powder (Figure 7.42b), the typical bands are observed, such as the C=O stretching vibration at 1654 cm⁻¹ of the amide group I, the N–H deformation vibration at 1521 cm⁻¹ of the amide group II, and N–H deformation vibration at 1284 cm⁻¹ of the amide group III band. Normally, the amide I band is strong, the amide II band is weak, and the amide III band is moderately intense.

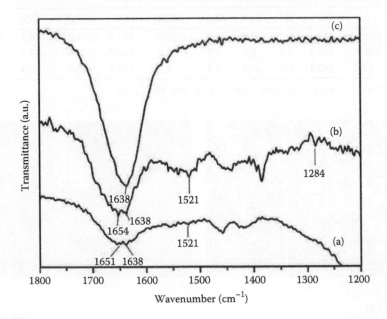

FIGURE 7.42
(a) IR spectra of composite coating, (b) a mechanical mixture of pure HA and collagen and (c) H_2O. (Reprinted from *Surf. Coat. Technol.*,201, Hu, K., Yang, X.J., Cai, Y.L., Cui, Z.D., Wei, Q., Preparation of bone-like composite coating using a modified simulated body fluid with high Ca and P concentrations, 1902, Copyright 2006, with permission from Elsevier.)

The band at 1638 cm^{-1}, which combines with the peak 1654 cm^{-1} of the amide I to form a shoulder structure, is the typical peak of H$_2$O (Figure 7.42c). The presence of H$_2$O might be due to the fact that the specimen is affected by moisture in the experiment. In the spectrum of the coating (Figure 7.42a), it is noticed that there is also a similar shoulder between 1600 and 1700 cm^{-1}. Besides the band at 1638 cm^{-1} for H$_2$O, the other band at 1651 cm^{-1} could be confirmed as being the typical peak of amide I of collagen. It is believed that the "red shift" of amide I is caused by a covalent bond formation with Ca^{2+} ions of HA crystals [78]. The band of amide II at 1521 cm^{-1} is very weak, and the amide III band almost disappears. It is implied that HA crystals grow on the self-assembled collagen fibrils and block the vibration of organic groups such as carboxyl and carbonyl groups [79]. The amide II and III bands not only correspond to the C=O stretching vibration but also include the N–H bending vibration. For this reason, the effects of mineralization on the amides II and III are greater than for amide I. Figure 7.43a shows the typical peaks of inorganic groups in the composite coating. There are PO$_4^{3-}$ $v1$ mode at 964 cm^{-1}, PO$_4^{3-}$ $v2$ mode at 474 cm^{-1}, PO$_4^{3-}$ $v4$ mode at 562 and 604 cm^{-1}, and –OH band at 632 cm^{-1}. These vibration bands could also be seen in the mechanical mixture of pure HA and collagen (Figure 7.43b). The 1037 cm^{-1} and 1096 cm^{-1} bands in Figure 7.43b are PO$_4^{3-}$ $v3$ mode and

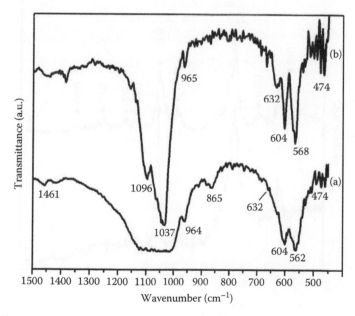

FIGURE 7.43
(a) IR spectra of composite coating, and (b) a mechanical mixture of pure HA and collagen. (Reprinted from *Surf. Coat. Technol.*,201, Hu, K., Yang, X.J., Cai, Y.L., Cui, Z.D., Wei, Q., Preparation of bone-like composite coating using a modified simulated body fluid with high Ca and P concentrations, 1902, Copyright 2006, with permission from Elsevier.)

asymmetric HA, respectively. However, both are very weak in the IR spectrum of composite coating (Figure 7.43a). Furthermore, we could see the CO_3^{2-} $v2$ mode at 865 cm^{-1} and the CO_3^{2-} $v3$ mode at 1461 cm^{-1} in the spectrum of composite coating. According to Chang and Tanaka [76], the CO_3^{2-} $v2$ mode at 865 cm^{-1} indicates the formation of carbonate apatite with B-type substitution in the HA/COL composite coating. This substitution leads to a distortion of the crystallographic lattice, i.e., *a*-axis contraction and *c*-axis extension [80].

The crystallographic structure of the coating was investigated by XRD. Figure 7.44 shows the XRD patterns of the coatings after the biomimetic growth in the MSBF for 4 days as well as in the SBF for 7 days. Both solutions contain identical concentrations of collagen as mentioned earlier. The phases of the coating obtained in the MSBF have broader diffraction peaks, indicating that the coatings are poorly crystallized with small particle size. The main diffraction peaks could be assigned to HA as shown in Figure 7.44a. An acute diffraction peak at 26.2° possibly corresponds to the most intense diffraction line of $CaCO_3$, which might be caused by carbonate substitution in HA/COL composite coating proved by IR results earlier. So it could be speculated that bone-like apatite is the main phase of the coating after the

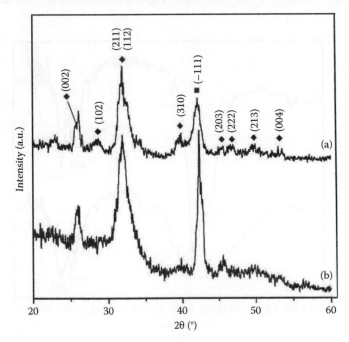

FIGURE 7.44
XRD patterns of coatings after biomimetic growth in the MSBF for 4 days (a), and SBF for 7 days (b). ◆, Hydroxyapatite; ■, NiTi. (Reprinted from *Surf. Coat. Technol.*, 201, Hu, K., Yang, X.J., Cai, Y.L., Cui, Z.D., Wei, Q., Preparation of bone-like composite coating using a modified simulated body fluid with high Ca and P concentrations, 1902, Copyright 2006, with permission from Elsevier.)

biomimetic growth in the MSBF. Compared with the coating formed in the SBF after 7 days (Figure 7.44b), the coating grown in the MSBF for 4 days has a higher degree of crystallinity, while high Ca and P concentrations cause the faster deposition of the coating in the solution.

Transmission electron microscopy (TEM) was employed to study the micromorphology and microstructure between HA and collagen fibrils. A plate-like particle grows in the middle of a collagen fibril of 800 nm in length and 20 nm in diameter (Figure 7.45a). At high magnification (Figure 7.45b), the particle shows a typical crystallographic structure, compared with the amorphous structure of uncalcified collagen fibril. This suggests that the current method could favor the nucleation of calcium phosphate crystal particles on the fibril at the initial stage of mineralization. To investigate the relative orientation between HA crystals and collagen fibril, a mineralized collagen fibril is analyzed by electron diffraction (Figure 7.46). In the electron diffraction pattern, the ring shaped diffraction is ascribed to (211) of HA, and the two pairs of bright diffraction points are ascribed to (002) and (004) of HA. This means that the *c* axis of HA crystals aligns from above-right to bottom-left, i.e., the *c*-axis is the preferential orientation and oriented parallel to the longitudinal direction of the collagen fibril. This special orientation could be observed more directly from high-resolution TEM images. As shown in Figure 7.47, HA crystals grown on the surface of a fully mineralized collagen fibril exhibit overall uniform lattice plane distribution. By virtue of the calculation of lattice plane spacing, the lattice plane of 0.2638 nm corresponds to the HA lattice plane 202 (0.2630 nm),

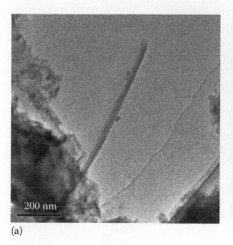

(a)   (b)

FIGURE 7.45
TEM micrographs of composite coating showing HA nuclei formed on the collagen fibril in the initial stage of mineralization at low (a) and high (b) magnifications. (Reprinted from *Surf. Coat. Technol.*,201, Hu, K., Yang, X.J., Cai, Y.L., Cui, Z.D., Wei, Q., Preparation of bone-like composite coating using a modified simulated body fluid with high Ca and P concentrations, 1902, Copyright 2006, with permission from Elsevier.)

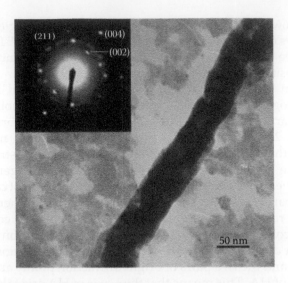

FIGURE 7.46

TEM micrographs of the mineralized collagen fibril in composite coating. The inset shows a selected area electron diffraction (SAED) pattern of the mineralized collagen fibril. (Reprinted from *Surf. Coat. Technol.*,201, Hu, K., Yang, X.J., Cai, Y.L., Cui, Z.D., Wei, Q., Preparation of bone-like composite coating using a modified simulated body fluid with high Ca and P concentrations, 1902, Copyright 2006, with permission from Elsevier.)

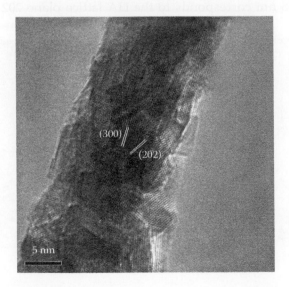

FIGURE 7.47

High-resolution TEM image of fully mineralized collagen fibril. (Reprinted from *Surf. Coat. Technol.*,201, Hu, K., Yang, X.J., Cai, Y.L., Cui, Z.D., Wei, Q., Preparation of bone-like composite coating using a modified simulated body fluid with high Ca and P concentrations, 1902, Copyright 2006, with permission from Elsevier.)

and the plane of 0.2734 nm corresponds to the HA lattice plane 300 (0.2720 nm). The (202) and (300) planes of HA crystal are ascribed to the ⟨010⟩ zone axis, thus the ⟨010⟩ direction is perpendicular to the paper plane. Because the ⟨001⟩ direction of the HA crystal is perpendicular to ⟨010⟩ direction and parallel to (300) lattice plane, the same conclusion is obtained that the c-axis of HA crystal aligns parallel to the longitudinal direction of the collagen fibril. Moreover, the measured lattice plane spacing is shown in Figure 7.45b. By comparison, the (202) plane of HA is basically perpendicular to the longitudinal direction of collagen fibril, which is different from the situation of Figure 7.47. It means that there is an angle between the c-axis of HA and the longitudinal direction of collagen fibril. It is believed that the individual size of crystallite is smaller during the onset of mineralization, and the calcium phosphate crystals nucleate on the collagen fibril oriented at random (Figure 7.45b). With the development of mineralization, the crystallites become larger and the c-axis of HA exhibits the preferential orientation (Figure 7.47). Gradually, the crystallite size along the c-axis of HA even grows beyond the size of the diameter of the collagen fibril. Therefore, the c-axis orients parallel to the longitudinal direction of the fibril in the later period of mineralization by means of structural self-organization between HA and collagen fibril [65]. The TEM results confirm that the relative orientation between HA crystals and collagen fibrils, similar to the nanostructure of calcified natural tissue [81,82], could be obtained in the composite coating by biomimetic growth in the MSBF. This meant that the composite coating has the microstructure similar to bone. Therefore, the coating is likely to possess bone-like properties and reaction at the implant site. The relative orientation maintained in the composite coating may benefit the chemically treated NiTi alloy to achieve better properties in hard tissue replacement.

Another method to shorten the formation period of apatite coating on the substrate is electrochemical deposition (ELD). In our studies, ELD was also employed to prepare HA/COL composite coating and compared with the biomimetic growth (BG) method. The first step of this work was preparing a HA/COL composite coating on chemically treated NiTi alloy by the ELD method. The electrolyte used was the MSBF with high Ca and P ion concentrations to maintain the proper concentrations and proportions of apatite and collagen in the solution. Then, the physical characteristics and in vitro biocompatibility of the two kinds of composite coatings were compared. To combine the two chemically prepared methods, the ELD coated samples were soaked in the MSBF again for further biomimetic growth for several days (it was called EBG method here), and the characteristics of this coating were also studied.

The different surface morphology for three methods is exhibited in Figure 7.48. The surface of ELD coating is relatively flat and smooth, which is composed of some strip-like crystals with 5–7 µm length and 1 µm width (Figure 7.48a). The BG coating exhibits a relatively rough surface composed

(a)   (b)

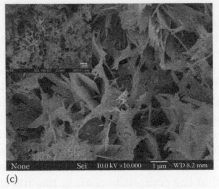

(c)

FIGURE 7.48
SEM micrographs of composite coating (a–c) are ELD (electrochemical deposition), BG (biomimetic growth), EBG (electrochemical deposition followed by biomimetic growth) coatings, respectively, and the insets are the morphology at low magnification. (With kind permission from Springer Science+Business Media: *Front. Mat. Sci. China,* Comparison of physical characteristics and cell culture test of hydroxyapatite/collagen composite coating on NiTi SMA: Electrochemical deposition and chemically biomimetic growth, 1, 2007, 229, Hu, K., Yang, X., Cai, Y., Cui, Z., Wei, Q.)

of plate-like crystals with about 3 µm in length (Figure 7.48b). The EBG coating shows a denser surface. Moreover, its morphology seems to be a transition from ELD coating to BG coating, and the length of EBG plate-like crystals also presents an intervenient size (Figure 7.48c). The XRD patterns of three coatings are shown in Figure 7.49. All the coatings have relatively typical diffraction peaks of apatite of which the peaks could match the major diffraction peaks of HA crystals. Except for the typical peaks of HA, the ELD coating also consists in the diffraction peaks of $Mg(OH)_2$ crystal. The phases of three coatings have broader peaks, indicating that the coatings are of low crystallinity with small particle sizes. The diffraction peak due to the NiTi substrate at 42° disappears in the EBG coating, which

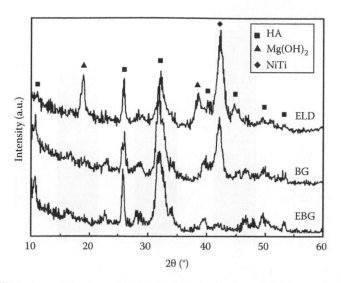

FIGURE 7.49
XRD patterns of composite coatings prepared by the three methods: ELD (electrochemical deposition), BG (biomimetic growth), and EBG (electrochemical deposition followed by biomimetic growth). (With kind permission from Springer Science+Business Media: *Front. Mat. Sci. China*, Comparison of physical characteristics and cell culture test of hydroxyapatite/collagen composite coating on NiTi SMA: Electrochemical deposition and chemically biomimetic growth, 1, 2007, 229, Hu, K., Yang, X., Cai, Y., Cui, Z., Wei, Q.)

proves that EBG coating is thicker than the others. Figure 7.50 shows the x-ray intensity ratios of (002) to (211) planes of the HA in the XRD patterns of three coatings. The standard intensity ratio of (002) to (211) planes is 0.4, according to random hydroxyapatite crystals (JCPDS No. 09-0432). The intensity ratios of the apatite coatings prepared are all larger than 0.4, indicating that the crystallographic lattice of HA along the c-axis is extended. This preferential orientation in the ELD coating is the most evident ($I(002)/I(211) = 0.955$).

The TEM images of EBG coating are shown in Figure 7.51. The collagen fibrils in EBG coating are about 500 nm in length and 20 nm in diameter. In addition, it is found that the collagen fibril is fully mineralized at the high-resolution TEM image (the middle image). The mineralized collagen fibril as shown exhibits overall uniform crystallographic lattice plane distribution compared with the amorphous structure of pure collagen fibril. The inset is the selected area electron diffraction (SAED) pattern of the mineralized collagen fibril. The c-axis of HA crystals aligns from above-left to bottom-right, according to the location of HA (002) and (004) diffraction spots. This means that the c-axis of HA is the preferential orientation in the mineralized process of composite coating and is oriented parallel to the longitudinal direction of the collagen fibril. Figure 7.51c shows a

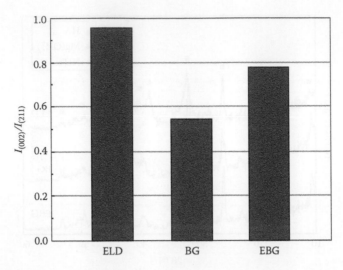

FIGURE 7.50

X-ray intensity ratios of the HA (002) to the (211) plane in the three coatings: ELD (electrochemical deposition), BG (biomimetic growth), and EBG (electrochemical deposition followed by biomimetic growth). (With kind permission from Springer Science+Business Media: *Front. Mat. Sci. China*, Comparison of physical characteristics and cell culture test of hydroxyapatite/collagen composite coating on NiTi SMA: Electrochemical deposition and chemically biomimetic growth, 1, 2007, 229, Hu K, Hu, K., Yang, X., Cai, Y., Cui, Z., Wei, Q.)

schematic diagram of the mineralized collagen fibril illustrating the relationship between HA crystals and collagen fibril as described. Similarly, TEM images of ELD coating are shown in Figure 7.52. The collagen fibrils in ELD coating exhibit shorter length and smaller diameter than those in EBG coating, indicating the difference in the self-organization of collagen fibrils (Figure 7.52a). The high-resolution image (Figure 7.52b) indicates that the mineralization occurs partly on the surface of collagen fibril. Some crystallographic lattice planes with different directions and also some unmineralized areas could be observed.

To study in vitro biocompatibility of HA/COL composite coating, OCT-1 osteoblast-like cells were employed to evaluate cellular responses on the composite coatings. Figure 7.53 shows the morphology of OCT-1 osteoblast-like cells attached on HA/COL composite coatings. On BG coating (Figure 7.53a), the cells are well attached on the coating with filopodia. At low magnification (Figure 7.53b), the cells seem to form bridges mutually and look more globose and compact. On ELD coating (Figure 7.53c), the cells spread dispersedly and are relatively rounded and flat. Moreover, the length of OCT-1 cells spread on ELD coating (about 10 μm) is less than that of the cells on BG coating (about 80 μm) after being cultured for the same period. The cell morphology on EBG coating is similar to that on ELD

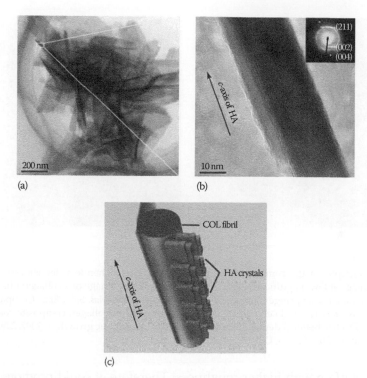

(a)　(b)

(c)

FIGURE 7.51
TEM micrographs of the composite coating for EBG (electrochemical deposition followed by biomimetic growth) method. (a) is the image at low magnification; (b) is a fully mineralized collagen fibril and the inset is its SAED pattern; (c) is a schematic diagram illustrating the relationship between collagen fibril and HA crystals. (With kind permission from Springer Science+Business Media: *Front. Mat. Sci. China*, Comparison of physical characteristics and cell culture test of hydroxyapatite/collagen composite coating on NiTi SMA: Electrochemical deposition and chemically biomimetic growth, 1, 2007, 229, Hu, K., Yang, X., Cai, Y., Cui, Z., Wei, Q.)

coating (Figure 7.53d). Figure 7.54 shows the number of cells after 4 and 8 days of culture. The percentages of cell numbers in 4 and 8 days for the three groups (ELD, BG, EBG) are 37.0%, 39.4%, and 33.9%, respectively, and that of the controls is 43.6%. Compared with the DMEM controls, OCT-1 cells cultured on BG coating show satisfactory results in cell proliferation and its growth trend is similar to that of the control group, both in the absolute value and the percentage. It indicates that the HA/COL composite coating prepared by BG method has better properties for OCT-1 cell adhesion and proliferation.

It is known that the bioactivity of cells is strongly influenced by surface composition, topography, roughness, and porosity [83–85]. In the present study, BG coating is composed of plate-like crystals and presents an open

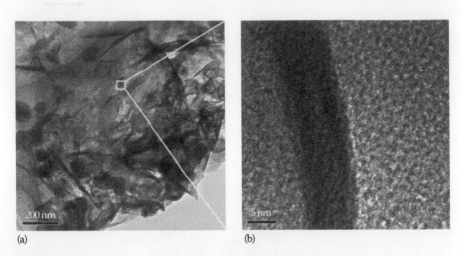

(a)   (b)

FIGURE 7.52
TEM micrographs of the composite coating for ELD (electrochemical deposition) method. (a) is the image at low magnification; (b) is the high-resolution image of a collagen fibril. (With kind permission from Springer Science+Business Media: *Front. Mat. Sci. China*, Comparison of physical characteristics and cell culture test of hydroxyapatite/collagen composite coating on NiTi SMA: Electrochemical deposition and chemically biomimetic growth, 1, 2007, 229, Hu, K., Yang, X., Cai, Y., Cui, Z., Wei, Q.)

porosity surface with higher roughness. Therefore, it could promote physicochemical surface reactions and cellular attachment behavior. On the other hand, some studies demonstrated that the apatite coating prepared by BG method dissolved faster than that prepared by ELD method, and the dissolution ability of calcium phosphate played an essential role in cell adhesion and bone-bonding [86,87]. Thus, it might be another reason of influencing the protein adsorption of HA/COL composite coating. It is found that the cell number on EBG coating is slightly larger than that on ELD coating, which could be ascribed to the formation of a more open surface structure to the cell culture medium after soaking the ELD samples in the MSBF for further BG. Our studies proved that BG is a better method than ELD in preparing an active, cell-friendly HA/COL composite coating in the acid MSBF.

7.4.2 Effect of Recombinant Human-Like Collagen*

For the consideration of bionics and biosafety, the recombinant human-like type I collagen produced in *Escherichia coli* containing human collagen's cDNA transcribed reversely from mRNA should be a good choice to provide

---

* *Source: Mater. Sci. Eng. C*, 29, Yang, X.J., Liang, C.Y., Cai, Y.L., Hu, K., Wei, Q., and Cui, Z.D., Recombinant human-like collagen modulated the growth of nano-hydroxyapatite on NiTi alloy, 25, Copyright 2009, with permission from Elsevier.

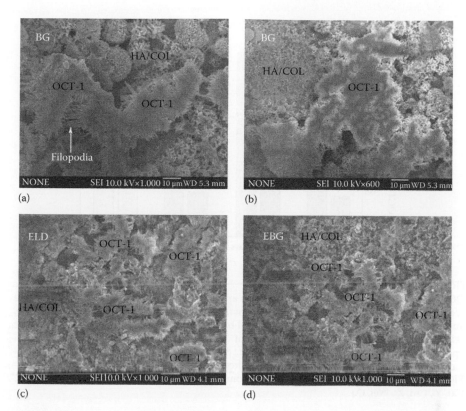

FIGURE 7.53
SEM micrographs of OCT-1 osteoblast-like cells attached on HA/COL composite coatings after 4 days culture: BG (biomimetic growth) coating (a) and (b); ELD (electrochemical deposition) coating (c); EBG (electrochemical deposition followed by biomimetic growth) coating (d). (With kind permission from Springer Science+Business Media: *Front. Mat. Sci. China*, Comparison of physical characteristics and cell culture test of hydroxyapatite/collagen composite coating on NiTi SMA: Electrochemical deposition and chemically biomimetic growth, 1, 2007, 229, Hu Hu, K., Yang, X., Cai, Y., Cui, Z., Wei, Q.)

a reliable source of purified human collagens. Moreover, recombinant human-like collagen has special characteristics that are significantly different from animal collagen, such as processability, virus-free, water solubility, little immunogenic reaction, and so on. This new material has the potential to be used as bone substitution grafts, but its relatively weak strength limits its application as load bearing bone grafts, where titanium and its alloys can exert their advantages of mechanical properties. In this work, recombinant human-like collagen was firstly introduced into the SBF to investigate its modulation effect on the growth of apatite coating and to achieve a combination of both excellent properties of NiTi alloy and the bionic function of bone-like coating.

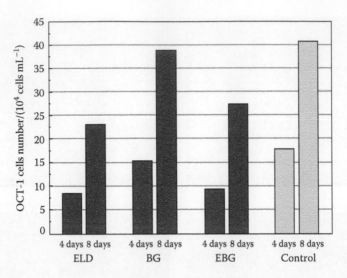

FIGURE 7.54

OCT-1 osteoblast-like cell proliferation after 4 and 8 days of culture on HA/COL composite coatings for the three groups: ELD (electrochemical deposition), BG (biomimetic growth), and EBG (electrochemical deposition followed by biomimetic growth), and the control (The cells were used at an initial concentration of 2.4×10^4/mL). (With kind permission from Springer Science+Business Media: *Front. Mat. Sci. China*, Comparison of physical characteristics and cell culture test of hydroxyapatite/collagen composite coating on NiTi SMA: Electrochemical deposition and chemically biomimetic growth, 1, 2007, 229, Hu, K., Yang, X., Cai, Y., Cui, Z., Wei, Q.)

XRD was used to confirm the crystallographic structure of the coating on NiTi alloy after biomimetic growth in 0.5 g/L 1.5CSBF for 3 days, as shown in Figure 7.55. The diffraction peak of NiTi substrate can be seen around 42°. A broadened peak between 31° and 33° should be the integrated peak of (211), (112), and (300) of HA with small size of crystals; the sharp peak at 26° could be (002) of HA; and other weak peaks are also assigned to the diffraction peaks of HA. In addition, the background of the pattern is relatively complex with many small peaks that couldn't be assigned to any certain calcium phosphates. It might be due to the presence of recombinant human-like collagen in the solution that influences the crystalline behaviors of calcium phosphates in the process of crystal nucleation and growth.

Figure 7.56 shows the surface morphology of the coating on NiTi alloy after soaking in 0.5 g/L 1.5CSBF for 3 days. Different from earlier network structure formed in animal collagen added SBF (Figure 7.36), uniform HA coating comprises numerous nanocrystal forms on the surface of NiTi alloy (Figure 7.56a), and some crystals have aggregated to form small clusters. Image-Pro Plus 5.0, a professional image analysis software, was used to measure the size of crystals in the high magnification of the

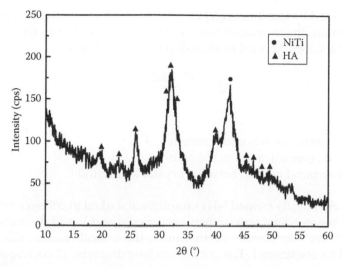

FIGURE 7.55

XRD pattern of the sample after soaking in 0.5 g/L 1.5CSBF for 3 days. (Reprinted from *Mat. Sci. Eng. C*, 29, Yang, X.J., Liang, C.Y., Cai, Y.L., Hu, K., Wei, Q., Cui, Z.D., Recombinant human-like collagen modulated the growth of nano-hydroxyapatite on NiTi alloy, 25, Copyright 2009, with permission from Elsevier.)

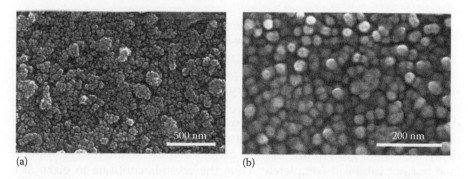

(a)   (b)

FIGURE 7.56

SEM micrographs of the coating on NiTi alloy after soaking in 0.5 g/L 1.5CSBF for 3 days (b is the high-magnification image of a). (Reprinted from *Mat. Sci. Eng. C*, 29, Yang, X.J., Liang, C.Y., Cai, Y.L., Hu, K., Wei, Q., Cui, Z.D., Recombinant human-like collagen modulated the growth of nano-hydroxyapatite on NiTi alloy, 25, Copyright 2009, with permission from Elsevier.)

nanophase coating (Figure 7.56b). The average diameter of the crystals is 45 nm with a standard deviation of 26 nm. The average size of crystals in this coating is relatively smaller compared with that of calcium phosphate particles formed on NiTi alloy in the SBF solution (Figure 7.36) without recombinant human-like collagen, which might be due to the addition of this kind of protein.

As discussed by Combes et al. [88], where adsorbent molecules are present in the solution, the interfacial tension will be altered due to the absorption phenomena. Gibbs' general relationship for absorption is

$$Q = \frac{-C/RT \mathrm{d}\gamma_{CL}}{\mathrm{d}C} \tag{7.13}$$

where
  Q is the quantity of adsorbed protein
  C the solute concentration in solution
  γ_{CL} the interfacial tension between crystal and liquid

When the chemically treated NiTi samples are soaked in collagen containing solution, the interfacial tension between crystal and liquid decreases due to the adsorption phenomena. Furthermore, the critical nucleus size of crystal should be considered. The critical nucleus diameter L^* corresponding to $\Delta G = 0$ is given next in the case of spherical nuclei:

$$L^* = \frac{4\gamma_{CL}}{(1-\alpha+v\alpha)C_c RT \ln C/C_0} \tag{7.14}$$

where
  α is the degree of dissociation of the nuclei
  v is the number of species in the precipitating phase
  C/C_0 the relative supersaturation ratio (C is the ionic product in the
    solution, and C_0 is the solubility product)
  γ_{CL} is the interficial tension at the crystal–liquid interface

As the interfacial tension is decreased as a result of collagen addition, the nucleus critical size would be decreased. At the same time, collagen may delay the crystal growth due to their presence at the solution–crystal interface but not inhibit it completely. Thus the crystals continue to grow at a slower rate, and numerous small nuclei would be stabilized. Consequently, nanophase HA crystals would be formed on the surface of chemically treated NiTi alloy. A similar phenomena could be found when bovine serum albumin (BSA) was added into the solution to study the influence of BSA at various concentrations on the kinetics of nucleation and growth of octacalcium phosphate (OCP). Higher concentrations of BSA led to bigger crystals than in the absence of protein, whereas smaller crystals were observed for lower BSA concentrations [88,89].

The possible structure and chemical interaction of HA and collagen in the coating were also characterized. Figure 7.57 exhibits IR spectrum of the coating on NiTi alloy after soaking in 0.5 g/L 1.5CSBF for 3 days. The bands at the position of 1027, 601, and 561 cm^{-1} are the typical peaks of phosphate. The peak at 1027 cm^{-1} is attributed to the stretching vibration of PO_4^{3-}, whereas

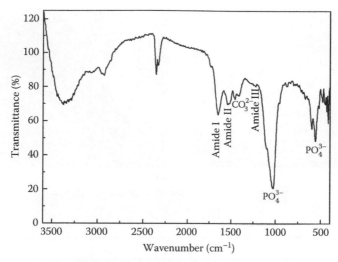

FIGURE 7.57

IR spectrum of the coating on NiTi alloy after soaking in 0.5 g/L 1.5CSBF for 3 days. (Reprinted from *Mat. Sci. Eng. C*, 29, Yang, X.J., Liang, C.Y., Cai, Y.L., Hu, K., Wei, Q., Cui, Z.D., Recombinant human-like collagen modulated the growth of nano-hydroxyapatite on NiTi alloy, 25, Copyright 2009, with permission from Elsevier.)

the other two peaks at 601 and 561 cm^{-1} can be ascribed to the deformation vibration of PO_4^{3-} [75]. Besides, the bands at the positions of 1650, 1524, and 1226 cm^{-1} represent typical amide bands of proteins, respectively: amide I band at 1650 cm^{-1} is ascribed to C=O stretching vibration, amide II band at 1524 cm^{-1} to C–N stretching vibration and N–H bending vibration, and amide III band of C–N stretching vibration and N–H bending vibration at 1226 cm^{-1}. Also, the band at 2930 cm^{-1} should be the typical band of CH_2 or CH_3 of the collagen [76]. Compared with the IR spectrum of pure recombinant human-like collagen in Ref. [90] that amide I band is at the position of 1658 cm^{-1}, red shift of amide I band to 1650 cm^{-1} is observed for the bonding between Ca^{2+} ions and C=O bands. This phenomenon suggests that the carbonyl groups on the surface of collagen molecules are the nucleation sites of calcium phosphate. Additionally, 1457 and 1414 cm^{-1} are carbonate $v3$ bands, which indicates the formation of carbonated apatite on NiTi alloy.

TEM analysis can provide us more detailed information about the micromorphology of coating and the orientation relationship between HA and recombinant human-like collagen. As can be seen in Figure 7.58a, bundles of mineralized fibrils are dispersed in the coating, and the length of fibrils is less than 100 nm, which is shorter than natural type I collagen fiber. It is because the molecular weight of recombinant human-like collagen used in the experiment is about 90,000 Da, much smaller than that of natural type I collagen (300,000 Da). EDX analysis of the fibril in Figure 7.58c presents the elements in the fibril, including C, O, Ca, P, Mg, N, S, and Cu. Cu element belongs to

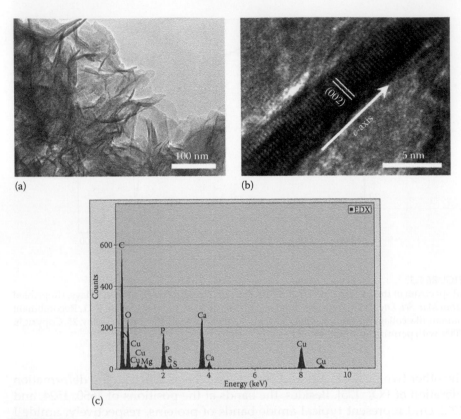

FIGURE 7.58
TEM images of the coating on NiTi alloy after soaking in 0.5 g/L 1.5CSBF for 3 days (a) TEM image of the coating on NiTi alloy; (b) HRTEM image of a mineralized collagen fibril; (c) EDX analysis of a mineralized collagen fibril). (Reprinted from *Mat. Sci. Eng. C*, 29, Yang, X.J., Liang, C.Y., Cai, Y.L., Hu, K., Wei, Q., Cui, Z.D., Recombinant human-like collagen modulated the growth of nano-hydroxyapatite on NiTi alloy, 25, Copyright 2009, with permission from Elsevier.)

the copper grids used as the supporting membrane. Ca, P, and O elements should be assigned to HA; a little amount of C, O, N, and S elements might be attributed to recombinant human-like collagen; and the presence of Mg element might be due to the substitution of Ca element in the formation process of HA. The high intensity of C might be partly attributed to collagen, partly due to the contaminants and carbonates in the calcium phosphate phase.

Figure 7.58b shows a HRTEM image of a collagen fibril after the mineralization. The diameter of the mineralization collagen fibril is about 5 nm, and numerous parallel lattice planes are aligned vertically to the longitudinal direction of the collagen fiber. The average lattice plane spacing of 0.34395 nm corresponds to the HA lattice plane (002) (0.3440 nm), and the orientation of (002), i.e., the *c*-axis of HA, should be vertical to the lattice plane and parallel to the longitudinal direction of the collagen fibril, which is the same as that in the natural bone.

The distribution of collagen in the coating directly influences the properties of the uncoated NiTi alloy; thus, it is examined by laser scanning confocal microscope (LSCM). The sample after biomimetic growth in 0.5 g/L 1.5CSBF for 3 days was washed with phosphate buffered saline (PBS), and then 50 μL of 10% BSA was added to block nonspecific antigen and incubated for 20 min at room temperature. About 1.5 mL of 1:200 rabbit anti-human collagen I was added as the primary antibody and incubated at 4°C for 12 h. After washing with PBS, 750 μL of 1:10 rhodamine (TRITC) labeled goat anti-rabbit antibody was added as the secondary antibody and then incubated at 4°C for 10 h. The distribution of collagen in the coating was observed at the excitation wavelength of 557 nm and the emission wavelength of 576 nm after the sample was further rinsed with PBS. As can be seen in Figure 7.59, the coating exhibits uniform red color, indicating the uniform distribution of collagen in the whole coating because the collagen was labeled by red colored TRITC.

The amount of collagen in the coating was measured using the Bradford method. The coating after biomimetic growth in 0.5 g/L 1.5CSBF for 3 days was detached from the substrate and dissolved in 0.1 M HCl. Bradford protein assay was used following the standard protocol. BSA was used to prepare the standard curve of protein concentration, and the collagen concentration was calculated according the measured OD_{595}. The measured results are shown in Figure 7.60. The collagen concentration in 0.1 M HCl is 10–20 μg/mL and 0.3–0.6 wt.% in the coating.

FIGURE 7.59
(See color insert.) Collagen distribution in the coating on NiTi alloy after soaking in 0.5 g/L 1.5CSBF for 3 days.

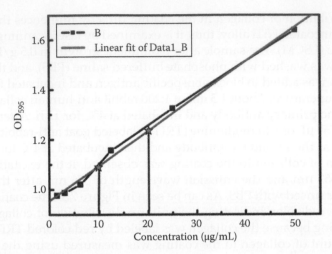

FIGURE 7.60
Collagen concentration in the coating on NiTi alloy after soaking in 0.5 g/L 1.5CSBF for 3 days.

In summary, apatite/collagen composite coating can be self-assembled on the chemically treated NiTi alloy. Collagen modulates the growth of HA crystals by means of influencing the nucleus critical size and the growth rate of crystals. The *c*-axis of HA is parallel to the longitudinal direction of the collagen fibril. These preferentially oriented HA nanocrystals on collagen fibril would give strength, while flexible collagen fiber itself gives ductility to implants [91]. Consequently, the hybrid of nano-HA crystals and recombinant human-like collagen fibers on NiTi alloy might greatly improve the mechanical properties of the coating, especially the toughness of the coating. Moreover, the interface between the coating and the substrate would become more ductile, and the whole implant material would be more bending resistant when used as fracture fixing materials in the orthopedic field to exhibit its shape memory effect and bone conduction function.

7.5 Conclusions

To make NiTi shape memory alloy bioactive and to use it as hard tissue implant, chemical treatment is employed to biofunctionalize NiTi alloy. The NiTi alloy is chemically treated by acid, alkali, and precalcification processes. Then biological apatite or apatite/collagen composite coatings are grown in simulated body fluids (SBF) without or with collagen. After the acid treatment, a passive titanium oxide layer forms on the surface of NiTi alloy.

An amorphous and fine crystalline sodium titanate layer produced in the alkali solution provides numerous nucleation sites for apatite crystals by absorbing Ca^{2+} and PO_4^{3-} ions on Ti-OH groups in the SBF. More Ca^{2+} and PO_4^{3-} ions are absorbed on the microscale network during the precalcification for accelerating the formation of apatite layer.

A biological apatite layer is grown in the SBF solution on the chemically treated NiTi alloy. The apatite-coated NiTi alloy facilitates the proliferation of 3T3 fibroblast cells and exhibits a better cytocompatibility than mechanically polished one. Direct bony coverage is observed on apatite coated NiTi implants after 6 weeks of implantation, the nature of osteo-bonding forms after 13 weeks, and a complete implant–bone interfacial osseointegration is achieved after 2 years.

To make the coating resemble natural bone more closely, apatite/collagen composite coating is self-assembled on the chemically treated NiTi alloy in the collagen containing SBF (CSBF). Apatite crystals and collagen fibrils coprecipitate in the CSBF with the *c*-axis of apatite crystals parallel to the longitudinal direction of the mineralized collagen fibrils. Collagen modulates the growth of apatite crystals by means of influencing the nucleus critical size and the growth rate of crystals. Moreover, faster cell proliferation takes place on the composite coating grown in the CSBF than that prepared by electrochemical deposition.

The biofunctionalization of NiTi alloy by the chemical treatment improves its biocompatibility and promotes its osseointegration to host bone, which makes the coated NiTi alloy more applicable as bone implant material.

Acknowledgments

The authors acknowledge the financial support from the National Natural Science Foundation of China (Project No. 59971032 and 50471048), Scientific and Technological Project of Tianjin (Project No. 043186311), and the Specialized Research Fund for the Doctoral Program of Higher Education (Project No. 20040056016).

References

1. Shabalovskaya SA. Surface, corrosion and biocompatibility aspects of nitinol as an implant material. *Bio-Medical Materials and Engineering* 2002;12:69.
2. Ryhänen J. Biocompatibility of nitinol. *Minimally Invasive Therapy and Allied Technologies* 2000;9:99.

3. Es-Souni M, Es-Souni M, Fischer-Brandies H. Assessing the biocompatibility of NiTi shape memory alloys used for medical applications. *Analytical and Bioanalytical Chemistry* 2005;381:557.

4. Nishiguchi S, Nakamura T, Kobayashi M, Kim HM, Miyaji F, Kokubo T. The effect of heat treatment on bone-bonding ability of alkali-treated titanium. *Biomaterials* 1999;20:491.

5. Yeung KWK, Chan RYL, Lam KO, Wu SL, Liu XM, Chung CY, Chu PK, Lu WW, Chan D, Luk KDK, Cheung KMC. In vitro and in vivo characterization of novel plasma treated nickel titanium shape memory alloy for orthopedic implantation. *Surface and Coatings Technology* 2007;202:1247.

6. Campbell AA. Bioceramics for implant coatings. *Materials Today* 2003;6:26.

7. Geesink RGT. Experimental and clinical experience with hydroxyapatitie-coated hip implants. *Orthopedics* 1989;12:1239.

8. Furlong RJ, Osborn JF. Fixation of hip prostheses by hydroxyapatite ceramic coatings. *Journal of Bone and Joint Surgery—Series B* 1991;73:741.

9. Bogdanski D, Epple M, Esenwein SA, Muhr G, Petzoldt V, Prymak O, Weinert K, Köller M. Biocompatibility of calcium phosphate-coated and of geometrically structured nickel-titanium (NiTi) by in vitro testing methods. *Materials Science and Engineering A* 2004;378:527.

10. Prymak O, Bogdansk D, Esenwein SA, Köller M, Epple M. NiTi shape memory alloys coated with calcium phosphate by plasma-spraying. Chemical and biological properties. *Materialwissenschaft und Werkstofftechnik* 2004;35:346.

11. Pichugin VF, Surmenev RA, Shesterikov EV, Ryabtseva MA, Eshenko EV, Tverdokhlebov SI, Prymak O, Epple M. The preparation of calcium phosphate coatings on titanium and nickel–titanium by rf-magnetron-sputtered deposition: Composition, structure and micromechanical properties. *Surface and Coatings Technology* 2008;202:3913.

12. Zhang JX, Guan RF, Zhang XP. Synthesis and characterization of sol–gel hydroxyapatite coatings deposited on porous NiTi alloys. *Journal of Alloys and Compounds* 2011;509:4643.

13. Lobo A, Otubo J, Matsushima J, Corat E. Rapid obtaining of nano-hydroxyapatite bioactive films on NiTi shape memory alloy by electrodeposition process. *Journal of Materials Engineering and Performance* 2011;20:793.

14. Chu CL, Pu YP, Yin LH, Chung CY, Yeung KWK, Chu PK. Biomimetic deposition process of an apatite coating on NiTi shape memory alloy. *Materials Letters* 2006;60:3002.

15. Chu CL, Hu T, Yin LH, Pu YP, Dong YS, Lin PH, Chung CY, Yeung KWK, Chu PK. Microstructural characteristics and biocompatibility of a Type-B carbonated hydroxyapatite coating deposited on NiTi shape memory alloy. *Bio-Medical Materials and Engineering* 2009;19:401.

16. Gu YW, Tay BY, Lim CS, Yong MS. Nanocrystallite apatite formation and its growth kinetics on chemically treated porous NiTi. *Nanotechnology* 2006;17:2212.

17. Kokubo T, Miyaji F, Kim HM, Nakamura T. Spontaneous formation of bonelike apatite layer on chemically treated titanium metals. *Journal of the American Ceramic Society* 1996;79:1127.

18. Wen HB, De Wijn JR, Liu Q, De Groot K, Cui FZ. A simple method to prepare calcium phosphate coatings on Ti6Al4V. *Journal of Materials Science: Materials in Medicine* 1997;8:765.

19. Wen HB, Liu Q, De Wijn JR, De Groot K, Cui FZ. Preparation of bioactive microporous titanium surface by a new two-step chemical treatment. *Journal of Materials Science: Materials in Medicine* 1998;9:121.
20. Miyazaki T, Kim HM, Kokubo T, Ohtsuki C, Kato H, Nakamura T. Mechanism of bonelike apatite formation on bioactive tantalum metal in a simulated body fluid. *Biomaterials* 2002;23:827.
21. Kokubo T, Kim HM, Kawashita M, Nakamura T. Bioactive metals: Preparation and properties. *Journal of Materials Science: Materials in Medicine* 2004;15:99.
22. Barrere F, Van Blitterswijk CA, De Groot K, Layrolle P. Influence of ionic strength and carbonate on the Ca-P coating formation from SBF×5 solution. *Biomaterials* 2002;23:1921.
23. De Andrade MC, Filgueiras MRT, Ogasawara T. Nucleation and growth of hydroxyapatite on titanium pretreated in NaOH solution: Experiments and thermodynamic explanation. *Journal of Biomedical Materials Research* 1999;46:441.
24. Kokubo T, Kim HM, Kawashita M. Novel bioactive materials with different mechanical properties. *Biomaterials* 2003;24:2161.
25. Ryhänen J, Kallioinen M, Tuukkanen J, Lehenkari P, Junila J, Niemelä E, Sandvik P, Serlo W. Bone modeling and cell-material interface responses induced by nickel-titanium shape memory alloy after periosteal implantation. *Biomaterials* 1999;20:1309.
26. Armitage DA, Parker TL, Grant DM. Biocompatibility and hemocompatibility of surface-modified NiTi alloys. *Journal of Biomedical Materials Research—Part A* 2003;66:129.
27. Chen MF, Yang XJ, Liu Y, Zhu SL, Cui ZD, Man HC. Study on the formation of an apatite layer on NiTi shape memory alloy using a chemical treatment method. *Surface and Coatings Technology* 2003;173:229.
28. Liu SX, Wang DH, Pan CH. *X-Ray Photoelectron Spectroscopy*. Peking, China: Science Publishing House, 1988.
29. Takadama H, Kim HM, Kokubo T, Nakamura T. XPS study of the process of apatite formation on bioactive Ti-6Al-4V alloy in simulated body fluid. *Science and Technology of Advanced Materials* 2001;2:389.
30. Kokubo T, Takadama H. How useful is SBF in predicting in vivo bone bioactivity? *Biomaterials* 2006;27:2907.
31. Yang XJ, Hu RX, Zhu SL, Li CY, Chen MF, Zhang LY, Cui ZD. Accelerating the formation of a calcium phosphate layer on NiTi alloy by chemical treatments. *Scripta Materialia* 2006;54:1457.
32. Jonášová L, Müller FA, Helebrant A, Strnad J, Greil P. Biomimetic apatite formation on chemically treated titanium. *Biomaterials* 2004;25:1187.
33. Milella E, Cosentino F, Licciulli A, Massaro C. Preparation and characterisation of titania/hydroxyapatite composite coatings obtained by sol-gel process. *Biomaterials* 2001;22:1425.
34. Kačiulis S, Mattogno G, Pandolfi L, Cavalli M, Gnappi G, Montenero A. XPS study of apatite-based coatings prepared by sol-gel technique. *Applied Surface Science* 1999;151:1.
35. Lu X, Leng Y. Theoretical analysis of calcium phosphate precipitation in simulated body fluid. *Biomaterials* 2005;26:1097.
36. Kim HM, Miyaji F, Kokubo T, Nakamura T. Effect of heat treatment on apatite-forming ability of Ti metal induced by alkali treatment. *Journal of Materials Science: Materials in Medicine* 1997;8:341.

37. Lee TM, Chang E, Yang CY. Attachment and proliferation of neonatal rat calvarial osteoblasts on Ti6Al4V: Effect of surface chemistries of the alloy. *Biomaterials* 2004;25:23.
38. Ponsonnet L, Reybier K, Jaffrezic N, Comte V, Lagneau C, Lissac M, Martelet C. Relationship between surface properties (roughness, wettability) of titanium and titanium alloys and cell behaviour. *Materials Science and Engineering C* 2003;23:551.
39. Ryhänen J, Niemi E, Serlo W, Niemelä E, Sandvik P, Pernu H, Salo T. Biocompatibility of nickel-titanium shape memory metal and its corrosion behavior in human cell cultures. *Journal of Biomedical Materials Research* 1997;35:451.
40. Cui ZD, Chen MF, Zhang LY, Hu RX, Zhu SL, Yang XJ. Improving the biocompatibility of NiTi alloy by chemical treatments: An in vitro evaluation in 3T3 human fibroblast cell. *Materials Science and Engineering C* 2008;28:1117.
41. Bogdanski D, Koller M, Muller D, Muhr G, Bram M, Buchkremer HP, Stover D, Choi J, Epple M. Easy assessment of the biocompatibility of Ni-Ti alloys by in vitro cell culture experiments on a functionally graded Ni-NiTi-Ti material. *Biomaterials* 2002;23:4549.
42. Manceur A, Chellat F, Merhi Y, Chumlyakov Y, Yahia L. In vitro cytotoxicity evaluation of a 50.8% NiTi single crystal. *Journal of Biomedical Materials Research—Part A* 2003;67:641.
43. Huang HH. In situ surface electrochemical characterizations of Ti and Ti-6Al-4V alloy cultured with osteoblast-like cells. *Biochemical and Biophysical Research Communications* 2004;314:787.
44. Bogdanski D, Esenwein SA, Prymak O, Epple M, Muhr G, Köller M. Inhibition of PMN apoptosis after adherence to dip-coated calcium phosphate surfaces on a NiTi shape memory alloy. *Biomaterials* 2004;25:4627.
45. Maitz MF, Pham MT, Matz W, Reuther H, Steiner G. Promoted calcium-phosphate precipitation from solution on titanium for improved biocompatibility by ion implantation. *Surface and Coatings Technology* 2002;158–159:151.
46. Choi J, Bogdanski D, Köller M, Esenwein SA, Müller D, Muhr G, Epple M. Calcium phosphate coating of nickel-titanium shape-memory alloys. Coating procedure and adherence of leukocytes and platelets. *Biomaterials* 2003;24:3689.
47. Kapanen A, Ilvesaro J, Danilov A, Ryhänen J, Lehenkari P, Tuukkanen J. Behaviour of Nitinol in osteoblast-like ROS-17 cell cultures. *Biomaterials* 2002;23:645.
48. Redey SA, Nardin M, Bernache-Assolant D, Rey C, Delannoy P, Sedel L, Marie PJ. Behavior of human osteoblastic cells on stoichiometric hydroxyapatite and type A carbonate apatite: Role of surface energy. *Journal of Biomedical Materials Research* 2000;50:353.
49. Yang B, Uchida M, Kim HM, Zhang X, Kokubo T. Preparation of bioactive titanium metal via anodic oxidation treatment. *Biomaterials* 2004;25:1003.
50. Chen MF, Yang XJ, Hu RX, Cui ZD, Man HC. Bioactive NiTi shape memory alloy used as bone bonding implants. *Materials Science and Engineering C* 2004;24:497.
51. Li CY, Yang XJ, Zhang LY, Chen MF, Cui ZD. In vivo histological evaluation of bioactive NiTi alloy after two years implantation. *Materials Science and Engineering C* 2007;27:122.

52. Berger-Gorbet M, Broxup B, Rivard C, Yahia LH. Biocompatibility testing of NiTi screws using immunohistochemistry on sections containing metallic implants. *Journal of Biomedical Materials Research* 1996;32:243.

53. Takeshita F, Takata H, Ayukawa Y, Suetsugu T. Histomorphometric analysis of the response of rat tibiae to shape memory alloy (nitinol). *Biomaterials* 1997;18:21.

54. Kapanen A, Ryhänen J, Danilov A, Tuukkanen J. Effect of nickel-titanium shape memory metal alloy on bone formation. *Biomaterials* 2001;22:2475.

55. Du C, Meijer GJ, Van de Valk C, Haan RE, Bezemer JM, Hesseling SC, Cui FZ, De Groot K, Layrolle P. Bone growth in biomimetic apatite coated porous Polyactive® 1000PEGT70PBT30 implants. *Biomaterials* 2002;23:4649.

56. Yuan H, Li Y, De Bruijn J, De Groot K, Zhang X. Tissue responses of calcium phosphate cement: A study in dogs. *Biomaterials* 2000;21:1283.

57. Ferris DM, Moodie GD, Dimond PM, Giorani CWD, Ehrlich MG, Valentini RF. RGD-coated titanium implants stimulate increased bone formation in vivo. *Biomaterials* 1999;20:2323.

58. Yan WQ, Nakamura T, Kawanabe K, Nishigochi S, Oka M, Kokubo T. Apatite layer-coated titanium for use as bone bonding implants. *Biomaterials* 1997;18:1185.

59. Ong JL, Bessho K, Cavin R, Carnes DL. Bone response to radio frequency sputtered calcium phosphate implants and titanium implants in vivo. *Journal of Biomedical Materials Research* 2002;59:184.

60. Wever DJ, Veldhuizen AG, De Vries J, Busscher HJ, Uges DRA, Van Horn JR. Electrochemical and surface characterization of a nickel-titanium alloy. *Biomaterials* 1998;19:761.

61. Giavaresi G, Fini M, Cigada A, Chiesa R, Rondelli G, Rimondini L, Torricelli P, Aldini NN, Giardino R. Mechanical and histomorphometric evaluations of titanium implants with different surface treatments inserted in sheep cortical bone. *Biomaterials* 2003;24:1583.

62. Weinlaender M. Bone growth around dental implants. *Dental Clinics of North America* 1991;35:585.

63. De Lange G, De Putter C. Structure of the bone interface to dental implants in vivo. *The Journal of Oral Implantology* 1993;19:123.

64. Sun L, Berndt CC, Gross KA, Kucuk A. Material fundamentals and clinical performance of plasma-sprayed hydroxyapatite coatings: A review. *Journal of Biomedical Materials Research* 2001;58:570.

65. Kikuchi M, Ikoma T, Itoh S, Matsumoto HN, Koyama Y, Takakuda K, Shinomiya K, Tanaka J. Biomimetic synthesis of bone-like nanocomposites using the self-organization mechanism of hydroxyapatite and collagen. *Composites Science and Technology* 2004;64:819.

66. Chang MC, Ikoma T, Kikuchi M, Tanaka J. The cross-linkage effect of hydroxyapatite/collagen nanocomposites on a self-organization phenomenon. *Journal of Materials Science: Materials in Medicine* 2002;13:993.

67. Roveri N, Falini G, Sidoti MC, Tampieri A, Landi E, Sandri M, Parma B. Biologically inspired growth of hydroxyapatite nanocrystals inside self-assembled collagen fibers. *Materials Science and Engineering C* 2003;23:441.

68. Wang Y, Yang C, Chen X, Zhao N. Biomimetic formation of hydroxyapatite/collagen matrix composite. *Advanced Engineering Materials* 2006;8:97.

69. Wahl DA, Czernuszka JT. Collagen-hydroxyapatite composites for hard tissue repair. *European Cells and Materials* 2006;11:43.

70. Cai Y, Liang C, Zhu S, Cui Z, Yang X. Formation of bonelike apatite-collagen composite coating on the surface of NiTi shape memory alloy. *Scripta Materialia* 2006;54:89.

71. Hu K, Yang XJ, Cai YL, Cui ZD, Wei Q. Preparation of bone-like composite coating using a modified simulated body fluid with high Ca and P concentrations. *Surface and Coatings Technology* 2006;201:1902.

72. Hu K, Yang X, Cai Y, Cui Z, Wei Q. Comparison of physical characteristics and cell culture test of hydroxyapatite/collagen composite coating on NiTi SMA: Electrochemical deposition and chemically biomimetic growth. *Frontiers of Materials Science in China* 2007;1:229.

73. Yang XJ, Liang CY, Cai YL, Hu K, Wei Q, Cui ZD. Recombinant human-like collagen modulated the growth of nano-hydroxyapatite on NiTi alloy. *Materials Science and Engineering C* 2009;29:25.

74. Zysset PK, Edward Guo X, Edward Hoffler C, Moore KE, Goldstein SA. Elastic modulus and hardness of cortical and trabecular bone lamellae measured by nanoindentation in the human femur. *Journal of Biomechanics* 1999;32:1005.

75. Zhang W, Liao SS, Cui FZ. Hierarchical self-assembly of nano-fibrils in mineralized collagen. *Chemistry of Materials* 2003;15:3221.

76. Chang MC, Tanaka J. FT-IR study for hydroxyapatite/collagen nanocomposite cross-linked by glutaraldehyde. *Biomaterials* 2002;23:4811.

77. Babini GN, Tampieri A. Towards biologically inspired materials. *British Ceramic Transactions* 2004;103:101.

78. Chang MC, Ko CC, Douglas WH. Preparation of hydroxyapatite-gelatin nanocomposite. *Biomaterials* 2003;24:2853.

79. Zhang W, Huang ZL, Liao SS, Cui FZ. Nucleation sites of calcium phosphate crystals during collagen mineralization. *Journal of the American Ceramic Society* 2003;86:1052.

80. Barralet J, Best S, Bonfield W. Carbonate substitution in precipitated hydroxyapatite: An investigation into the effects of reaction temperature and bicarbonate ion concentration. *Journal of Biomedical Materials Research* 1998;41:79.

81. Traub W, Arad T, Weiner S. Three-dimensional ordered distribution of crystals in turkey tendon collagen fibers. *Proceedings of the National Academy of Sciences of the United States of America* 1989;86:9822.

82. Sasaki N, Sudoh Y. X-ray pole figure analysis of apatite crystals and collagen molecules in bone. *Calcified Tissue International* 1997;60:361.

83. Deligianni DD, Katsala ND, Koutsoukos PG, Missirlis YF. Effect of surface roughness of hydroxyapatite on human bone marrow cell adhesion, proliferation, differentiation and detachment strength. *Biomaterials* 2001;22:87.

84. Montanaro L, Arciola CR, Campoccia D, Cervellati M. In vitro effects on MG63 osteoblast-like cells following contact with two roughness-differing fluorohydroxyapatite-coated titanium alloys. *Biomaterials* 2002;23:3651.

85. Borsari V, Giavaresi G, Fini M, Torricelli P, Salito A, Chiesa R, Chiusoli L, Volpert A, Rimondini L, Giardino R. Physical characterization of different-roughness titanium surfaces, with and without hydroxyapatite coating, and their effect on human osteoblast-like cells. *Journal of Biomedical Materials Research—Part B Applied Biomaterials* 2005;75:359.

86. Mayr-Wohlfart U, Fiedler J, Gnther KP, Puhl W, Kessler S. Proliferation and differentiation rates of a human osteoblast-like cell line (SaOS-2) in contact with different bone substitute materials. *Journal of Biomedical Materials Research* 2001;57:132.

87. Wang J, Layrolle P, Stigter M, De Groot K. Biomimetic and electrolytic calcium phosphate coatings on titanium alloy: Physicochemical characteristics and cell attachment. *Biomaterials* 2004;25:583.
88. Combes C, Rey C. Adsorption of proteins and calcium phosphate materials bioactivity. *Biomaterials* 2002;23:2817.
89. Combes C, Rey C, Freche M. In vitro crystallization of octacalcium phosphate on type I collagen: Influence of serum albumin. *Journal of Materials Science: Materials in Medicine* 1999;10:153.
90. Zhai Y, Cui FZ. Recombinant human-like collagen directed growth of hydroxyapatite nanocrystals. *Journal of Crystal Growth* 2006;291:202.
91. Rhee SH, Suetsugu Y, Tanaka J. Biomimetic configurational arrays of hydroxyapatite nanocrystals on bio-organics. *Biomaterials* 2001;22:2843.

8

Investigation and Application of HA Composite Coating on the Ti Alloy

Wei Qiang, Zhao Jin, Zhang Lijun, Liu Shimin, and Liang Yanqin

CONTENTS

Titanium (Ti)-based alloys have been widely used in biomedical fields for many years, especially commercially pure (CP) Ti, Ti-6Al-4V alloys, and TiNi alloys [1], due to their excellent mechanical property and biocompatibility. However, extensive medical application of Ti alloy has been hindered for a long time owing to its lack of surface biocompatibility [2]. Therefore, it is crucial to improve surface biocompatibility and anticorrosion properties of titanium and its alloys by activation surface technology for clinical applications in the biomedical field.

Bioactive hydroxylapatite (HA) coatings are commonly used to improve the biocompatibility and osteoinductivity of the titanium alloy. Titanium alloy coated with HA has the advantages of metal and ceramics and has been considered as one of the most promising bone replacement material. Recently, there has been a tendency to develop HA-contained composite coatings to obtain new type of coatings that satisfy the different requirements of clinical applications. One type of composite coatings is to incorporate biomacromolecules, and the other is to add inorganic materials. Different composites have different functions and applications. The addition of carbon nanotubes (CNTs) into HA coatings can improve the mechanical properties. The preparation of HA/TiO$_2$ composite coating is helpful to the augment of corrosion resistance and bioactive properties of Ti alloy substrate. The introduction of biomacromolecules, such as collagen, gelatin, polyamidoamine, and chitosan, can promote the biological performance and exhibit the antibacterial, osteoblastic, and inorganic crystallization functions. In this chapter, the recent investigation and application of HA composite coatings, including biomacromolecules and inorganic components, on Ti alloy are summarized and introduced.

8.1 HA/CNTs Composite Coating

Hydroxyapatite (HA, Ca$_{10}$ (PO$_4$)$_6$(OH)$_2$) is a very important biomaterial due to its structural and chemical similarities to the mineral components of natural bones. Because of the poor mechanical properties of HA, however, it is difficult to apply the load-bearing parts of the body. Therefore, HA coating on metallic substrates is one of the most promising ways to develop new biological bone substitute materials.

In order to improve the mechanical properties of HA, a lot of materials are introduced to HA composite material as a strengthening phase, with CNT being one of the materials. CNT is a new family of crystalline carbon that has special electromagnetic, mechanical, thermal, and other properties. They have now become a hot topic since they were found especially in composite fields.

The mechanical properties of CNTs are quite prominent, such as high elastic modulus and high plasticity. Ti alloy coated with HA/CNTs appear to have good biocompatibility and can be used as a hard tissue implant material. The HA/CNTs composite coating on the titanium substrate surface combined with the advantages of Ti/HA/CNTs must obtain excellent mechanical properties and biological properties of complex biological materials and is expected as a novel high-performance embedded body material that can meet the requirements of weight-bearing parts of the body.

8.1.1 Structure and Properties of CNTs

CNTs were found in 1991 when S. Iijima of NEC Corporation of Japan was preparing C_{60} by arc evaporation of graphite electrodes [3]. Since then, its special structure and excellent performance have greatly attracted the interest of scientists from various countries. CNTs are typical one-dimensional nanostructures, as shown in Figure 8.1. According to the number of graphite layers, CNTs can be divided into multiwalled carbon nanotubes (MWCNTs) and single-walled carbon nanotubes (SWNTs). Their superior mechanical properties, large diameter ratio (typically greater than 1000), excellent chemical and thermal stability, and good optical performance enhance their application prospects in biosensors, hydrogen storage containers, super capacitors, electromechanical actuators, structural reinforcements, and other aspects.

CNTs made by a single layer of graphite are called single-walled carbon nanotubes (referred to as SWNTs), with a typical diameter and length of 0.75–3 μm and 1–50 μm, respectively. SWNTs have a high uniform consistency. As an important nanoelectronics material, they can be used as conductors, semiconductor materials, as well as microelectronic devices. The ones coiled by multilayer graphite are called multiwalled carbon nanotubes (referred to as MWNTs), with the number of layers ranging from 2 to 50. The stacking is close to ABAB, and the layer spacing is 0.34 ± 0.01 nm, which is equal to the graphite layer. At the beginning of MWNTs formation, it is easy to capture a variety of defects between layers; for this reason, MWNTs are usually covered with small holes on the wall-like defects, which result in a large surface area. In addition, MWNTs have many other merits such as high crystallinity, good conductivity, and holes diameter concentration [4].

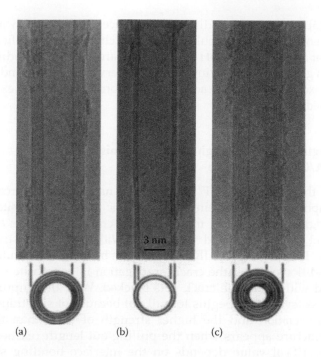

FIGURE 8.1
Electron micrographs of microtubules of graphitic carbon. (From Iijima, S., *Nature*, 354, 56, 1991.) Parallel dark lines correspond to the (002) lattice images of graphite. A cross section of each tubule is illustrated. (a) Tube consisting of five graphic sheets, diameter 6.7 nm. (b) Two-sheet tube diameter 5.5 nm. (c) Severn-sheet tube diameter 6.5 nm, which has the smallest hollow diameter (2.2 nm).

CNTs usually appear as a dark gray or blue powder-like substance. The excellent mechanical properties of CNTs are closely related to their microstructure. CNTs are composed of carbon atoms combined with strong covalent bonds. The four outermost electrons generate three tracks with same energy level by sp2 hybridization, which incorporate with other carbon atoms to form a strong binding of σ key. This microstructure makes allows CNTs good stability and high modulus of elasticity. According to theoretical estimates, the Young's modulus and shear modulus are similar to those of diamond. The theoretical estimate of Young's modulus of SWNTs is up to 5 TPa; according to test measurements, the Young's modulus of MWNT is 1.8 Tpa, flexural strength is 14.2 Gpa, tensile strength is about 100 times that of steel, and density is 1/7–1/6 of steel. Although the tensile strength of CNTs is high, their brittleness is not as high as that of carbon fiber. The carbon fiber will break when the deformation is about 1%, while the carbon will break in about 18% of deformation. The inter-laminar shear strength of CNTs reached 500 Mpa, which is higher by an order of magnitude than that of traditional carbon fiber reinforced epoxy composite materials.

Hui Hu [5] and others modified the surface of CNTs with a chemical method. Then, neurons were cultivated using chemically modified CNTs as a substrate at a physiological pH of 7.35. Compared with nonmodified CNTs, the neurons grow well according to SEM observation, which indicates that CNTs have excellent biological activity. Therefore, we can try to explore their use in biomedical materials.

8.1.2 Strengthening and Toughening Mechanisms of HA/CNTs Composite

In terms of the nature of CNT/HA, they are nanofiber-reinforced ceramic matrix composites. In CNT-reinforced ceramic matrix composites, pulling out the bridging and steering mechanism of crack contributes a lot to the improvement of strength and toughness. With high axial strength and stiffness, CNT can be repeatedly bent and not be destroyed under extreme twisted conditions. When the crack propagation in composite materials is encountered with CNT, the crack gets blocked. With the improvement of applied stress level, CNT begins to pull out because of substrate and CNT interface dissociation and the higher strength of CNT than that of the substrate. Fracture appears when the pulling out length reaches a critical value. (The critical value depends on the interface bonding strength of their own strength and CNT.) Therefore, crack propagation must overcome the CNT pulling out power and the breaking power, resulting in high-fracture toughness. In the fracture process, the main crack turns to different directions according to the different positions of the fiber breakage, which will increase the crack growth resistance (due to crack propagation path twists resulting in an increase in the crack surface area) and further enhance the toughness [6].

In addition to the increase in CNTs in nanoceramic matrix composites, CNTs can suppress nanoceramic grain growth to some extent and promote the improvement of ceramic density so that they further improve the mechanical properties.

8.1.3 Preparation and Characterization of HA/CNTs Composite Coating

Recently, the formation of CNTs–HA composite coatings has been reported using a variety of different techniques, including plasma spraying, laser deposition, and electrophoretic deposition (EPD).

8.1.3.1 Plasma Spraying

Plasma spraying is an existing commercially viable technique and has been used by researchers to economically coat HA composite coating for real-life implants. In recent years, numerous investigations have been carried out to fabricate HA coating or HA composite coating using a plasma spraying technique.

With plasma spraying, the powder is sent to plasma flame, heated to melt or semimolten state, and then sprayed to the surface from the nozzle at the speed approaching or exceeding the speed of sound. The sprayed particles deformed after violent collision with the substrate and deposited on the substrate surface. Meanwhile, the coating layer is formed while quench cooling and rapid solidification.

The combination of plasma spray coating and the substrate is mainly based on mechanical combination. Melting powder particles enter into the surface separation and cool down quickly to combine with the substrate by pinning. At the same time, HA at high temperatures can react with titanium dioxide [7]:

$$Ca_{10}(PO_4)_6(OH)_2 + xTiO_2 \rightarrow$$

$$xCaTiO_3 + (1-x)Ca_{10}(PO_4)_6(OH)_2 + 3xCa_3(PO_4)_2 + xH_2O \, (1000°C)$$

$$2TiO_2 \rightarrow (Ti^{4+})(TiO_4^{4-}) \rightarrow HA$$

To some extent, it can improve the bonding force between the apatite coating and Ti substrates.

During the formation of HA composite coating, addition of CNTs aids the nucleation sites in the crystallization of HA, preserving its inherent crystal structure.

There are various advantages of plasma-sprayed HA coating such as a short time to obtain the coating, the high strength (up to 60 MPa) between coating and the substrate, etc. So plasma spraying has been the most extensively studied method of preparation of HA coating. Although plasma-sprayed HA coatings have resulted in successful improvement in promoting bone attachment and integration of the implants, the long-term stability of these coatings is still a very challenging issue since these coatings tend to have uncontrollable dissolution and sometimes exhibit insufficient bond strength to the metal substrate. In Balani's work [8], the coatings bonded well to the substrate and appeared uniform in thickness. CNTs are not damaged and are evenly distributed in the coating, as shown in Figure 8.2a. The cell growth on HA–CNT surface is observed in Figure 8.2b, indicating biocompatibility of this coating. It is important to note that biocompatibility and toxicity of CNTs have been debated in terms of (i) cell growth and proliferation and (ii) free CNTs in the blood/body.

8.1.3.2 Laser Deposition

In this method, HA + CNT powder is used as preset materials on metallic substrates. Then, a laser device is utilized to synthesize composite coating with bioactive ceramic under a cladding process (ventilate Ar as protection gas) chosen by orthogonal design.

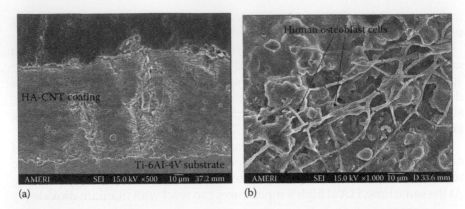

FIGURE 8.2
(a) Cross-sectional image of plasma-sprayed HA-CNT coating, and (b) cell growth, adhesion, and proliferation of human osteoblasts on plasma-sprayed HA-CNT coating. (From Balani, K. et al., *Acta Biomater.*, 3, 944, 2007.)

The microstructure characteristics, hardness, and elastic modulus of the laser-surface-alloyed MWNT-reinforced hydroxyapatite composite coating have been reported in Chen's research work [9,10], and the results showed that the addition of CNTs has a notable effect on the increase in hardness but no strong effect on the increase in elastic modulus, which contributes to a decrease in the modulus mismatch relative to the living bone issues under the condition of guaranteeing higher strength.

Meanwhile, metallurgical bonding between the HA and the metallic substrate can be easily obtained using laser surface processing. This is very useful to improve the mechanical fastening between coating material and human bone tissue.

Although CNTs in mixed powders (CNTs mixed with hydroxyapatite powder) can form TiC with Ti component in substrates, there are still CNTs that are unique tubular or multiparietal structures in the coating.

The nanoindentation technique is used to test the hardness and elastic modulus of HA/CNTs. Under the same pressing conditions, compared with HA coating, the plastic deformation of HA/CNTs composite coating is improved.

Also, with the increase in CNT in the preset powder, the hardness of HA/CNTs is significantly improved because of the residual CNTs and TiC in situ formation, especially the CNTs with unique structure in the coating. However, the amplitude of increase in elastic modulus is much lower than that of the hardness. This is mainly on account of a number of structural defects existing in MWNTs, which will result in a decrease in the elastic modulus.

In [9], high-quality CNT-reinforced hydroxyapatite composite coatings have been successfully deposited on the surface of Ti-6Al-4V substrate using laser surface alloying. The bonding characteristics of the coating/substrate system

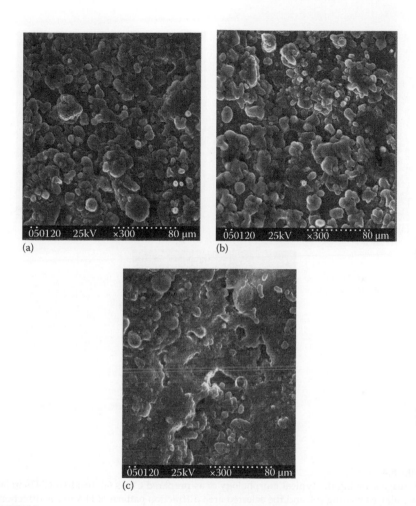

FIGURE 8.3
SEM micrographs showing the surface morphology of as-alloyed with a powder mixture of HA–5% CNTs (a), HA–10% CNTs (b), and HA–20% CNTs (c). (From Chen, Y. et al., *Carbon*, 44(1), 37, 2006.)

are all of metallurgical fusion bonding. In Figure 8.3, SEM images showed that the coatings have a rough surface with interlinking of pores, and TEM observation showed that a large amount of CNTs can be found with their original tubular morphology, even though some CNTs react with titanium element in the substrate during laser irradiation, as shown in Figure 8.4.

8.1.3.3 Electrophoretic Deposition (EPD)

In plasma spray and laser cladding, high temperature leads to a large number of HA decomposing to CaO and TCP, which affects the biological

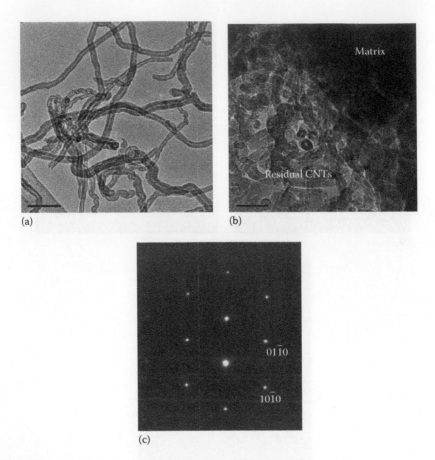

FIGURE 8.4

TEM image showing the typical morphology of as-prepared CNTs (a), residual CNTs in laser surface alloyed coating (b), and the selected area diffraction pattern of HA in the direction of [0001] (c). (From Chen, Y. et al., *Carbon*, 44(1), 37, 2006.)

activity. Thus, it is necessary to prepare HA/CNT composite coating under mild conditions.

The EPD process is promising for producing composite coatings on different substrates for several reasons: simplicity, low equipment costs, lack of restrictions on substrate shape, and lack of requirements for binding materials. Most importantly, EPD technology facilitates the fabrication of a homogeneous microstructure and porous coating surface, the latter of which is beneficial to successful biointegration by the penetration of bone tissue into a coating.

To date, there have been several publications that reported the coating formation of CNTs/hydroxyapatite by EPD from suspensions containing CNTs and HA nanoparticles. Figure 8.5 shows the preparation and EPD process for CNTs–HA nanocomposites.

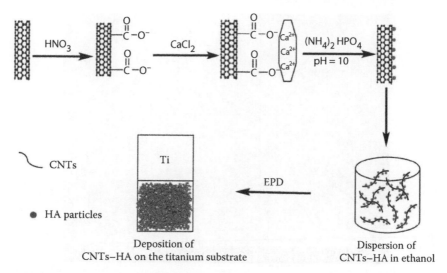

FIGURE 8.5
Schematic representation of preparation and EPD process of CNTs–HA nanocomposites. (From Bai, Y. et al., *Mater. Sci. Eng. C*, C30, 1043, 2010.)

According to Yu Bai's research [11], CNTs and HA were successfully preassembled and deposited onto a titanium substrate by electrophoretic deposition, as shown in Figures 8.6 and 8.7.

8.1.3.4 Biomimetic Growth

Simulated body fluids (SBF) with ion concentrations similar to those of the inorganic constituents of human blood plasma were able to act as a medium for the development of a poorly crystallized HA similar to that of bone on substrates of different materials. The MWNTs/HA composite material can be prepared through chemical biomimetic growth at 37°C chemistry formation of the composite material. The possible mechanism of the reaction is shown in Figure 8.8.

MWNTs with negative charge were achieved by refluxing MWNTs with nitric acid. After dispersion in SBF (37°C), MWNT-COOH was found to be efficient for the growth of HA from SBF. The high surface area of CNTs makes for the reaction of Ca^{2+}, PO_4^{3-}, and carboxyl-functionalized CNTs in SBF and promotes the formation of MWNTs/HA composites.

In Aryal's work [12], synthesis of hydroxyapatite (HA) using carboxylated CNTs was investigated using SBF similar to physiological conditions and products tailored to have chemistry found to mimic natural bone. The result showed that the carboxylated CNTs were capable of nucleating HA from SBF, which can be used as a biomaterial for the modification of implant materials (Figure 8.9).

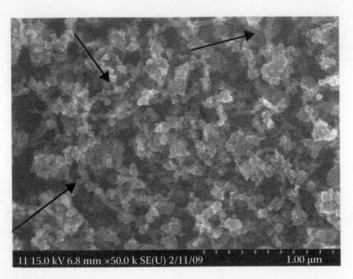

FIGURE 8.6
FE-SEM image of CNTs–HA coatings, deposited at a deposition voltage of 20 V, with a 2 min deposition time; arrow shows HA particles loaded to the CNTs. (From Bai, Y. et al., *Mater. Sci. Eng. C*, C30, 1043, 2010.)

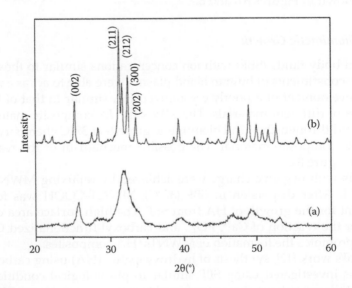

FIGURE 8.7
XRD patterns of (a) as-prepared CNTs–HA nanocomposites and (b) CNTs–HA nanocomposites after hydrothermal treatment. (From Bai, Y. et al., *Mater. Sci. Eng. C*, C30, 1043, 2010.)

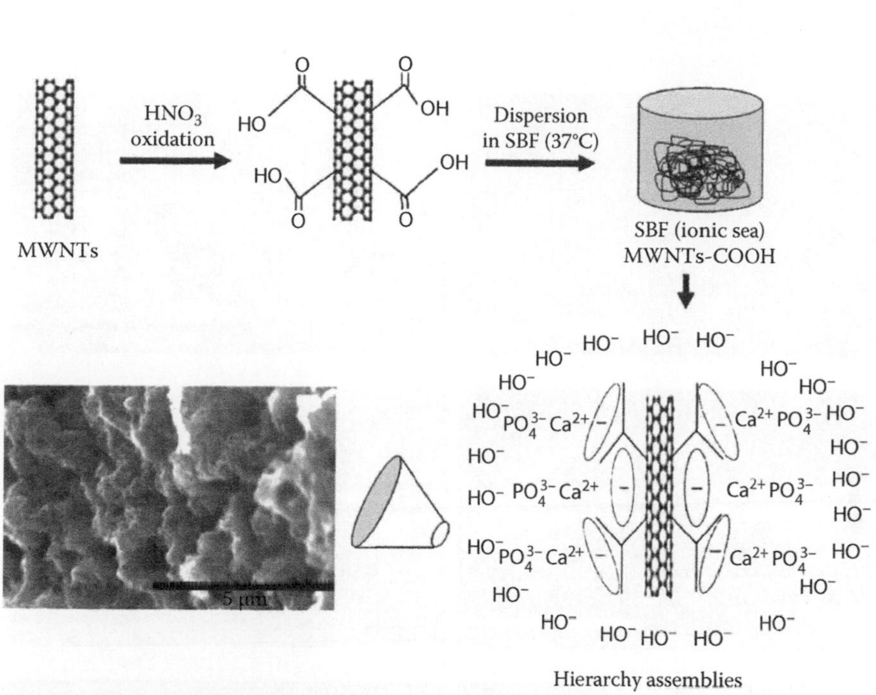

FIGURE 8.8
Schematic representation of hydroxyapatite growth on MWNTs-COOH matrix. (From Aryal, S. et al., *Mater. Sci. Eng. A*, 426, 202, 2006.)

8.1.3.5 Electrochemical Deposition

The preparation of ceramic materials by electrochemical deposition started late; we have not seen any related research literature until 1991. For the purpose of simulating the environment of the human body, the SBF is adopted for electrochemical deposition. The biologically active apatite is prepared, and the formation mechanism of apatite is discussed further.

The basic principle of electrochemical deposition of calcium phosphate compounds is as follows: Ca^{2+}, PO_4^{3-} in electrolyte near the cathode will precipitate due to a certain degree of supersaturation in the higher pH environment. A number of factors have an effect on the crystallization process such as the electrode materials, electrolyte composition, deposition model, and its current and voltage parameters, as well as the deposition temperature, the electrode surface mass quality conditions, and deposition time. The initial crystal charge of calcium phosphate obtained by biomimetic deposition in SBF provides sites for the nucleation of HA. At the same time, under the load current, Ca^{2+}, PO_4^{3-} ions will attract the MWNT in solution to promote self-assembly of MWNTs to form a mesh structure on the substrate surface, which will combine with the initial formation of calcium phosphate.

FIGURE 8.9

SEM micrographs of MWNTs-HA composites recorded after (a) 7, (b) 14, and (c) 21 days of reaction under ambient conditions. Micrographs (d) 7, (e) 14, and (f and g) 21 days are of MWNTs-HA composites synthesized at 37°C. (From Aryal, S. et al., *Mater. Sci. Eng. A*, 426, 202, 2006.)

Meanwhile, the mineralization reactions in solution are engaged and are beneficial for the formation of calcium phosphate. Calcium phosphate will grow up on the initial crystal grain on the substrate surface. There is a mutual promotion between MWNTs self-assembly and the formation of calcium phosphate. They are both deposited on the substrate surface, and the calcium phosphate turns to the most stable structure of hydroxyapatite in a suitable current density and temperature condition. Finally, we get the HA/MWNTs composite coating.

Bone-HA or high degree of crystallinity of pure HA is prepared by electrochemical methods under mild conditions. The mechanism of electrochemical methods is inconclusive because the electrochemical process and ion reaction are more complex. Whether there is a precursor or not in the electrodeposition process is the focus of the debate. But one thing is certain: the pH directly affected the solution supersaturation of calcium and phosphorus, and thus the nucleation and growth of HA are affected.

In our research group, the HA/MWNTs composite coating is prepared on the surface of NiTi alloy by the electrochemical deposition method, as shown in Figures 8.10 and 8.11. We can conclude that the coating is uniform and dense, morphology is lamellar structure, and the crystallinity is good.

8.1.4 Property Evaluation of HA/CNTs Composite Coating

The main properties of HA/CNT composites are mechanical and biological. The mechanical properties determine materials bearing effect and durability, while the biological properties determine the compatibility, bone conduction, and toxic reactions of implanted materials. These properties play a decisive role in determining the success of implanted materials.

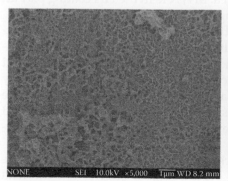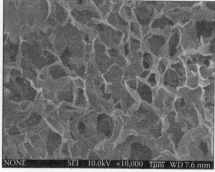

FIGURE 8.10
Morphology of HA/MWNTs composite coating on the surface of NiTi alloy. (From Yu, W., Preparation of HA/MWNTs composite coating on the surface of NiTi alloy, Master degree thesis, Tianjin University, Tianjin, China, vol. 33, pp. 37–38, 2008.)

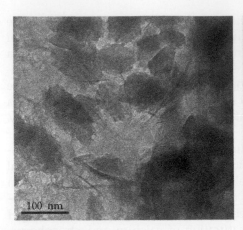
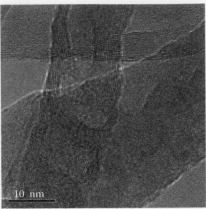

FIGURE 8.11
TEM images of HA/MWNTs composite coating on the surface of NiTi alloy. (From Yu, W., Preparation of HA/MWNTs composite coating on the surface of NiTi alloy, Master degree thesis, Tianjin University, Tianjin, China, vol. 33, pp. 37–38, 2008.)

8.1.4.1 Mechanical Properties

The mechanical properties involved in CNT/HA are very extensive, but most current studies focus on toughness, wear resistance, and coating bond strength. Han Huijuan et al. [14] have studied the bonding strength of MWNTs/HA composite coatings by bonding tensile test, whose results show that the bonding strength of MWNTs/HA composite coating and the substrate obtained by electrophoresis deposition increased significantly. Compared with pure HA, the bonding force of the composite coating and the substrate increased from 20.6 to 35.4 MPa. MWNTs in the HA coating play an extremely important role in toughening and strengthening.

Balani et al. [8] investigated the friction properties of the CNTs/HA coating in physiological solution. The results show that the wear rate of the composite coating and the substrate significantly reduced than the pure HA and wearing capacity was low. As time increased, the amount of wear capacity increased very little. This indicates that HA-CNTs composite coatings have excellent wear resistance. The wear resistance was improved and attributed to not only the deformation of CNTs and self-lubrication but also the function of bridging and stretching.

Chen used the nanoindentation technique to study the variation of the mechanical properties, as shown in Figure 8.12. It shows that the load increases with increasing content of CNT in the powder mixtures, indicating that the higher the content of CNTs, the higher the hardness and the elastic modulus. After removal of the indenter tip, the plastic deformation of CNT-free coating is about 1480 nm, while it is 1290 nm for the HA–20%

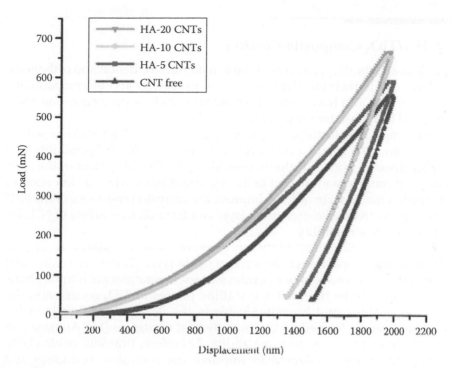

FIGURE 8.12
Typical load–displacement curves of laser surface alloyed coating with different content of CNT. (From Chen, Y. et al., *Carbon*, 44(1), 37, 2006.)

CNTs coating. This means the CNT-free hydroxyapatite coating undergoes a larger plastic deformation during nanoindentation experiments.

8.1.4.2 Biological Properties

The evaluation of biological properties commonly includes two stages of the in vivo and in vitro studies. At present, the studies of HA/CNT composite coating biological properties are still in in vitro stage, while there is less investigation of long-term performance (including biological and mechanical properties) for HA/CNT composite coating materials implanted in animals.

Han Huijuan et al. [14] with others conducted an in vitro cell culture experiment. The results show that cells can adhere well to the surface. Balani [8] simulated artificial bone cells in vitro of the CNTs/HA composite coating. They found that the osteoblasts grew limitlessly along the surface of CNTs and then embedded in the substrate, which indicates that the composite is nontoxic. HA precipitated and mineralized on the surface, and crystallinity increased from 53.7% to 80.4%.

8.2 HA/TiO₂ Composite Coating

Titanium and its alloys are very suitable materials for dental and orthopedic implants due to their excellent mechanical properties and biocompatibility. However, they also have some disadvantages, such as poor corrosion resistance and bad bioactive properties.

For example, despite the satisfactory clinical use of NiTi alloy in biomedical applications, NiTi is still a controversial biomaterial for the possibility that Ni may dissolve from NiTi due to corrosion [15]. The Ni content in the alloy is of great concern with regard to its biocompatibility when its ion releases into the human body. In order to improve the corrosion resistance of the NiTi alloy, a protective nickel-free oxide layer can form on the surface of NiTi for improved biocompatibility.

In fact, the biocompatibility of titanium and its alloys is closely related to the properties of the surface oxide layer. The native oxide that spontaneously forms on the titanium surface in ambient air is amorphous and is known to be bio-inert. Crystalline phases of TiO_2 occur naturally as rutile, anatase, and brookite. Rutile is the thermodynamically stable phase and can be formed on titanium by heat treatment [16]. Anatase and rutile have good apatite-forming ability. Therefore, titanium oxide (TiO_2) coating has been considered to improve the corrosion resistance and restrain Ni release.

TiO_2 coating on titanium alloys has recently shown promising in vivo corrosion behavior, acting as a chemical barrier against release of metal ions from the implant. However, it is difficult to achieve chemical bond with bone and form new bone on the implant with only TiO_2 surface at an early stage. In order to further improve bioactivity of the implant, it is necessary to prepare hydroxyapatite (HA) on the TiO_2 coated surface.

A double-layer HA–TiO_2 coating on titanium or its alloys with HA as the top layer and a TiO_2 coating as the inner layer should show a very good combination of good corrosion resistance and bioactive properties.

8.2.1 Duplex Treatments for Preparation HA/TiO₂ Composite Coating

TiO_2 film can be prepared by methods such as heat treatment, chemical treatment, chemical vapor deposition, and sol–gel dip coating technique. Several coating methods such as plasma spray, electrophoretic, sol–gel, and electrochemical deposition are usually used to prepare HA coating on the implant surfaces.

8.2.1.1 MAO Combined with Other Treatments

Microarc oxidation (MAO), also called microplasma oxidation, has attracted more and more attention by virtue of its convenience and effectiveness in preparing oxide ceramic coatings on the surface of titanium or its alloys [15]. By applying a positive voltage to Ti or its alloy substrate immersed in an electrolyte, a microarc occurs, and a TiO_2 coating on the surface is formed when the applied voltage is increased to a certain point. The formed TiO_2 film is porous and beneficial for the biological performance of the implants. An advantage of MAO process is that the formed TiO_2 coating is in situ formed on titanium surface and thus bonds to titanium substrate firmly. The MAO process also has the potential to prepare coatings on metal surfaces with complex geometry [17].

During the early beginning of the MAO process, a compact film forms in the electrolyte. With the increment of the voltage, microarcing appears due to electrical breakdown of the dielectric layer, resulting in a porous microstructure on the titanium surface. The surface morphology of the Ti after MAO in electrolyte containing calcium and phosphate is shown in Figure 8.13. Microspores, protuberant areas, and concave areas were observed on the film, which made the film rough. The pores are channels of microarc discharge in electrolyte. The MAO coating should act as a barrier to the release of metal ions to human body and avoid deleterious effects.

The titanic coating is mainly composed of anatase structure containing rutile (as shown in Figure 8.14). With the applied voltage increasing, some anatase may translate to rutile.

Another advantage of the MAO process is the possibility of incorporating element in the electrolyte into the coating by controlling the composition and concentration of the electrolyte. Titanium oxide coating containing Ca and P can form on titanium substrate by MAO in electrolyte containing $(CH_3COO)_2Ca \cdot H_2O$ and $NaH_2PO_4 \cdot 2H_2O$. The XPS spectra of a sample subjected to MAO treatment reveal that the outer layer of the MAO sample

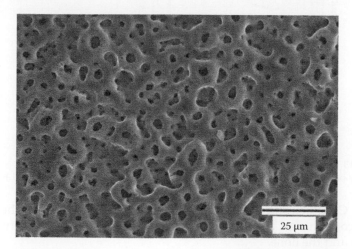

FIGURE 8.13
Surface morphology of the MAO sample. (From Liu, S.M. et al., *Mater. Lett.*, 65(6), 1041, 2011.)

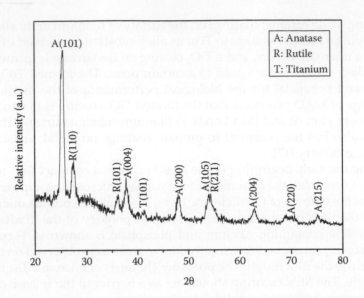

FIGURE 8.14
XRD pattern of the MAO samples. (From Liu, S.M. et al., *Mater. Lett.*, 65(6), 1041, 2011.)

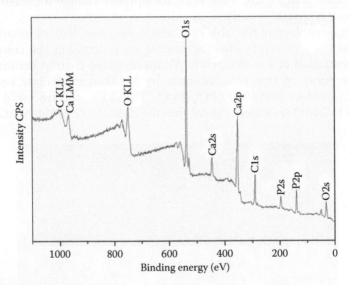

FIGURE 8.15
XPS survey of the MAO coating. (From Liu, S.M. et al., *Mater. Lett.*, 65(6), 1041, 2011.)

contains Ca, P, C, and O elements (as shown in Figure 8.15). The porous coating containing Ca and P ions can enhance the bonding between the implants and the bone.

HA/TiO$_2$ composite coating on titanium and titanium alloy can be prepared by MAO combined with subsequent treatment.

8.2.1.1.1 MAO and Electrophoretic Deposition

X. Nie made efforts to improve implant biocompatibility and durability. Ti-6Al-4V was used as the substrate. A relatively thick and hard TiO$_2$ coating can be produced by MAO treatment, and then a pure HA coating with an intermediate HA/TiO$_2$ composite layer deposited on top of the TiO$_2$ layer can easily be formed by hybrid MAO treatment and electrophoretic deposition. The porous TiO$_2$ surface is helpful in anchoring the HA powder to the surface to allow sintering to occur [19].

The adhesive strength between the coating and substrate was assessed using scratch adhesion testing. The result indicated that the adhesion upper critical loads of the TiO$_2$ coating and HA/TiO$_2$ coating were 8 and 10 N, respectively. Relatively high adhesive strength was achieved.

8.2.1.1.2 MAO and Electron Beam Evaporation

For deposition, anodized surface and calcium phosphate deposition by electron beam evaporation methods were combined. Han Li prepared the titanium samples of microarc oxidized surface (anodized Ti) and anodized surface deposited with calcium phosphate thin film (coated Ti) by this way [20]. The porous oxide layer of an anodized Ti surface is shown in Figure 8.16a. After deposition of calcium phosphate, no obvious differences of morphology were observed (as shown in Figure 8.16b).

The film consisted of HA, b-TCP, and calcium oxide (CaO) before heat treatment. After heat treatment, the particle grows larger and only HA was detected.

8.2.1.2 Heat Treatment and Biomimetic Deposition

In order to transform titanium surface into rutile, commercially pure titanium plates were oxidized in a furnace. Small faceted crystals that describe the topography constitute the surface (Figure 8.17). X-ray diffraction analysis

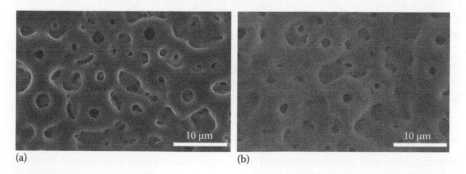

(a)   (b)

FIGURE 8.16
Surface morphology of samples: (a) anodized Ti and (b) coated Ti. (Li, Y. et al., *Biomaterials*, 29, 2025, 2008.)

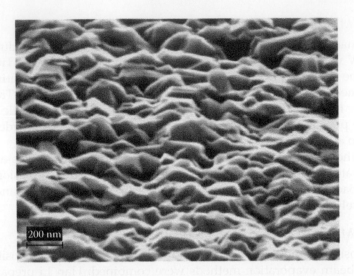

FIGURE 8.17
Surface morphology of the oxidized sample; the sample is tilted 70°. (From Forsgren, J. et al., *Acta Biomater.*, 3, 980, 2007.)

shows that the titanium surface has been transformed into rutile after the heat treatment (see Figure 8.18).

After cleaning the samples with agent and ethanol, they were exposed to phosphate-buffered saline (PBS) to deposit HA. PBS is a solution with pH and ion concentrations similar to those of human blood plasma. HA coating had formed on the rutile surface after immersion for 7 days (see Figures 8.19 and 8.20). When the rutile surface is exposed to body fluids, OH groups adsorb to the titanium ions in the oxide and negative Ti–O groups are formed over it. These negative sites attract Ca^{2+} ions from the body fluid. A layer of amorphous calcium titanate is formed, and the surface becomes slightly positively charged. It will then attract negatively charged phosphate ions and calcium phosphate is formed. This layer is then crystallized into bone-like apatite because it is thermodynamically more favorable in a wet environment [16].

A scratch test was used to evaluate the adhesion of the coating/substrate composite. The critical pressure for the coating to detach from the substrate was estimated to be 2.4 ± 0.1 GPa. When the critical pressure was reached, large pieces of the coating started to detach from the substrate.

8.2.1.3 Anodization and Electrodepositon

Anodization and electrochemical technique have been developed to form HA/TiO_2 composite coating on a titanium substrate [21]. Initially, Ti substrate was anodized in an acidified fluoride solution to prepare TiO_2 nanotubes (Figure 8.21a). Vertically oriented TiO_2 nanotubular arrays acted as templates

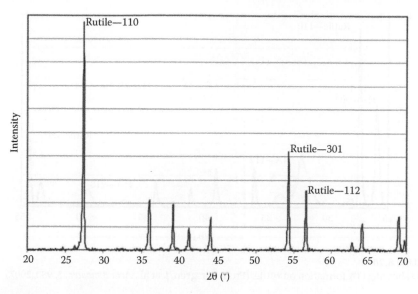

FIGURE 8.18
XRD of oxidized titanium where all peaks correspond to rutile TiO_2. (From Forsgren, J. et al., *Acta Biomater.*, 3, 980, 2007.)

FIGURE 8.19
Surface morphology of the HA deposited on rutile TiO_2; the sample is tilted 70°. (From Forsgren, J. et al., *Acta Biomater.*, 3, 980, 2007.)

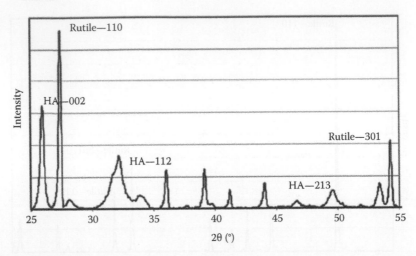

FIGURE 8.20
XRD showing HA formation on rutile. (From Forsgren, J. et al., *Acta Biomater.*, 3, 980, 2007.)

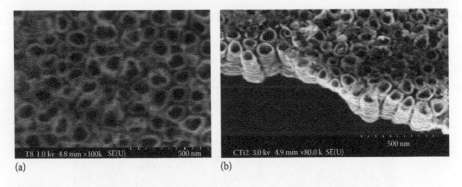

(a)　　　　　　　　　　　　　(b)

FIGURE 8.21
Morphology of the TiO_2 nanotubes: (a) top and (b) side views of the nanotubes. (From Kar, A. et al., *Surf. Coat. Technol.*, 201, 3723, 2006.)

and anchorage for growth of the apatite nanocrystals during subsequent electrodeposition process (as shown in Figure 8.21b).

During the anodization process, phosphate ions can incorporate in the TiO_2 lattice, which is proved by the peak at 133.8 eV (as shown in Figure 8.22). The presence of phosphate ions could facilitate nucleation of calcium phosphate within the nanotubes. Growth of HA inside the nanotubes would give anchoring effect to the electrodeposition, which will enhance the interfacial bond strength between TiO_2 and HA coating.

Anodized sample was pretreated with NaOH solution subsequently. The result showed that a "ring-like" structure was observed at the neck of the nanotubes. The EDAX results suggested the formation of sodium titanate

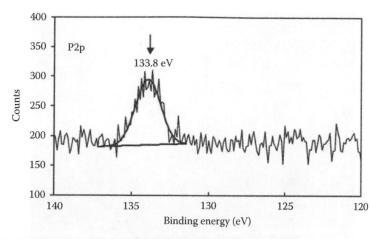

FIGURE 8.22
P2p XPS spectrum of the TiO₂ nanotubular sample anodized in phosphate solution. (From Kar, A. et al., *Surf. Coat. Technol.*, 201, 3723, 2006.)

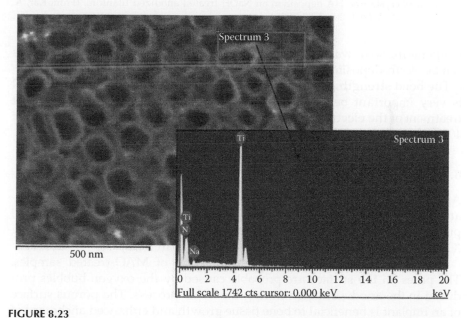

FIGURE 8.23
FESEM micrograph and EDAX analysis of the sodium titanate rings on the nanotubular TiO₂. (From Kar, A. et al., *Surf. Coat. Technol.*, 201, 3723, 2006.)

rings (as shown in Figure 8.23). The sodium titanate was benefited to enhance the formation of the HA coating.

The morphology of crystalline HA on NaOH-treated anodized titanium after electrodeposition process is shown in Figure 8.24. The electrodeposition process has some advantages. For example, the process operates at low

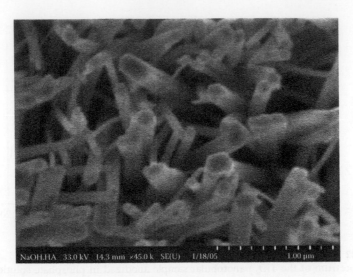

FIGURE 8.24
Side view of crystalline HA deposition on NaOH treated anodized titanium. (From Kar, A. et al., *Surf. Coat. Technol.*, 201, 3723, 2006.)

temperature, so unwanted phase changes could be avoided. Complex shapes can be electrodeposited, and this process is relatively inexpensive.

The bond strength of the hydroxyapatite coating on the implant material is very important because it has to withstand bone growth stress. Heat treatment of the electrodeposited HA coating resulted in a bond strength of 40 MPa.

8.2.2 One-Step Synthesis

Apatite/titania coating was prepared on commercial pure titanium by micro-arc oxidation in electrolyte containing calcium and phosphate. Commercial pure titanium was used as the substrate material. The electrolyte containing $(CH_3COO)_2Ca \cdot H_2O$ and $NaH_2PO_4 \cdot 2H_2O$ was prepared [18].

Figure 8.25a shows the surface morphology of MAO-treated samples. The pores on the surface are probably caused by the oxygen bubbles produced in the anodic reaction during the MAO process. The porous surface of an implant is beneficial to bone tissue growth and enhanced anchorage of implant to bone. Furthermore, the microporous structure may be valuable as a depot for bioactive constituents and has the function of an enhanced cell proliferation. The cross section of the coated sample exhibits apparently three layers, which in turn are apatite, TiO_2, and Ti substrate from the outside to inside (as shown in Figure 8.25b). No obvious discontinuity is observed among the three layers, indicating that the composite coating can be tightly adhered to the substrate. Such MAO films are expected to have significant applications as artificial bone joints and dental implants [22].

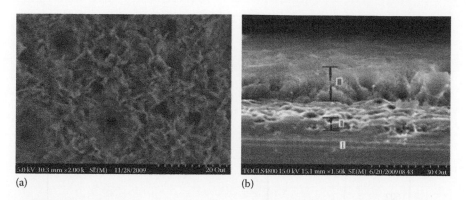

(a)   (b)

FIGURE 8.25
(a) Surface morphology and (b) cross-sectional view of the MAO sample. (From Liu, S.M. et al., *Mater. Lett.*, 65(6), 1041, 2011.)

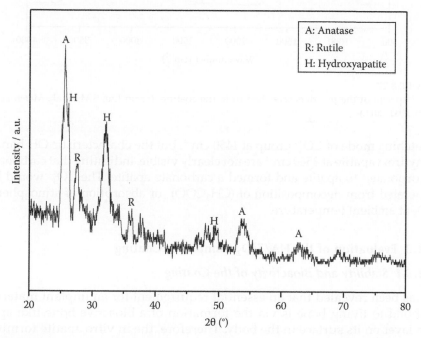

FIGURE 8.26
XRD pattern of the MAO sample. (From Liu, S.M. et al., *Mater. Lett.*, 65(6), 1041, 2011.)

The pattern of sample subjected to MAO treatment exhibits diffraction peaks of anatase and hydroxyapatite. Three peaks at 26°, 32°, and 49° are indexed as (002), (211), and (213) plane of hydroxyapatite (as shown in Figure 8.26).

FT-IR spectra of the powders scratched from the coating are shown in Figure 8.27. The CO_3^{2-} absorption band is detected by the characteristic

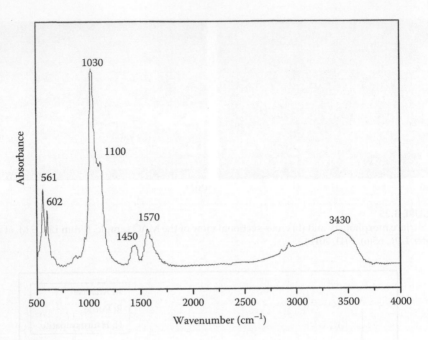

FIGURE 8.27
FT-IR spectra of the powders scratched from the coating. (From Liu, S.M. et al., *Mater. Lett.*, 65(6), 1041, 2011.)

stretching mode of CO_3^{2-} group at 1450 cm^{-1}, but the characteristic OH bands of hydroxyapatite at 3560 cm^{-1} are not clearly visible, indicating that carbonate incorporated to apatite and formed a carbonate apatite. The CO_3^{2-} would be generated from decomposition of $(CH_3COO)^-$ or absorption of atmospheric CO_2 at ambient temperature.

8.2.3 Evaluation of the HA/TiO$_2$ Composite Coating

8.2.3.1 Stability and Bioactivity of the Coating

It has been revealed that an essential requirement for an implant material to bond to living bone is via the formation of a bioactive bone-like apatite layer on its surface in the body. Therefore, the in vitro apatite forming ability of the material surface is an important assessment to evaluate the bioactivity of the implants. In addition, the layer may be scratched during operation accidentally or dissolved after implantation by contacting with the body fluids. So a bioactive and stable coating for an implant material is prerequisite.

In order to evaluate the stability and bioactivity of coating, much efforts have been made to investigate the immersion behavior of coating in solutions, such as Hank's balanced salt solution (HBSS), phosphate buffered saline (PBS) solution, and Kokubo's SBF. These solutions can well reproduce in vivo surface changes due to the same ion concentrations as those of the human blood plasma. For example, the SBF is usually prepared by dissolving reagent-grade mixtures of $CaCl_2$, $K_2HPO_4 \cdot 3H_2O$, KCl, NaCl, $MgCl_2 \cdot H_2O$, $NaHCO_3$, and Na_2SO_4 in distilled water and buffering at pH 7.40 with tris-hydroxymethyl aminomethane and hydrochloric acid (HCl) at 37°C. In order to accelerate the rate of apatite formation, 1.5 SBF is usually prepared. These in vitro studies are important in a practical sense in that they could give some indication of the in vivo behavior of the coating.

Han Li had immersed the anodized Ti and anodized Ti coated with calcium phosphate (coated Ti) in DPBS solution for determined intervals [20]. The DPBS solution was prepared by dissolving Dulbecco's PBS and reagent-grade $CaCl_2$ in ultra-pure water. The coated surface exhibited apatite formation as shown in Figure 8.28. A newly formed homogeneous layer was observed on the coated surface after incubation in DPBS solution after only 1 h and the bigger flake-like crystals appeared after 24 h of incubation, while no apparent changes appeared on the anodized surface of Ti until 24 h of immersion in DPBS solution. The XRD pattern also demonstrated the new peaks for bone-like apatite.

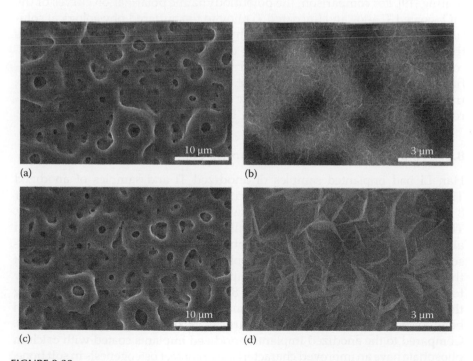

FIGURE 8.28
Surface morphologies of sample anodized Ti after incubation in DPBS solution for (a) 1 h and (c) 24 h, and sample coated Ti after incubation in DPBS solution for (b) 1 h and (d) 24. (From Li, Y. et al., *Biomaterials*, 29, 2025, 2008.)

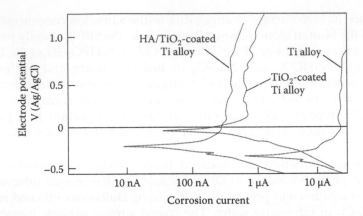

FIGURE 8.29
Potentiodynamic polarization curves of samples. (From Nie, X. et al., *Surf. Coat. Technol.*, 125, 407, 2000.)

8.2.3.2 Corrosion Resistance

Nie investigated the corrosion resistance of the Ti alloy with HA/TiO_2 coating [19]. For comparison, the potentiodynamic polarization curves of the TiO_2-coated Ti alloy and uncoated Ti alloy in the buffered physiological solution are also plotted in Figure 8.29. The HA/TiO_2-coated Ti alloy possesses the lowest corrosion current. The TiO_2-coated Ti alloy shows the highest corrosion potential, while the uncoated Ti alloy gave the lowest corrosion potential and the highest corrosion current. This result showed that the duplex treatment of MAO and electrophoretic deposition could provide an anticorrosive HA/TiO_2 composite coating on Ti alloys.

8.2.4 Application of HA/TiO_2 Composite Coating

Han Li had implanted samples of anodized Ti and samples of anodized Ti coated with calcium phosphate (coated Ti) in four male minipigs for 8 weeks [20]. Two groups of implants were placed on each side, respectively. Throughout the course of the study, all animals were healthy and tolerated all procedures well. At necropsy, no abnormal findings were found macroscopically. Radiographs of each mandible taken immediately after sacrifice of the animals were used to locate the implants. The implants contacted closely to the host bone (as shown in Figure 8.30). The typical sections comprising the implant and surrounding tissues are shown in Figure 8.31. Appositional bone formation occurs around the surfaces of the two groups of implants. Compared to the anodized implants, anodized implants coated with calcium phosphate have an improved characteristic of contact osteogenesis in soft bone. The calcium phosphate coating did not separate from the implant surfaces.

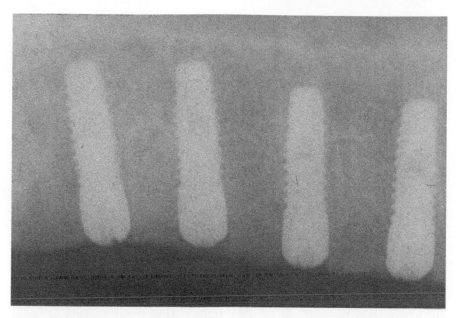

FIGURE 8.30
Radiograph of implant sites (From Li, Y. et al., *Biomaterials*, 29, 2025, 2008.)

8.3 HA/Collagen Composite Coating

Human bone is a bioceramic composite that is mainly composed of calcium phosphate (69%), collagen (20%), and water (9%). Other organics, such as proteins, polysaccharides, and lipids, only account for a small percentage. Calcium phosphate exists in the form of a certain degree of crystallization HA, which can ensure the rigidity of the bone.

The main ingredient in the organic part is collagen fibers, which make the bone flexible and elastic. From the view of bionics, it is the best when the synthetic materials are designed on the basis of natural bone composition and structure. There is no doubt that nanohydroxyapatite/collagen composite materials must have good mechanical properties and biological characteristics because the structure and chemical composition are similar to the natural bone. Compared to a pure HA layer, this organic–inorganic composite layer has a better application prospect. Therefore, in order to develop better surface treatment technologies for implant materials, apatite/collagen composite coatings formed on the surface of titanium and its alloys have attracted many researchers' attention.

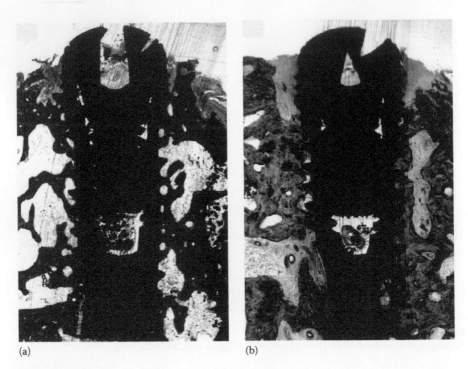

(a)   (b)

FIGURE 8.31
Light micrographs of the implant sites after 8 weeks of healing: (a) anodized Ti; (b) coated Ti (original magnification × 10). (From Li, Y. et al., *Biomaterials*, 29, 2025, 2008.)

8.3.1 Collagen and Its Characteristics

Collagen is a type of protein, mainly existing in animal skin, bone, cartilage, teeth, tendons, ligaments, and blood vessels. As a very important structural protein of connective tissue, it supports organs such as the heart, kidneys, liver, and lungs and protects organisms. Collagen is the most abundant protein present in mammals, composing 25%–30% by weight of body protein tissue [23].

Like other proteins, collagen has a quaternary structure, which plays a decisive role in determining its molecular size, shape, chemical reactivity, biological functions, and so on. Figure 8.32 gives the hierarchical design of collagen. The primary structure points to the arrangement of amino acids in the protein molecules. There is peptide bond between amino acids. It is the basic structure of protein, while the secondary structure, tertiary structure, and quaternary structure are three-dimensional structures of protein molecules as the high-level structure of protein molecules.

The structural features of collagen range from the amino acid sequence, tropocollagen molecules, collagen fibrils to collagen fibers.

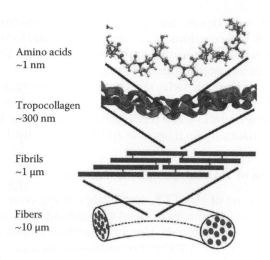

FIGURE 8.32
The hierarchical design of collagen. (From Buehler, M.J., *Curr. Appl. Phys.*, 8, 440, 2008.) The structural features of collagen ranges from the amino acid sequence, tropocollagen molecules, collagen fibrils to collagen fibers.

Collagen exists in the human body in the form of collagen fibers. The smallest unit of collagen structure is called tropocollagen molecules with diameters of about 1.5 nm, lengths of approximately 280 nm, and relative molecular mass about 300,000. Each procollagen molecule consists of three α-peptide chains, which have α-helix structure. Three α-peptide chains are entangled into a rope-like structure in parallel, right-handed forms and then formed collagen fibers by multistage polymerization in molecules. The three forces to stable collagen triple helix are the van der Waals force and the hydrogen bonds between spirals and the covalent cross-links [25].

Some of the basic physical and chemical properties of collagen are as follows: collagen is a poly-ampholyte, which is due to each peptide chain of collagen having a lot of acid or alkaline lateral groups, and both ends of each peptide chain are determined by α-carboxyl and α-amino. These groups have the ability to accept or give protons. In a certain pH, collagen solution has no charge, in the state of zwitterionic, meaning that it has the same positive charge and negative charge (zero charge). We defined the PH as the isoelectric point of collagen. The isoelectric point of collagen is pH 7.5–7.8, showing alkaline. The isoelectric point of collagen is quite different when it is prepared by alkaline and acid methods.

Collagen has colloidal properties in aqueous solution combined with water molecules by hydrogen bonding. There is a water membrane around the collagen molecules, and then hydration occurs. If the temperature is increased and the water immersion time is extended, the collagen can further absorb water to form a hydrophilic colloid. When external conditions change, the collagen solution may freeze and precipitate. The denatured reaction is called collagen gel. The reactions of collagen with nitrite, formaldehyde, and ninhydrin can be used to determine amino acids. The ionic bond of collagen in salt solution will be opened, thus swelling.

The collagen-specific triple helix structure is responsible for many useful features of amino acids, such as high tensile strength, biodegradability, low antigen activity, low irritation, low cell toxicity, and promotion of cell growth. All these characteristics make it become an ideal biomedical material.

8.3.2 Preparation of HA/Col Composite Coating

The aim of preparation of HA/COL composite coating on metal substrate is trying to combine the advantages of metal and coating to create a more useful hard tissue replacement materials. However, because collagen is a low melting point organic polymer, a lot of the traditional methods to prepare HA coating are no longer applicable (such as plasma spraying and laser coating). At present, the main methods of preparation of HA/COL composite coating on metal substrate includes biomimetic coating procedures, the electrochemical co-deposition method, and electrospray deposition (ESD) process.

8.3.2.1 Biomimetic Growth Method

Biomineralization is a process involving nucleation, growth, and phase transformation to obtain a biomineral material with a highly ordered structure. This process occurs when it is under certain conditions, in different parts of the organism, with a variety of different ways, and under the influence of the organic substance. The interaction of organic molecules and inorganic mineral ions at the interface controls the precipitation of inorganic minerals at the molecular level, which gives the mineral a special high-level biological structure and assembly methods. Biomineralization has a special reaction medium.

The growth process of HA/COL composite coating on titanium surface is similar to the mineralization process in vivo, including the preorganization of organic molecules, the recognition of interface molecular, growth modulation, and cell processing these four stages. Specifically, it is the accumulation of calcium and phosphorus ions in intracellular and extracellular matrix. When it is up to a specific saturation, it will become a nucleation in the matrix protein and then grow and form a hard tissue together with the substrate.

The collagen-specific three-dimensional configuration and certain chemical groups (lysine, hydroxyproline groups) can combine with phosphate to induce the crystallization of calcium phosphate so as to promote the calcification process. Collagen plays the role of nuclei in the inorganic deposition process. It is a site of starting the crystallization. It has been proven that collagen can covalently combine with a small amount of phosphate and the serine residues or carbonyl take the priority.

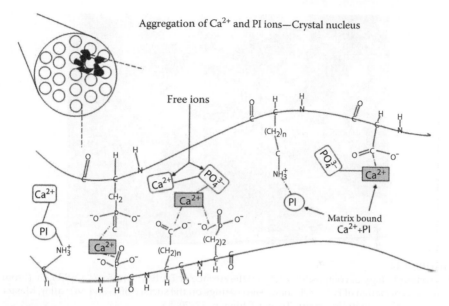

FIGURE 8.33
Schematic map of collagen compartments localizing a microenvironment which serves to nucleate the mineral phase heterogeneously. (From Qiyi, Z., The fabrication and bioactivity study of Ca-P coatings on titanium surface, Doctoral dissertation, Sichuan University, Sichuan, China, p. 110, 2003.)

In the process of bone formation, millions of collagen molecules self-assemble into collagen microfibers through the interaction of side chains and the connection of end to end. The diameter is about 20–500 nm. The process of bone formation in hydroxyapatite nucleation sites on the collagen fibers is very similar to the cyclical nature of its mechanism of nucleation, which can be explained in Figure 8.33. Controlled nucleation and growth are accomplished within the fiber cell formed in the collagen matrix microstructure. Collagen molecules self-assemble into microfibrils. There are interspaces or holes with sizes of 40 nm at the terminals of two molecules. The microenvironment in holes contains free mineral ions and side chain groups among which there is a certain molecule period to promote heterogeneous nucleation of mineralization phase. From Figure 8.34, obvious collagen fibers with a diameter of about 200–300 nm can be seen. The calcium and phosphorus compounds grow on and around collagen fibers whose diameter is about 50 nm. The particles on collagen fibers grow periodically. And the interval is approximately 600–700 nm.

Cai Yanli et al. [28] prepared the HA/COL composite layer using the biomimetic method on titanium surface, as shown in Figure 8.35. The morphology of the composite coating was uniform and compact calcium phosphate particles with some colloidal and elliptical particulates, and the micromorphology of that was bone-like lamellar structure.

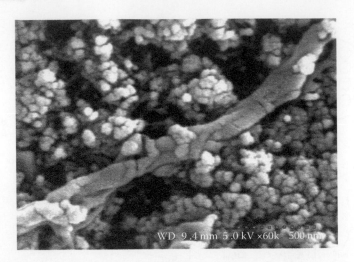

FIGURE 8.34
Micromorphology of composite coating on the surface of Ti alloy: (a) 0.5 g/L CSBF 12 h. (From Yanli, C., Formation of HA/COL composite coatings on the surface of Ti and NiTi alloy, Master degree thesis, Tianjin University, Tianjin, China, p. 22, 2005.)

The formation mechanism is as follows. After acid–alkali treatment, there is a negatively charged microstructure on the sample surface. After precalcification process, the negatively charged surface gathered a large amount of Ca^{2+} and PO_4^{3-} ions for the sake of static electricity, and the initial nucleation of calcium phosphate salt formed. Then under the control of the collagen, apatite crystals form nucleation on the organic/inorganic interface in the CSBF solution. With extension of growth time, the nuclei ultimately change into a HA/collagen composite layer under the influence of the transport rate and association rate of ions in solution.

8.3.2.2 Electrochemical Deposition

In the past researches, several different methods have been employed to deposit bone-like apatite coating onto metallic implant surface. Electrochemical deposition (ELD) is one of the most attractive methods because with it irregularly shaped substrates can be coated relatively quickly. Additionally, the corresponding devices for ELD are simple and convenient.

The possible mechanism for the present electrolytic deposition of HA/collagen coating is understood as follows. First of all, the electroreduction of water molecules at the cathode surface leads to the pH increase due to the generation of OH^- ions around Ti alloy and produced hydrogen gas, as shown in the following electrochemical half reaction:

$$2H_2O + 2e^- \rightarrow H_2 + 2OH^-$$

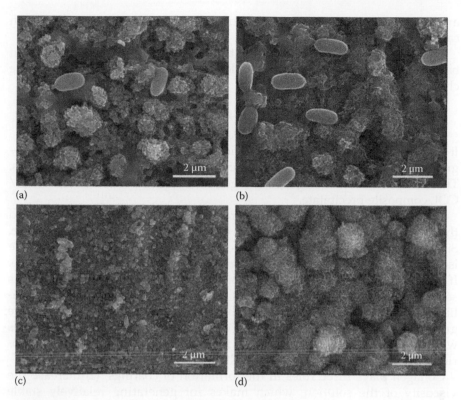

FIGURE 8.35
SEM micrographs of the coatings deposited in CSBF (a, b for 3 and 7 days) and SBF (c, d for 3 and 7 days). (From Yanli, C. et al., *Scripta Mater.*, 54(1), 89, 2006.)

The hydroxide ions generated at the cathode surface then reacted with dihydrogen phosphate in MSBF to form PO_4^{3-} in succession:

$$H_2PO_4^- + OH^- \rightarrow H_2PO_4^{2-} + H_2O$$

$$HPO_4^{2-} + OH^- \rightarrow PO_4^{3-} + H_2O$$

The following reaction resulted in the formation of HA:

$$10Ca^{2+} + 6PO_4^{3-} + 2OH^- \rightarrow Ca_{10}(PO_4)_6(OH)_2$$

At the beginning of the ELD process, Ca^{2+} ions are first attached to the cathode as the nucleation of CaP. When the titanium surface is covered with Ca^{2+} ions, the surface tends to attract negative ions, and this hierarchical organization is growth of CaP. As the pH of the electrolyte is adjusted to the isoelectric point of collagen or even lower, positively charged collagen molecules are drifted toward the titanium substrate and are deposited together with calcium phosphate.

The COL molecules experience a transformation from amphoteric ion balance to the electronegative molecules with the pH increase (P is peptide chain):

$$-OOC-P-NH^{3+} + OH^- \rightarrow -OOC-P-NH^{2+} + H_2O$$

It was believed that there were chemical interactions between the $-COO-$ group in the COL molecules and HA crystal in the co-precipitation process. Thus, the formation of negative charge ion $-COO-$ could induce the HA/COL composite precipitate.

With the proceeding of the cathode reaction, more OH^- ions gather around the Ti substrate, thus a pH gradient occurs around the titanium surface. Simultaneously, collagen is negatively charged when the local pH rises to the isoelectric point of collagen or even higher and is drifted away from cathode. If pH arrives at the isoelectric point of collagen, collagen fibril comes to its self-assembling process, which means the collagen fibril is fixed at the equilibrium position accompanying the absorption of Ca^{2+} ions and phosphate group ions on itself. The mineralized collagen and the CaP layer combines with the ionic bond, which means the formation of the coating. The equilibrium position is determined by the voltage.

Collagen plays three roles in the formation of coatings: (1) increase the viscosity of the solution, which makes for generating relatively stable hydrogen bubbles to form a "bubble template," promoting the growth of HA whiskers along the bubble wall. Finally, porous coating with nanomicron secondary structure was prepared. (2) In the process of HA crystallization, it can play a part in inducing HA nucleation formation and growth and influencing the growth orientation of HA crystal and grain geometric scale. Fan Yuwei et al. [29] consider that the presence of calcium ions in solution will also promote collagen deposition at the cathode interface. (3) The protonized collagen and hydrogen phosphate may have chemical bonds and also participate in chemical co-deposition reaction on electrode surface to make the organic–inorganic sediments cross-link with each other so as to help improve the mechanical strength in the co-deposition layer and the bonding strength between the deposition layer and metal substrate. In addition, the addition of collagen may greatly increase the degradation of hydroxyapatite.

In view of this, electrolytic deposition (ELD) has been used to generate homogeneous collagen-CaP (Col-CaP) composite coatings onto metallic implants [30–32]. Composite coating shows uniform and orderly distribution of the porous structure. It is a very obvious secondary structure, that is, nano-HA whiskers knot together in a certain orientation to constitute a relatively uniform micron-scale pore structure. Pore size is several microns, while the diameter of whiskers is about several tens of nanometers, which

FIGURE 8.36
SEM images of the HA/collagen coatings by electrochemical deposition. (From Sun, T. et al., *Mater. Lett.*, 65(17–18), 2575, 2011.)

looks thin and soft and is similar to natural bone microstructure. Composite coating consists of completely through holes as shown in Figure 8.36.

8.3.2.3 Spin Coating

Spin coating is a simple and effective technique for controlling the thickness of thin coatings. The spin coating process used to produce the HA/col composite coating on Ti substrates is described in Figure 8.37.

Ultrafine HA particles were first prepared by the simultaneous titration of $Ca(OH)_2$ and H_3PO_4 solutions. After the repeated centrifugation/washing process, the HA precipitates so obtained were further centrifuged at high speed to remove as much of the water as possible. Subsequently, the gel-like HA precipitates, without any dry treatment, were directly re-dispersed in the collagen/HFP (1,1,1,3,3,3-hexafluoro-2-propanol [HFP]) solution.

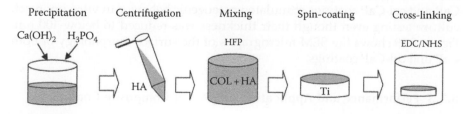

FIGURE 8.37
Schematic illustration of the processes used to produce the collagen/HA composite coating on the Ti substrates. (From Teng, S.-H. et al., *J. Mater. Sci.: Mater. Med.*, 19, 2453, 2008.)

A viscous, homogeneous composite sol was obtained by stirring the mixture for 24 h at room temperature. After the spin coating, the thin film that was obtained was cross-linked in EDC/NHS solution (ethanol containing N-(3-dimethylaminopropyl)-N0-ethylcarbodiimide (EDC) hydrochloride and N-hydroxysuccinimide (NHS) solutions) to ensure its chemical and structural stability, as presented in Figure 8.37.

In Teng's study [34], a homogeneous collagen/HA nanocomposite sol was obtained by the biomimetic preparation and centrifugation of nanosized HA precipitates followed by mixing with collagen without any dry treatment. Subsequently, collagen/HA composite thin coatings with an approximate thickness of 7.5 μm were uniformly formed on Ti substrates through a spin coating process. The SEM results showed that the nanosized HA particles were well distributed within the collagen matrix, without any obvious aggregation, as shown in Figure 8.38.

8.3.2.4 Electrospray Deposition (ESD)

Using ELD or biomimetic, synthesis of Col-CaP coatings were time consuming and control over coating thickness was poor. The ESD process, on the other hand, is among the most promising techniques for generating organic–inorganic composite coatings on implant materials since its low processing temperatures allow for simultaneous deposition of both biomolecules and CaP. Using ESD, faster deposition rates and exact control over coating thickness can be achieved as compared to both ELD and biomimetic coating synthesis. As a result, the ESD technique offers the possibility to deposit coatings of nanometer thickness onto metallic materials.

Using the ESD technique, nanometer-thick col/CaP coatings were successfully deposited at a collagen/CaP ratio comparable to that of native bone tissue [35], of which the fibrous nanostructure resembled the structure of mineralized collagen in bone tissue. With increasing ESD times, the CaP particles and collagen became more tightly packed within the coating as compared to the relatively thicker composite coatings for the short spraying times. The mechanical properties of the electrosprayed CaP coatings improved significantly upon co-deposition of collagen. Furthermore, both CaP and col-CaP coatings stimulated osteogenic behavior in vitro upon cell culture testing even though their thickness was reduced to below 100 nm. Figure 8.39 shows the SEM micrographs of the surface morphology of both CaP and col-CaP coatings.

8.3.3 Performance and Application of HA/Col Composite Coating

To comprehend that bone is apatite/collagen composite material in micro scale is the foundation of the preparation of bioactive ceramic/polymer bone biomimetic composite materials. The preparation of bone repair

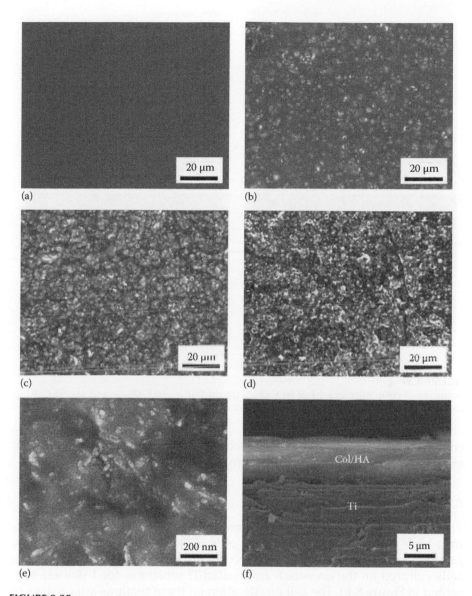

FIGURE 8.38
SEM morphologies of the coatings with different compositions on Ti substrates: (a) collagen; (b) COL-10%HA; (c) COL-20%HA; (d) COL-30%HA; (e) COL-30%HA at high magnification; and (f) its cross section. (From Teng, S.-H. et al., *J. Mater. Sci.: Mater. Med.*, 19, 2453, 2008.)

and replacement materials by compositing hydroxyapatite has become one of the most useful methods of biomaterials investigation. The specific nano/microstructure hydroxyapatite/collagen composite coating on titanium surface is prepared under mild conditions. Using the combination of biological macromolecules and hydroxyapatite, an organic–inorganic

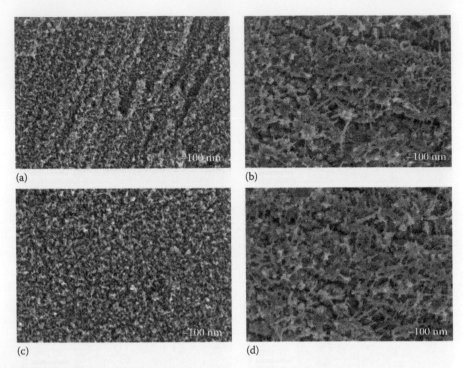

FIGURE 8.39
Scanning electron micrographs of (a) CaP$_{thin}$, (b) col-CaP$_{thin}$, (c) CaP$_{thick}$, and (d) col-CaP$_{thick}$-coated Ti disks. (From de Jonge, L.T. et al., *Biomaterials*, 31, 2461, 2010.)

composite coating was prepared to obtain bone substitute material with excellent biological and mechanical properties.

8.3.3.1 Biological Performance

The osteoblast responses to the collagen and collagen/HA composite coatings were assessed using MC3T3-E1 cells in terms of the cell attachment, growth, and morphology [34]. The electron microscopy morphologies of the cells cultured on the uncoated- and coated-Ti substrates for 1 day are presented in Figure 8.40. The cell proliferation on these coatings after 3 days of incubation is presented in Figure 8.41.

As compared to both the bare Ti substrates and the conventional composite coatings with equivalent compositions, the collagen/HA nanocomposite coatings on the Ti substrates obtained with the present method showed much better cell proliferation behaviors and osteogenic potentials, thus confirming the improved activity of the cell functions that they afford. In conclusion, collagen/HA nanocomposites have great potential for use as a coating material for future medical applications in hard and soft tissue replacements.

FIGURE 8.40
The morphology of the MC3T3-E1 cells cultured on (a) titanium substrate, (b) collagen, and (c) COL-30%HA composite coatings. (From Teng, S.-H. et al., *J. Mater. Sci.: Mater. Med.*, 19, 2453, 2008.)

8.3.3.2 Mechanical Performance

Generally, the strengthening phenomena of collagen are obvious for hydroxyapatite coatings. Xu Yanli et al. [36] made a comparison of the pure HA and HA/col mechanical properties of composite coatings through bond-tensile test. Experiments show that under the same conditions, the adhesion between HA/collagen composite coating and titanium substrate is doubled that of pure HA coating and substrate (5.02 MPa), approximately to 10.31 MPa. In addition, the turn-off time of the coating with collagen is about 60s, which is longer than the stretching time (about 37 s) of the pure coating indicating that both the coating adhesion and the flexibility are enhanced after adding collagen in the coating. Similarly, Keng-Liang Ou et al. [37] reported that the maximum indention depth under the same load decreased for the organic–inorganic composites obtained by introduced collagen into hydroxyapatite coatings, which are similarities between these HA/collagen and natural composite materials (Figure 8.42).

In Hu Kai study [38], the scratch method is used to evaluate the bonding properties of the coating and the substrate. The coating was prepared by electrochemical deposition and biomimetic growth.

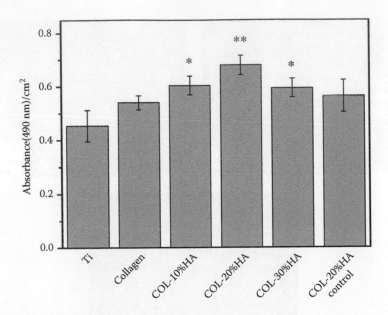

FIGURE 8.41
Growth activity of the MC3T3-E1 cells on different coatings after 3 days. Pure Ti plates were used as a control, and differences at $*P < 0.05$ and $**P < 0.01$ with respect to pure Ti were considered statistically significant. (From Teng, S.-H. et al., *J. Mater. Sci.: Mater. Med.*, 19, 2453, 2008.)

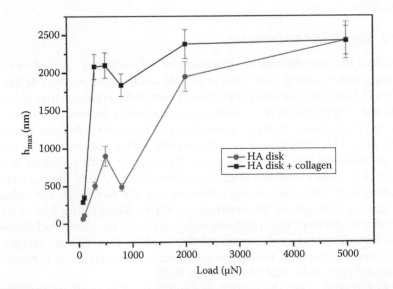

FIGURE 8.42
Relationship between the maximum indention depth and the loads for the sintered HA with or without collagen treatment. (Ou, K.-L. et al., *J. Mech. Behav. Biomed. Mater.*, 4(4), 618, 2011.)

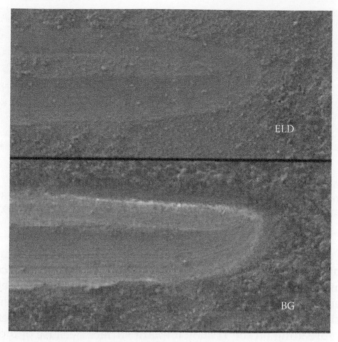

FIGURE 8.43
Montage of SEM micrographs of the scratch traces for the composite coatings. (From Kai, H., Study on preparing HA/COL composite coating by chemically biomimetic growth and electrochemical deposition, Master degree thesis, Tianjin University, Tianjin, China, p. 38, 2007.)

The scratch method is based on the use of a diamond stylus under a constant or continuous positive pressure, which engraves the coating surface with a certain speed till the damage of coating bonding ability. The critical load L_C is the measurement of bond strength of the substrate/coating. According to the scratch morphology in Figure 8.43, it can be concluded that the coating surface roughness and crystal arrangement have a certain impact on the scratches. BG coating has a larger crystal spacing, is loosely arranged, has more apparent scratches, and the critical load is smaller (40N). The surface roughness of ELD coating is relatively small, resulting in a small resistance along the direction of scratches. The critical load is 50N. The results indicate that the combination performance of HA/COL prepared by electrochemical method is better than the coating prepared by chemical biomimetic growth method.

8.3.3.3 Clinical Applications

The water-soluble recombinant human-like collagen is applied to modulate the growth of hydroxyapatite on NiTi alloy so as to better biofunctionalize the alloy in X.J. Yang' report [39], as shown in Figure 8.44. The co-precipitation

FIGURE 8.44
SEM micrographs of the coating on NiTi alloy after soaking in 0.5 g/L 1.5CSBF for 3 days ([b] is the high-magnification image of [a]). (Yang, X.J. et al., *Mater. Sci. Eng. C*, 29, 25, 2009.)

coating of hydroxyapatite and recombinant human-like collagen on NiTi alloy might improve the toughness of the coating and make the composite materials more applicable in the orthopedic field. Xiaolong Zhu et al. gave a characterization on hydroxyapatite/collage surfaces and described the cellular behaviors. The studies showed that the nano-HA/collagen surface favors cell spreading on the Ti alloy implant surface because of the collagen attraction to cells and cell motility on the surface by inducing well-developed cellular filopodia through nano HA (Figure 8.45), which enhanced cell adhesion to nano HA.

Jin Qiong et al. [41] investigated the efficacy of a new collagen-hydroxyapatite (COL-HA) composite membrane on bone regeneration of rat cranial defects. The research shows that COL-HA composite membranes can guide bone regeneration of rat cranial defects. As an absorbed membrane that can guide bone regeneration, the COL-HA composite film has a good prospect for clinical application. The efficacy of bone regeneration of COL-HA double-layer

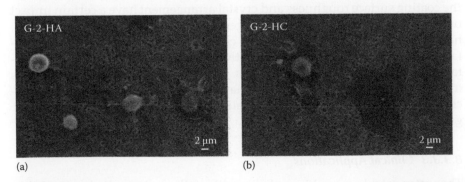

FIGURE 8.45
SEM of osteoblasts growing on nano HA (a) and nano-HA/collagen (b) films for 1 h. (From Zhu, X. et al., *J. Biomed. Mater. Res. Part A*, 79(1), 114, 2006.)

membrane is superior to COL-HA single-layer dense membrane, because its property is more propitious to the adherence and proliferation of osteoblasts.

8.4 HA/Gelatin Composite Coating

It is well known that an implant with a structure and composition similar to natural hard tissue can improve bone repair or substitute. Coatings on Ti and Ti alloy are managed to be designed as similarly as possible to natural bone structure. HA deposited on Ti alloy via various traditional methods has, in fact, little similarity to bone mineral in morphology, crystallinity, composition, and trace element content [42]. Bone-like apatite is actually a needle-like nanocrystal with poor crystallinity containing CO_3^{2-}. The nucleation and growth of these crystals are modulated by collagen fibrils, the other major component of natural bone [43]. Thereafter, composite coatings combining collagen or collagen-like materials with HA are considered to modify the metal surface to either bridge the gap in HA coatings with poor mechanical properties or to simulate natural hard tissue [35]. On the other hand, thanks to the good cell adhesive property and osteointegration ability [44], collagen-like macromolecules can also function as carriers for releasing drugs, such as antibiotics and osteoinductive proteins, to enhance bone healing around the implant, and therefore implant fixation and osseointegration. Although collagen possesses many good properties as mentioned earlier, it is predominantly derived from natural sources, so there is some risk of allergy [45]. The drawbacks of high price and immune antigenicity limit its application [46]. Thus, there is a need for alternatives of collagen to fabricate HA composite coatings. Gelatin serves as one of the best choices.

8.4.1 Property of Gelatin

Gelatin is the denatured state of collagen, sharing much similarity to collagen molecules, and is always used instead of collagen because of the lower immune antigenicity and lower cost [46]. This macromolecule possesses arginine-glycine-aspartic acid (RGD) sequences reported to stimulate the adhesion of osteoblasts and improve the osteointegration [44].

Gelatin is actually a heterogeneous mixture of water-soluble proteins of high average molecular weights present in collagen. The proteins are extracted by boiling skin, tendons, ligaments, bones, etc., in water. There are two typical form of gelatin, Type A and Type B. Type A gelatin is mainly derived from acid-cured tissue, while Type B is derived from lime-cured tissue. The structure of gelatin molecules is shown in Figure 8.46. Gelatin has 70–120 mmol of free carboxyl groups per 100 g of protein. The charge on a

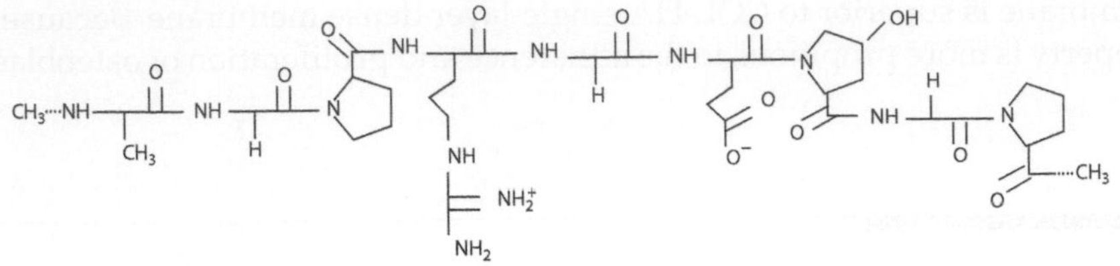

FIGURE 8.46
Chemical structure of gelatin. (Gaihre, B. et al., *J. Mater. Sci.: Mater. Med.*, 20(2), 573, 2009.)

gelatin molecule is primarily determined by the ratio of free carboxyl groups to free amino groups on the side chains.

Gelatin is often used to improve cell attachment for a variety of cell types. Additionally, as a biocompatible polymer, gelatin has been used as a delivery vehicle for the release of bioactive molecules and in the generation of scaffolds for tissue engineering applications.

8.4.2 Preparation of HA/Gelatin Coatings

There are several methods to prepare HA/gelatin composite coatings on Ti and Ti alloy surfaces, including the dip coating method, biomimetic processes, and electrochemical deposition.

8.4.2.1 Dip Coating Method

Dip coating methods are classified into two types: dip/spin coating and dip/phase inversion process.

Rohanizadeh et al. used dip/spin coating method to obtain an oxidized Ti alloy surface, which was further immersed in a supersaturated calcium phosphate solution to allow apatite deposition [48]. Briefly, insoluble TiO_2 (rutile or anatase) powder was added into a gelatin solution with an adequate viscosity for dip/spin coating to prepare TiO_2 slurries. After that, the rutile and anatase slurries were stirred at 45°C to make them workable and ready for application to the Ti disks. The slurries were then placed onto the Ti disk, followed by spinning the slurry dipped disks at a fast speed. During the coating procedure, the slurry was stirred and kept in a water bath at 45°C. Because of centrifugal forces during the spinning stage, the excess of slurry was removed from the surface. As a result, a homogeneous TiO_2 film was achieved, and the disks were sequentially cooled to room temperature to evaporate the slurry. The treated Ti disks were incubated in a supersaturated calcium phosphate solution containing $CaCl_2$ and NaH_2PO_4 at pH 7.2 for the formation of apatite.

Recently, Xiao et al. developed an asymmetric coating method that involves a dip coating and a phase inversion process [44]. In their method, HA nanoparticles were first fabricated through chemical precipitation. The dipping solution was prepared by homogenously dispersing hydroxyapatite nanoparticles into deionized water by using ultrasonic cell disrupter and then dissolving gelatin in this solution and sonicating. Ti-6Al-4V alloy plates as substrate were etched to form a fresh titanium oxide surface. This substrate was immersed into the dipping solution for some time, followed by immersing into the quenching solution. Finally, asymmetric coatings composed of gelatin and hydroxyapatite were achieved by crosslinking in glutaraldehyde.

8.4.2.2 Biomimetic Processes

This method often involves SBF, which is prepared according to the inorganic ion concentration of human blood plasma. SBF solution can mimic the ion components and their concentration of actual human body fluid to some extent. HA formed in SBF solution under human body temperature has the features of small size, low crystallinity, and high surface activity, very similar to the natural HA.

Bian et al. built a gelatin and HA composite coating on porous Ti alloy [49]. A titanium substrate treated by alkali and heat treatment was immersed in Na_2HPO_4 and $Ca(OH)_2$ in advance. After that, it was put into SBF solution at 36.5°C, followed by immersing in a gelatin-containing die. However, the actual human body fluid contains not only the inorganic components but also some organic components, such as proteins, which have essential effects on the nucleation, polymorphism, growth, chemical composition, shape, and dimensions of apatite. Recently, Pan et al. employed a biomimetic process to form a HA/collagen-like protein coating on the alkali-heat treated Ti substrate in 1.5 times SBF (1.5 × SBF) with the addition of a recombinant collagen-like protein [45]. The composition of 1.5 × SBF is listed in Table 8.1 [45]. Considering gelatin is also a kind of collagen-like protein, HA/gelatin may also be fabricated via this method.

8.4.2.3 Electrochemical Deposition

Low-temperature electrochemical deposition technique was also used to prepare HA/collagen composite coating on NiTi shape memory alloy [33]. It is expected in the near future that this technique could also be used to form HA/gelatin composite coatings as well.

Uncoated NiTi SMA disks were mechanically polished to a mirror finish, followed by ultrasonic cleaning in acetone and ethanol separately. Electrochemical deposition of coatings on metal disk was carried out at 37°C in order not to denature protein. The metal disk served as the working electrode, and a pure platinum plate was used as the counter electrode. A double-strength simulated body fluid solution (2SBF) containing soluble collagen was used as the electrolyte. Composite-coated samples were gently rinsed with distilled water and dried after electrochemical deposition.

TABLE 8.1

Composition of 1.5 × SBF (pH 7.4)
and Order of Addition of the
Reagents to Water with
Recombinant Collagen-Like Protein

Reagent	Weight (g/L)
NaCl	12.0540
NaHCO$_3$	0.5280
KCl	0.3375
K$_2$HPO$_4$·3H$_2$O	0.3450
MgCl$_2$·6H$_2$O	0.4665
1 M HCl	15 mL
CaCl$_2$	0.4440
Na$_2$SO$_4$	0.1080
C$_4$H$_{11}$NO$_3$ (Tris)	9.0945
1 M HCl	50 mL
Collagen-like protein	0.5

Source: Pan, M. et al., *Mater. Chem.
Phys.*, 126, 811, 2011.

8.4.3 Characterization of HA/Gelatin Coatings

The characteristics of HA/gelatin coatings formed on titanium vary in terms of the fabrication method employed. That is, the morphology, crystallinity, size, and chemical composition of HA, as well as the mechanical properties, adhesion strength to alloy substrate, and bioactive properties of the composite coatings, are related to the formation condition and mechanisms.

The fabrication process involving an SBF medium often leads to the formation of nanosized calcium-deficient carbonated HA with a poorly crystalline and flake-like structure [33,45]. As shown in Figure 8.47, nanosized apatite with the flake-like structure was obtained and was combined by the collagen fibrils. The broadened peak of the XRD pattern indicated the low crystallinity of HA. This characteristic of apatite may be due to the biomimetic chemistry mechanism resembling the biomineralization in vivo in the physiological condition. A treated Ti or Ti alloy surface is able to bond OH$^-$ and form Ti–OH groups [50]. The Ti–OH groups were negatively charged and could selectively combine with the positively charged Ca^{2+} in SBF to form calcium titanate, as was reported by Kokubo et al. [51]. With the accumulation of more Ca^{2+} on the surface, the overall positive surface further combined with negatively charged PO$_4^{3-}$ to induce apatite nucleation. Then, they could spontaneously grow by consuming the Ca^{2+} and PO$_4^{3-}$ ions from the SBF, which were supersaturated with respect to apatite [45].

It is speculated that a high proportion of the amino acids with charged groups, such as aspartic acid residues with carboxyl group, which also exist

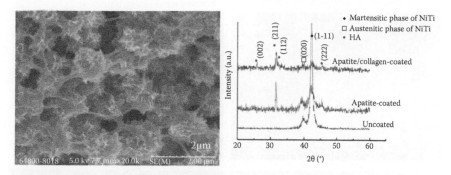

FIGURE 8.47
Surface morphology and XRD patterns of apatite/collagen-coated NiTiSMA samples 2SBF containing soluble collagen was used as the electrolyte. (From Sun, T. et al., *Mater. Lett.*, 65(17–18), 2575, 2011.)

in gelatin, promoted the nucleation of HA [45]. With the addition of such proteins, the nucleation and growth of HA in SBF on the Ti substrate were found to be promoted effectively, and the morphology of apatite turns from lamellar flake to needle-like crystals (Figure 8.48).

XRD result showed that more apatite formed and a larger aspect ratio of HA with protein molecules existed (Figure 8.49), which suggested that the presence of collagen-like protein led to the preferential growth of HA crystal along the c-axis, similar to the natural bone.

All the difference in characteristics lies in the modulation of collagen-like protein on the nucleation and growth of HA. As mentioned earlier, the positive charge surface due to the accumulation of Ca^{2+} absorbs negatively charged groups. When protein was introduced, $-COO^-$ of special amino acid groups (aspartic acid residues) competed with PO_4^{3-} to combine with Ca^{2+}, which changed the nucleation and growth of HA on the metal surface. As a result, the presence of collagen-like protein accelerates the apatite deposition and HA formation on the pretreated Ti substrate.

On the other hand, the fabrication process not involving an SBF medium always leads to obtain HA comparatively more different from natural mineral. For example, HA prepared via chemical precipitation during preparing a HA/gelatin composite coating on Ti alloy surface exhibited a nanorod morphology with a length of 30–50 nm and diameter of 10–15 nm and with a higher crystallinity (Figure 8.50). This phenomenon is caused by the distinction between the HA synthesis condition and the physiological environment.

Further addition of gelatin in this synthetic HA seemed to no longer influence the HA crystals. Gelatin just serves as joining of HA and affects the compactness of the composite coatings. Moreover, through different treatments, gelatin can be prepared to various morphologies, which influence the mechanical property and bioactive property of the composite coatings.

FIGURE 8.48
SEM images of the surface of pretreated Ti substrates after soaking in 1.5 × SBF (a, c) or 1.5 × SBF with addition of a recombinant collagen-like protein (b, d) for 8 days. (From Pan, M. et al., *Mater. Chem. Phys.*, 126, 811, 2011.)

During a dip/phase inversion process, gelatin is soluble in warm water as solvent and insoluble in anhydrous ethanol as nonsolvent (quenching solution). An asymmetric HA/gelatin coating composed of a thin dense outer layer and a thick porous inner layer was formed. The porous layer is due to the phase inversion by water–ethanol exchange in the quench solution, while the dense layer is caused by evaporation of the solvent in the air. Hence, long immersion time in the quenching solution obviously increases the porosity of the coating (Figure 8.51).

8.4.4 Application of HA/Gelatin Coatings

8.4.4.1 Drug Release

Metal ions and infection were always detected in tissues close to Ti alloy implants and became one of the major problems of these implants. Controlled antibiotics release systems in combination with implants have been developed to solve these issues. Composite coatings composed of gelatin and hydroxyapatite on Ti-6Al-4V alloy implant were developed using the dip/phase inversion method for the delivery of water-insoluble drug ibuprofen to reduce the infection rate [44]. Ibuprofen was added in the quenching solution.

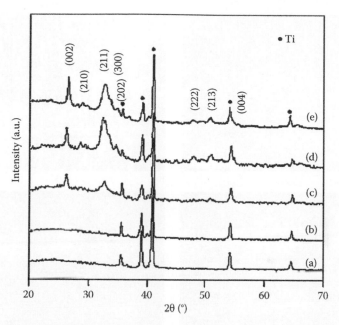

FIGURE 8.49

XRD patterns of pretreated Ti substrate (a), Ti substrates after soaking in 1.5 × SBF for 8d (b) and for 14d (d), Ti substrates after soaking in 1.5 × SBF with addition of a recombinant collagen-like protein for 8d (c) and for 14d (e). (From Pan, M. et al., *Mater. Chem. Phys.*, 126, 811, 2011.)

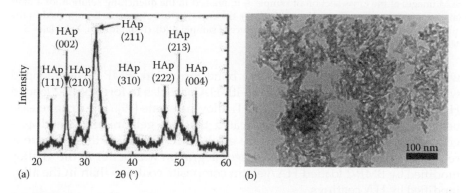

FIGURE 8.50

XRD pattern (a) and TEM image (b) of hydroxyapatite particles obtained through chemical precipitation. (From Junwu, X, *J. Mater. Sci.: Mater. Med.*, 2009, 20, 889, 2009.)

Drug loading depended on the immersion time and drug concentration in the quenching solution. The in vitro release from this coating was always at an approximately zero-order rate and at least lasted for 30 days.

HA/gelatin coatings were also used as vehicles for bone morphogenetic protein (BMP). BMP-2 possesses osteoinductivity and is considered a protein

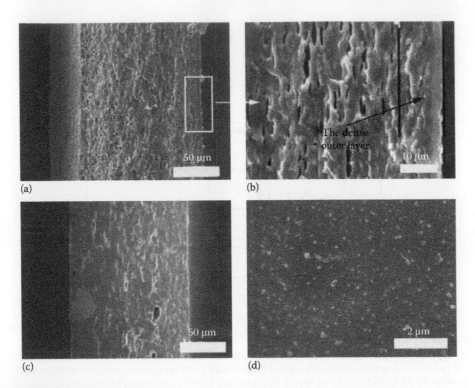

FIGURE 8.51
SEM images of the cross section of sample A immersed in the quenching solution for 4 days (a), the thin dense outer layer in sample A (b), the cross section of sample E immersed in the quenching solution for 2 days (c), and the surface of sample A (d). (From Junwu, X, *J. Mater. Sci.: Mater. Med.*, 2009, 20, 889, 2009.)

with the highest ability to improve bone formation. BMP-2 was blended with gelatin and further combined with HA particles to form BMP-2 loaded HA/gelatin composite coatings on a Ti alloy substrate [49]. The drug release behavior was studied. Results indicated that the gelatin functioned as a good delivery system to prolong the effect of BMP-2. With the continuous release of this bioactive drug, the osseointegration was better in the Ti alloy modified by BMP-2 loaded HA/gelatin composite coatings than in the alloy modified by HA coatings.

8.4.4.2 Tissue Engineering

Biomimeic HA/collagen-like protein composite coatings on Ti substrate were prepared by Pan et al. Osteoblastic MG-63 cells was cultured on this coating. It was found that the existence of collagen-like protein in coating could improve the cell adhesion, proliferation, and differentiation (Figure 8.52). The authors believed that this superiority is due to the RGD sequence in collagen-like proteins, since RGD can increase the overall attachment of the surface

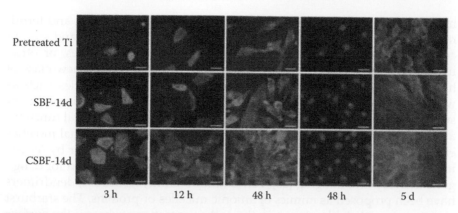

FIGURE 8.52
Fluorescence images of cellular actin cytoskeleton and cell nucleus on the different substrates after 3 h, 12 h, 48 h, and 5 d of cultivation. Arrows indicate the pseudopodia. (SBF represent samples with HA coating on pretreated Ti surface; CSBF represent samples with HA/collagen-like protein composite coating). (From Pan, M. et al., *Mater. Chem. Phys.*, 126, 811, 2011.)

for osteoblasts and enhance the rate of cell spreading through forming the integrin-binding domain [45]. Moreover, HA in this composite coating is modulated by the collagen-like protein. That is to say, hydrophilicity and surface roughness, which favors cell attachment and proliferation, can be modulated by the collagen-like protein. Hence, Ti alloy with HA/collagen-like composite coatings may serve as a good implant candidate for bone regeneration.

As mentioned in Section 1.2.5, the drug release behavior of coatings composed of gelatin and hydroxyapatite on Ti-6Al-4V alloy implant has been investigated. In addition, in vitro calcification of this implant was also studied. More apatite was found to form on the composite coatings than on the pure gelatin coatings and bare Ti-6Al-4V implant as controls. According to some former literature, the apatite formation ability on the implant played critical roles on facilitating the chemical fixation of biomaterials to bone tissue, and ultimately affecting the in vivo success of the bone grafting materials. Therefore, it is reasonable to believe that HA/gelatin composite coating may significantly improve the bond with the surrounding bone, in vivo, compared with bare Ti-6Al-4V alloy implants [44].

8.5 HA/PAMAM (Polyamidoamine) Composite Coating

Dendrimers are used for modifying HAP coatings due to their relative low toxicity, intrinsic biocompatibility, and ability to interact with HAP surfaces. The potential biomedical applications of dendrimers have become an active area of research. Unlike linear polymers, dendrimers are composed of a core molecule, hyperbranches, which regularly extend from the core, and terminal groups. Thus, dendrimers have a definite molecular weight and size. In addition, they can encapsulate mental complexes, nanoparticles, or other inorganic and organic guest molecules [52]. Dendrimers are a new class of hyperbranched macromolecules possessing distinctive properties such as well-defined globular architecture, narrow polydispersity, and tunability of surface functionalities. Their tunable nanometric size and chemical functionality offer versatility for incorporating a wide variety of functional moieties either through encapsulation in the interior of the dendrimer or by tethering onto the periphery via covalent modification or physisorption for drug/gene delivery and imaging [53]. Polies (amidoamine) (PAMAM) dendrimers have been proposed as mimics of anionic micelles or proteins. The starburst structures are disk-like shapes in the early generations, whereas the surface branch cell becomes substantially more rigid and the structures are spheres [54]. Because of the unique and well-defined secondary structures of the PAMAM, it should be a good candidate for studying inorganic crystallization.

8.5.1 Property of PAMAM

Poly(amidoamine) dendrimers (PAMAM) are arguably the most extensively studied dendrimers for biomedical applications, especially as carriers of biologically active agents. As early as 1984, PAMAM dendrimers were the first complete dendrimer family (G=0–7) to be synthesized and characterized, followed by commercialization in 1990 [55]. They are synthesized by the divergent method, involving a two-step iterative reaction sequence that produces concentric shells of branch cells (generations) around a central initiator core. This PAMAM core–shell architecture grows linearly in diameter as a function of added generations, while the surface groups amplify exponentially at each generation (Figure 8.53). For the PAMAM dendrimer family initiated from an ethylenediamine core with a branch cell multiplicity of two, the expected mass values double, approximately, from generation to generation. [56]

These polymers have been called artificial proteins; they can be used as protein biomimics because the terminal amino groups can be functionalized and the dendrimers covalently linked together. PAMAM dendrimers with amino terminal groups have low toxicity to eukaryotic cells. Modification of the amino groups with poly(ethylene glycol) (PEG) or lauroyl chains can further improve their biocompatibility [57] (see Figure 8.54).

8.5.2 Preparation of HA/PAMAM Composite Coating

Many methods have been developed for HA/PAMAM composites. The composites can be prepared by different methods, including microarc oxidation, hydrothermal biomineralization and crystallization, and self-precipitation in SBF adding PAMAM.

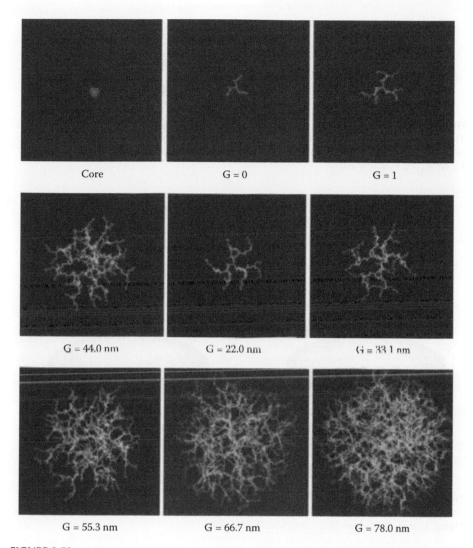

Core G = 0 G = 1

G = 44.0 nm G = 22.0 nm G = 33.1 nm

G = 55.3 nm G = 66.7 nm G = 78.0 nm

FIGURE 8.53
Graphical presentation of PAMAM dendrimers from core to generation G=7 showing the linear increase in diameter and exponential growth of the number of surface groups. (From Sönke, S. and Tomalia, D.A., *Adv. Drug Deliver. Rev.*, 57(15), 2106, 2005.)

8.5.2.1 Microarc Oxidation

MAO method is usually used to prepare TiO_2 coatings in different acid or alkaline electrolytes [58]. Some researchers also tried to prepare HA coating containing Ca and P ions in calcium and phosphate electrolytes, The MAO process was started by applying higher voltage and finished in a very short time. Such MAO coatings can provide good corrosion resistance and

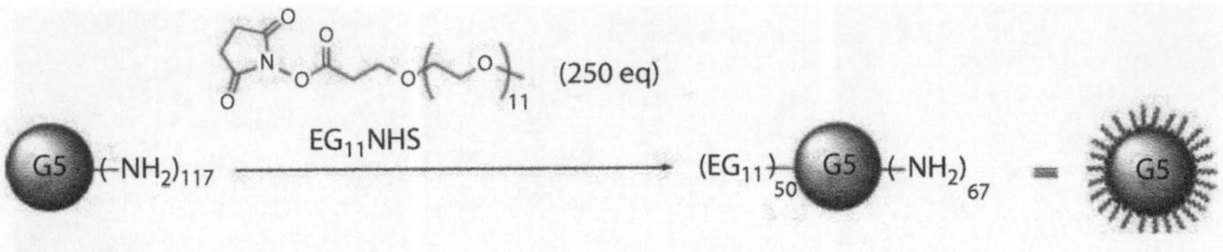

FIGURE 8.54
The reaction for modifying G5 PAMAM dendrimer, possessing 117 amino groups at its periphery, with PEG consisting of 11 ethylene glycol units (EG11). (From Jevprasesphant, R. et al., *Int. J. Pharm.*, 252(1–2), 263, 2003.)

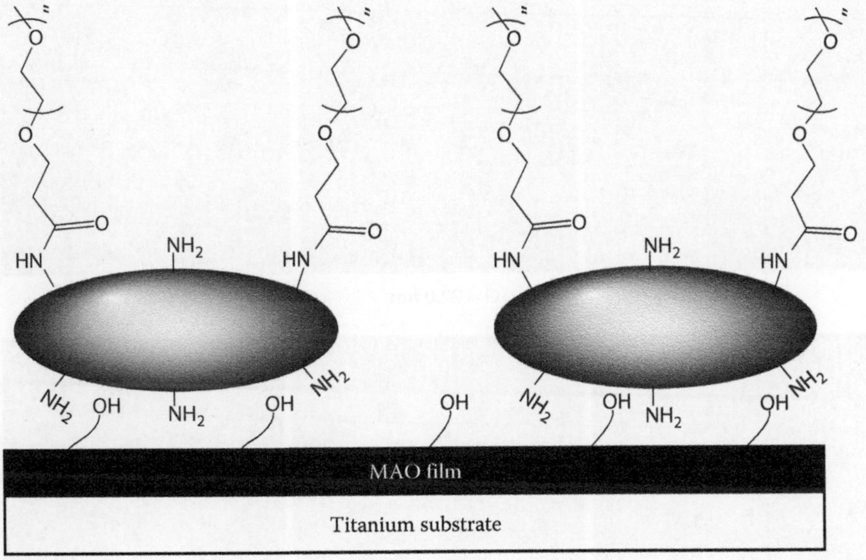

FIGURE 8.55
Schematic diagram of the PEGylated PAMAM dendrimer film on MAO substrate. (From Wang, L. et al., *ACS Appl. Mater. Interfaces*, 3, 2885, 2011.)

biochemical stability in the complex environment of the human body. It is expected to be applied in artificial bone joints and dental implants. The MAO substrate was immersed into the dendrimer solution and incubated with shaking. A schematic diagram of HA/PAMAM on Ti substrate is shown in Figure 8.55.

8.5.2.2 Hydrothermal Biomineralization and Crystallization

The formation of stable HAP nanoparticles with controlled size and shape can be achieved by hydrothermal crystallization using a mixture of supersaturated solutions of Ca^{2+} and PAMAM. These conditions ensured fast homogeneous nucleation and effective growth inhibition by adsorption of PAMAM predominantly onto Ca-rich nanoparticles during the initial stages of crystallization. Hydroxyapatite nanoparticles were synthesized in the presence of various generations and concentrations of PAMAM by hydrothermal treatment of a calcium phosphate precipitate produced by the addition of aqueous ammonium dihydrogen phosphate to a solution containing calcium nitrate and PAMAM at different molar ratios.

8.5.2.3 Self-Precipitation

Calcium hydroxyapatite particles as cores were produced by self-precipitation in the presence of PAMAM dendrimers from SBF [60]. SBF was prepared by dissolving reagent-grade chemicals NaCl, NaHCO$_3$, KCl, Na$_2$HPO$_4$·7H$_2$O, MgCl$_2$·6H$_2$O, CaCl$_2$, and Na$_2$SO$_4$ into deionized water and buffered at pH = 7.4 with tris-hydroxy-methy aminomethane ((CH$_2$OH)$_3$CNH$_2$) and 1 M HCl at 37°C. PAMAM dendrimers were dissolved in SBF (pH 7.4). HA/PAMAM composites were allowed to self-precipitate from SBF solutions containing dendrimer after nucleation and crystal growth.

8.5.3 Characterization of HA/PAMAM Composite Coating

The various morphologies of HA/PAMAM composites can be obtained, such as HA/PAMAM plates, nanorods, elliptical particles, and spherical cores. Figure 8.56 shows the morphology and size of the composites prepared by different methods. The HA/PAMAM plates [61] are shown in Figure 8.56a with an average length of 80 nm and width of ca. 30 nm. The HA/PAMAM plates were prepared by hydrothermal method in Ca$_{10}$(PO$_4$)$_6$(OH)$_2$ and G1.0 PAMAM dendrimer mixtures. Figure 8.56b shows that the samples prepared with carboxylic terminated PAMAM are composed by well-dispersed nanorods [62] (length ca. 65 nm and diameter ca. 26 nm). With the concentration of PAMAM to 40 g, the morphology of hydroxyapatite crystals was nearly ellipsoid-like, as shown in Figure 8.56c. The elliptical particles have an average grain size of ca. 30 nm. The particle sizes of hydroxyapatite crystals depended on the concentration of the PAMAM dendrimers. The particles obtained by self-precipitation were spherical sub-micrometer-sized hydroxyapatite particles [63] (Figure 8.56d). The particles appear hollow, and the size ranged from less than 50 to approximately 150 nm.

The dendrimers affected morphology of HAP crystals as growth inhibitors. PAMAM has a stronger control on crystallographic faces of HAP crystals, which leads to a different crystal habit of HAP crystallization. PAMAM dendrimer might serve as a nucleation site due to calcium binding on the carboxylic acid groups present on the surface and the amphipathic half-generation dendrimer acting as a supramolecular (micellar) aggregate, which might serve as a nucleation site [64].

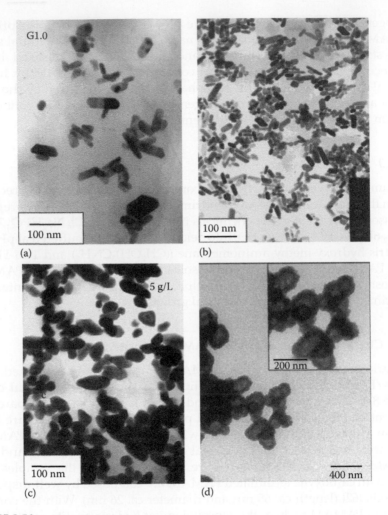

FIGURE 8.56
TEM of images of HA/PAMAM composites. (a) HA/PAMAM plates, (b) HA/PAMAM nano-rods, (c) elliptical HA/PAMAM particles, (d) hollow HA/PAMAM cores. (From Yan, S.-J. et al., *Mater. Chem. Phys.*, 99(1), 164, 2006.)

8.5.4 Application of HA/PAMAM

8.5.4.1 Therapeutic of PAMAM

A thin layer of PAMAM/HA coated on titanium-based substrates has many applications on biochemical fields. The most active area in dendrimer-based therapeutics is gene transfection by dendrimers as nonviral vectors. Dendrimers are very actively under investigation for the delivery of DNA and small organic molecule drugs, especially for cancer therapy. To gain a

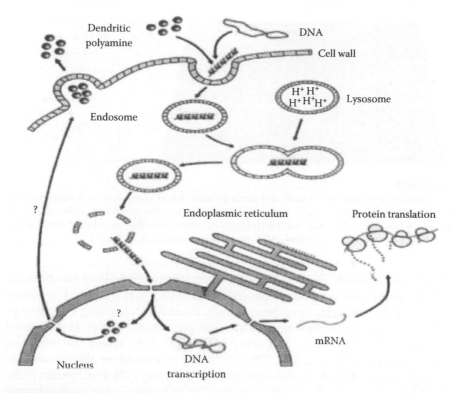

Dendritic polyamine

DNA

Cell wall

Endosome

H⁺ H⁺
H⁺ H⁺H⁺ Lysosome

?

Endoplasmic reticulum

Protein translation

mRNA

?

Nucleus

DNA transcription

FIGURE 8.57
Hypothetical mechanism of DNA transfection with cationic dendrimers. (From Oliveira, J.M. et al., *Biomaterials*, 30(5), 804, 2009.)

better understanding of the rules that govern dendrimer-based gene delivery, PolyFect, obtained by partial chemical hydrolysis of PAMAM dendrimers [64], based on the rigid diphenylethyne core was synthesized and their activity as transfection agents described. Under physiological conditions, the branches bear positively charged ammonium end groups, which can interact with negatively charged phosphate groups of nucleic acids. The PolyFect reagent assembles DNA into compact toroidal structures, thus optimizing the entry of the DNA into cells. Once inside the cell, the dendrimers buffer the lysosomes after fusion with the endosomes [65] (Figure 8.57).

8.5.4.2 Dendrimers in Antimicrobial Application

Dendrimers exhibit low toxicity and are nonimmunogenic. They are widely used in biomedical applications. Based on their dimensional length scaling, narrow size distribution, and other biomimetic properties, dendrimers are often referred to as bartificial proteins. Several measures of antibacterial

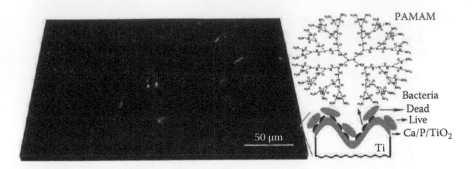

FIGURE 8.58
3D fluorescence images of the indicated surfaces after 3 h of incubation in a suspension of PA01 in PBS buffer containing PI as the viability indicator. The green channels indicate all the bacteria on the surface and the red channels indicate those that were killed (membrane-compromised). (From Wang, L. et al., *ACS Appl. Mater. Interfaces*, 3, 2885, 2011.)

activity, particularly the concentration needed to kill half of the bacteria and the minimum inhibitory concentration (MIC) that inhibits the visible growth of the bacteria, have been widely used for evaluating the susceptibility of potential drugs against specific pathogens. Cai et al. [59] used a standard, more efficient, and reproducible assay to assess the activities of the dendrimer derivatives. The bactericidal activity of the dendrimer coatings was maintained after storage for 30 days in PBS. Furthermore, the PAMMA/HA coating had a low cytotoxicity to human bone mesenchymal stem cells and did not alter the osteogene expression of the cells. It is also demonstrated that simple modification of calcium phosphate coated titanium substrates with PAMAM dendrimers furnished the materials with antibacterial activity without apparently decreasing their bioactivity, as shown in Figure 8.58.

8.5.4.3 Dendrimers in Artificial Oxygen Carrying System

PAMAM dendrimers have recently attracted attention for studying drug delivery [66]. As known to all, one of the important functions of the blood is oxygen transport to the tissues. An immediate resuscitative solution to prevent excessive blood loss is to deliver oxygen-carrying fluid to the patient. Hemoglobin-based oxygen carriers are being developed as resuscitative agents, and many of them are in the final phase of clinical trials [67]. Khopade et al. reported that the aquasomes carried by PAMAM/HA could load hemoglobin approximately and retain oxygen-affinity, cooperativity, and stability for at least 30 days. Figure 8.59 shows the oxygen-binding characteristics of fresh blood, hemoglobin solution, and hemoglobin aquasome formulation (HAF) [60]. This shows that the properties of hemoglobin and its ability to carry oxygen are retained by the HAF.

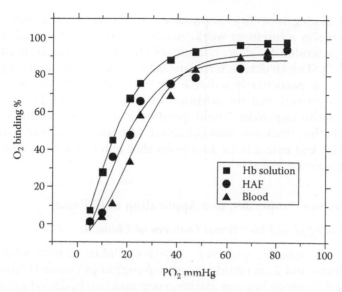

FIGURE 8.59

Oxygen-binding characteristics (oxygen saturation curve) of fresh blood, hemoglobin solution and hemoglobin aquasome formulation. (From Khopade, A.J. et al., *Int. J. Pharm.,* 241(1), 145, 2002.)

8.6 HA/Chitosan Composite Coating

Although high bioactive HA is the main inorganic composition of bone tissue, it is brittle and hard. The cracking of HA coatings can result from the stage of heating treatments such as deposit sintering because of the difference in the thermal expansion coefficients of the sintered coating and the substrate, and additional cracking during cooling can further damage the coatings. Other problems are related to the phase transformations and oxidation of the substrates and the chemical reactions between the deposits and substrates during sintering at elevated temperatures. In addition, due to the favorable bioactive property and the proper body temperature and nutrition, HA is usually in favor of the bacteria adherence and colonization on the implant, which consequently causes the failure of the implant.

Therefore, it is important to modify HA or synthesize novel composites of polymer and HA to enhance the mechanical and antibacterial properties. It is natural to think of adding antibacterial materials into HA coating to effectively reduce the postoperative infection; however, it would generally lead to cytotoxicity and be detrimental for bone formation and growth, while bone in-growth at early stage is much critical for implant/bone fixation and patient recuperation. Seeking a kind of material possessing combined biodegradability, nontoxicity, and antibacterial and hemostasis properties is a foremost work. Among all the antibacterial materials, chitosan, generated by deacetylation of chitin, is a widely used natural polymer [68]. The structure with predictable pore sizes and degradation rates makes it particularly suitable for bone and cartilage regeneration. It has been proved that the addition of chitosan to the HA suspension was effective in improving the deposition rate of the coating and that a uniform coating thickness was achieved. Moreover, chitosan exhibits the antibacterial and osteoblastic functions that make it a good candidate in tissue engineering [69].

8.6.1 Structure, Properties, and Application of Chitosan

8.6.1.1 Structural and Functional Features of Chitosan

Chitosan is a linear copolymer of β-(1–4) linked 2-amino-2-deoxy-β-D-glycopyranose and 2-acetamido-2-deoxy-β-D-glucopyranose (Figure 8.60). It is polycationic, which has one amino group and two hydroxyl groups in the repeating glucosidic residue.

The primary amino groups on the molecule are active and provide reactive sites for a variety of side group attachments employing mild reaction conditions. Thus, the facile derivatization makes chitosan an ideal candidate for biofabrication. In addition, as it is cationic, hemostatic, and insoluble at high pH, chitosan can be reversed by protonation of the amine, which makes the molecule anionic and water-soluble. The attached side groups on chitosan make it a versatile material with specific functionality, altered biological properties, or modified physical properties.

8.6.1.2 Biodegradation and Toxicity of Chitosan

Chitosan is known to be degraded in vertebrates predominantly by lysozyme and by certain bacterial enzymes in the colon. This biodegradation plays a major role in the metabolic fate of chitosan in the body and makes it a useful polymer used in drug delivery systems and scaffolds in tissue engineering.

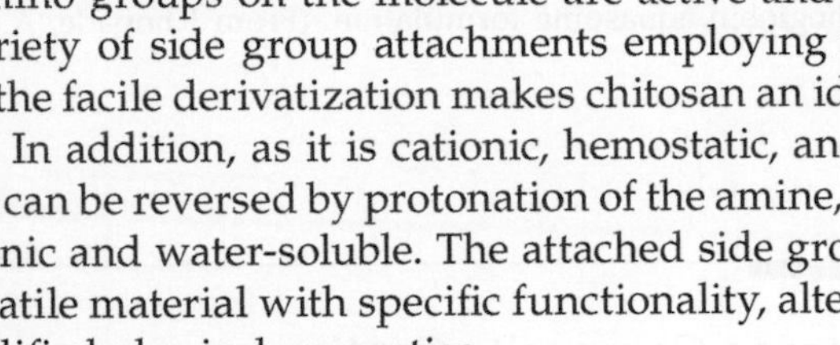

FIGURE 8.60
Chitosan structure. (From Agrawal, P. et al., *Adv. Drug Deliver. Rev.*, 62(1), 42, 2010.)

As chitosan is a nontoxic, biologically compatible polymer, it is approved for dietary applications in Japan, Finland, and Italy for many years, and it has been approved by the FDA for use in wound dressings [71]. However, chemical modifications implemented on chitosan could make it more or less toxic. In drug formulations, chitosan has been proved to be a safe excipient over the last decades. Interestingly, some studies [72] have proved that chitosan and its derivatives seem to be toxic to several bacteria, fungi, and parasites. This pathogen-related toxicity is very important as it could be beneficial in the control of infectious disease, especially in implant engineering.

8.6.1.3 Application of Chitosan in Tissue Engineering

Chitosan offers several advantages, such as its ability to control the release of active agents and avoid the use of hazardous organic solvents while fabricating particles, since it is soluble in aqueous acidic solution. Owing to these properties, chitosan is extensively used in developing drug delivery systems. Also, hydrogels obtained from chitosan are widely used in bioapplications and play a crucial role in current strategies to remedy malfunctions in and injuries to living systems. Apart from these applications, chitosan has been extensively used in bone tissue engineering, since it was shown to promote cell growth and mineral rich matrix deposition by osteoblasts cells in culture. By the biocompatibility of chitosan, the additional local inflammation can be minimized, and the porous structures that design to allow osteoconduction can be easily molded. Several studies have focused on chitosan–calcium phosphate (CP) composites for this purpose with a 3D macroporous CP bioceramic. It is revealed that the nested chitosan sponge enhances the mechanical strength of the ceramic phase through matrix reinforcement and preserves the osteoblast phenotype. Incorporating hydroxyapatite (HA) or CP glass, the macroporous chitosan scaffolds had an interconnected porosity of approximately 100 μm and are used for clinical applications. Chitosan–HA nanocomposite with high strength and bending modulus renders the material suitable for possible application for internal fixation of long bone fractures.

In tissue engineering approaches, especially cartilage repair, the ideal cell-carrier substance should mimic the natural environment in the articular cartilage matrix. The cartilage-specific extracellular matrix (ECM) components such as type II collagen and glycosamine glycans (GAGs) play a critical role in regulating the expression of the chondrocytic phenotype and in supporting chondrogenesis in vitro and in vivo. The hybrid polymer fibers based on chitosan increased cell attachment and proliferation, delivered growth factors in a controlled fashion to promote the in-growth and biosynthetic ability of chondrocytes, and showed increased tensile strength, implying a possible use in developing a 3D load-bearing scaffold for cartilage regeneration.

In liver tissue engineering, for the structural similarity of chitosan to GAGs and since GAGs are components of the liver ECM, chitosan was selected as a scaffold material for hepatocytes culture. The cross-linked matrix based on chitosan had moderate mechanical strength, good hepatocyte compatibility, as well as excellent blood compatibility.

In nerve tissue engineering, chitosan is suitable for nerve regeneration based on its biocompatibility and biodegradability. Many researchers reported that neurons cultured on the chitosan membrane can grow well and that chitosan tube or fibers can promote repair of the peripheral nervous system [73]. Chitosan-based microfibers provide very good scaffolds for nerve tissue engineering applications with the advantages of ease of fabrication, simplicity, and cost effectiveness.

In conclusion, chitosan, almost the only cationic polysaccharide in nature with such great innate medical potential with its exciting properties, is one of the most promising bio-based polymers for drug delivery, tissue engineering, gene therapy, and theranostics.

8.6.2 Preparation of HA/Chitosan Composite Coating on Ti Alloy

8.6.2.1 Plasma Spraying Process

Various methods have been used to deposit calcium phosphate ceramic coatings on metallic implants, such as plasma spraying, sol-gel derived coating, pulsed laser deposition, electrophoretic deposition, and biomimetic coating. Among them, plasma spraying is the most widely accepted method for producing a calcium phosphate, especially HA-chitosan coating on metallic implants. On the basis of the difference in feedstock, the process is classified as atmospheric plasma spraying (APS) [74] and liquid precursor plasma spraying (LPPS) [75]. In the former process, powder materials are used as the feedstock, and they undergo rapid heating, melting, and rapid cooling processes. In the later process, the liquid precursor is directly used as the feedstock, thus avoiding a series of additional preparation processes for the powder feedstock involved in the APS process. Thus, the later is a simple, low-cost, efficient, and potentially an alternative plasma spraying technique for commercially synthesizing HA-chitosan coatings. The main difference between the two processes is the preparation of feedstock.

Step 1, disposing substrates: Choosing a Ti alloy such as Ti-6Al-4 V disk with 14 mm diameter and 1.5 mm thickness as the substrate, its surface is grit-blasted with alumina abrasives (380 μm particle size, 0.4–0.7 MPa pressure) prior to the plasma spraying and ultrasonically cleaned in acetone and ethanol.

Step 2, preparation of feedstock: For APS process, a wet chemical synthesis is used to prepare HA powders. First, both $Ca(OH)_2$ (95 wt%) and H_3PO_4 (85 wt%) are diluted in water, and the later aqueous solution is added dropwise to the stirred dispersion of the former. The mixture is stirred for 3–5 h at 35°C–95°C, depending on the desired crystallinity degree. Thirdly, the reaction mixture is ripened for 2 or 24 h to give HA powders with the crystallinity degree between 20% and 80%. At last, the precipitates are washed with distilled water, filtered three times, and finally freeze-dried. The obtained powders were sieved at 400 and 150 μm. *For LPPS process*, 2.03 M $(NH_4)_2HPO_4$ aqueous solution is added into a stirred 1.69 M $Ca(NO_3)_2$ aqueous solution based on the Ca/P atomic ratio of 1.67. The pH value of the solution is adjusted to 11 with ammonia solution (NH_4OH, 30 wt%), and the mixture is stirred 48 h at 70°C to afford a 25 wt% HA solid content of liquid precursor. More porous microstructure can be obtained from lowering the HA solid content by adding a certain amount of deionized water into the former liquid precursor.

Step 3, spraying HA coatings: With Metco MN air plasma spraying system (Metco Ltd., USA) and AR2000 thermal spraying robot, the feedstock, acetic acid solution of HA powders or liquid HA precursor, is transported through a peristaltic pump and injected into the plasma flume through an atomizing nozzle (Bete Inc). Nitrogen and hydrogen are used as the primary gas and the secondary gas, respectively. Porous HA coatings are deposited onto the Ti alloy substrates with the thicknesses of about 100 μm.

Step 4, loading with chitosan: The chitosan with about 85% degree of deacetylation is dissolved in ultrapure water with concentrations of 10 30 g/L. Then, the HA-coated Ti alloy sample is immersed into the chitosan solution for 24 h in sterilized condition. Next, the sample is kept in a laminar flow cabinet for air drying, and HA/chitosan composite coating on Ti alloy is obtained.

8.6.2.2 Electrophoretic Deposition Process (EPD)

In the EPD process [76], the composite coating is achieved via the motion of charged particles toward an electrode under an applied electric field. Its main advantage is the ability to control the deposit stoichiometry, that is, the degree of stoichiometry of electrophoretic deposits can be controlled by the composition of the powder used. In its procedure, Ammonium phosphate solution (0.6 M) is slowly added into 1.0 M calcium nitrate solution to perform precipitation at 70°C, and the pH of the mixture is then adjusted to 11 with NH_4OH. Stirring is kept for 8 h at 70°C and 24 h at room temperature. The precipitate is washed with water and finally with isopropyl alcohol to afford the stoichiometric HA nanoparticles. Chitosan (MW = 200,000) with a degree of deacetylation of about 85% is dissolved in 1% acetic acid solution. Electrodeposition is performed from suspensions of 0–8 g/L HA nanoparticles in a mixed ethanol–water solvent, containing 0–0.5 g/L chitosan. Cathodic deposits were obtained by galvanostatic method on alloy plates (10 mm × 50 mm × 0.1 mm) at current densities ranging from 0.1 to 3 mA/cm². The electrochemical cell for deposition included a cathodic substrate centered between two parallel platinum counter electrodes.

8.6.2.3 Aerosol Deposition (AD)

This process is an alternative coating method for depositing dense ceramic coatings on various substrates such as metals, ceramics, and plastics. It is a kind of spray coating technique that can produce a dense and well-adherent ceramic coating at room temperature (RT). Solid powder particles are used as the starting materials, and the formation of the coating layer is achieved by the collision of the particles onto a substrate.

In the process [77], HA–chitosan powder mixture is first prepared. HA nanocrystalline powder having a volumetric mean diameter (d50) of 15 nm and chitosan powder with a degree of deacetylation of about 85% were used as the starting materials. The HA powder is heated at 1050°C for 2 h, and the chitosan powder is dry ball-milled for 12 h in a planetary mill using ZrO_2 balls and a jar. Thereafter, the heat-treated HA powder is mechanically mixed with the ball-milled chitosan powder by dry ball milling, resulting in the preparation of the HA–chitosan powder mixture.

The alloy plate with a size of 20 mm × 20 mm × 1 mm is used as the substrate, which is ground with 2000 grit SiC paper and ultrasonically rinsed in distilled water for 5 min. The apparatus of AD is shown in Figure 8.61. It consists of a carrier gas supply system with mass flow control, a powder chamber containing the above composite powder, and a deposition chamber with motored X–Y stage and nozzle evacuated by rotary vacuum pump with mechanical booster. Teflon tube with 6.35 mm in diameter connects the three parts. Tape is used to fix commercial purity Ti substrate on a steel plate, which is fixed to the motored X–KY stage using magnets. About 50 g of the powder is poured into the powder chamber that is shaken during coating, making

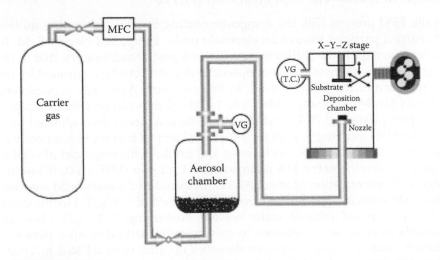

FIGURE 8.61
Schematic diagram for apparatus of aerosol deposition. VG, vacuum gauge. (From Hahn, B.-D. et al., *J. Am. Ceram. Soc.*, 92, 683, 2009.)

fine particle suspension in the upper portion of the chamber. The deposition chamber is evacuated to 0.65 Pa. Oxygen (99% purity) is used as a carrier gas and flows at 5×10^{-4} m³/s for carrying fine composite particles from the powder chamber, spraying them vertically onto the substrate through a nozzle with a slit type opening (0.5 mm × 25.4 mm) in the deposition chamber. At the same time, a motor for the X–Y stage was turned on, moving the substrate at 1 mm/s. During deposition, pressure of the deposition chamber is 920 Pa. The coated sample is heated to 400°C–500°C in air, cleaned in an ultrasonic bath. and washed using ethanol, to give the desired material.

8.6.3 Characterization and Functional Test

8.6.3.1 Composition Analysis

The HA–chitosan composite coatings can be characterized by X-ray diffraction (XRD) using monochromatized CuK_α radiation at a scanning speed of 4°/min. The HA peaks in the XRD patterns of the coatings may be somewhat reduced and broadened, compared to those of the corresponding HA–chitosan composite powders. In XRD, the peaks for calcium phosphates, such as tricalcium phosphate and tetracalcium phosphate, were not always detected, which suggested that no significant phase change of HA occurred during the coating process [77].

8.6.3.2 Thermogravimetric Analysis (TGA)

To quantify the amount of chitosan in the HA–chitosan composite coating, the composite coating layer can be scraped from the substrate and subjected to TGA [79]. By applying temperatures of up to 800°C at a rate of 5°C/min, the relative amounts of HA and chitosan in the composite coating can be calculated from the weight loss of the coating layer. The TGA curve of the pure HA coating shows a small weight loss upon heating up to 800°C, indicating its high thermal stability. The HA–chitosan composite coatings exhibited a remarkable weight loss in the temperature range of 200°C–600°C, which was attributed to the decomposition of the chitosan. At 800°C, the total weight losses of the HA–chitosan composite coatings can be obtained from the HA–chitosan powder mixtures curves. Therefore, the amount of chitosan in the HA–chitosan composite coating can be calculated from the weight loss profiles of the TGA curves.

8.6.3.3 Morphology and Roughness

The microstructures of the HA–chitosan composite coatings can be investigated using scanning electron microscopy (SEM) [76]. The surface morphology of the coatings shows different microstructures according to the amount of chitosan added. The coating surface becomes smoother than the pure HA coatings, and the later often has a rough surface with a network structure. With adding chitosan, the surface morphology of the coating layer shows fairly dense microstructures without pores or cracks, implying that chitosan did not change significantly the dense nature of the HA coating layer.

8.6.3.4 Adhesive Strength Test

The tensile adhesion strength between the coating and the substrate can be measured using a universal testing machine, such as Instron series IX automated materials testing system, Instron Corp., MA, USA, according to ISO 13779-4. An alloy disk with 25 mm in diameter and 1 mm in thickness is coated with HA–chitosan on one side, and both its top and bottom surfaces are bonded to the supporting fixture of the universal tensile testing machine with thermal cure, high-strength epoxy, such as 3 M Scotch-Weld Epoxy Adhesive 2214, 3 M Corp., MN, USA. Perform three to five adhesion tests for each sample. It is proved that the adhesion strength values of the coatings exhibit a tendency to decrease slightly with the increasing amount of chitosan in the composite coating. However, even the lowest adhesion strength value obtained from the composite coatings is higher than those of HA coatings, which indicates the helpfulness of chitosan in the formation of composite layers [79].

8.6.3.5 Bacteria Inhibition

The antibacterial properties of HA/chitosan complex coatings can be tested using *Staphylococcus aureus* (ATCC 25923) [80]. A suspension of *S. aureus* bacteria with a concentration of $1\sim2\times10^6$ CFUs/mL is added to 24-well plates and cultured for 6, 12, and 24 h, respectively. The bacterial inhibition behavior is investigated with MTT assay. Optical density (OD) is detected at 490 nm, and the proliferation of *S. aureus* is determined by the OD value. SEM is used to analyze the bacteria adhesion on the coating surface after 24 h culture time. It is proved that no significant difference is found between the HA/chitosan complex coating and the control at 6 h. At 12 h, porous HA/chitosan complex coatings with four different concentrations all demonstrated significant antibacterial activity, while significant bacterial growth was observed for the control. At 24 h, the antibacterial properties of the HA/chitosan complex coatings showed distinct difference. While the HA coatings loaded with 20, 50, and 100 g/L chitosan still possessed obvious antibacterial properties, significant growth of bacteria has been observed for coating loaded with 10 g/L chitosan.

8.6.3.6 Cytotoxicity on Osteoblastic Cell

Suspension of MG63 cells with number of 1×10^4/mL is added to 24-well plates and cultured for 2, 4, and 6 days, respectively. MTT assay is selected and the OD value at 570 nm is measured [80]. The result shows that the OD value of HA coating loaded with chitosan is significantly lower than that of the control coated with pure HA. It is attributed to the intrinsic antibacterial property of chitosan, which is also expected to have an adverse effect on osteoblastic cell viability. It is also shown that the loading of chitosan is even beneficial for the proliferation of osteoblastic cells at low concentrations. It mainly relies on its positive charge to attract and break the negatively charged bacterial cell wall, which would also work adversely on the osteoblastic cells. However, at certain concentrations and contact time, the negatively charged osteoblastic cells may be just attracted to the HA/chitosan coating surface without undergoing further impairment, which leads to enhanced cell adhesion and proliferation. In addition, the electrostatic interaction of chitosan with the negatively charged molecules would also likely help the adsorption of certain proteins, which would promote the adhesion of the osteoblast cells.

Therefore, chitosan can selectively do harm to the bacteria but not the osteoblastic cells. It also appears that chitosan sticks very well to the HA coating surface, even during the in vitro tests. By tuning the chitosan concentration, combined antibacterial activity and enhanced osteoblast function can be achieved simultaneously in the porous-HA/chitosan complex coatings. Therefore, the physical and chemical properties of the chitosan can be controlled and optimized to allow the coexistence of antibacterial activity and enhanced osteoblastic cell response.

8.6.4 Applications of Hydroxyapatite/Chitosan Composite Coatings

Hydroxyapatite/chitosan composite coatings can be extensively used in bone tissue engineering, since it was shown to promote cell growth and mineral rich matrix deposition by osteoblasts cells. The biocompatibility of chitosan minimizes additional local inflammation, and it can be molded into porous structures to allow osteoconduction. A study [81] in vivo has confirmed that 1 week after repairing the bone defect with the composite in the experimental group, new bone appeared around the composite and matured gradually. At 24 weeks after surgery, little collagenous tissues were observed between the material and surrounding bones. At 40 weeks after surgery, new bone had grown into the mature bone and total osseointegration had occurred. In the control group, however, no bone defect healing was observed at 40 weeks after surgery. All these results suggest that the composite has a good hard tissue biocompatibility and an excellent osteoconductivity. It is suitable for artificial bone implants and frame materials of tissue engineering.

In hydroxyapatite/chitosan composite, the nested chitosan sponge enhances the mechanical strength of the ceramic phase through matrix reinforcement and preserves the osteoblast phenotype. A hydroxyapatite/chitosan multilayer nanocomposite with high strength and bending modulus rendering the material suitable for possible application for internal fixation of long bone fractures has been prepared. In vitro osteoblast cell culture studies revealed the biocompatibility of the scaffolds with cells exhibiting healthy morphology and avid proliferation throughout the culture period. As a new synthesized material, hydroxyapatite/chitosan composite is promising in tissue engineering.

References

1. Grosgogeat B, Reclaru L, Lissac M, Dalard F. Measurement and evaluation of galvanic corrosion between titanium/Ti6A14V implants and dental alloys by electrochemical techniques and auger spectrometry. *Biomaterials*, 1999, 20(10): 933–941.
2. Fu Q, Hong Y, Liu X, Fan H, Zhang X. A hierarchically graded bioactive scaffold bonded to titanium substrates for attachment to bone. *Biomaterials*, 2011, 32: 7333–7346.
3. Iijima S. Helical microtubes of graphitic carbon. *Nature*, 1991, 354: 56–58.
4. Zeng Q, Li Z, Zhou Y. Synthesis and application of carbon nanotubes. *Journal of Natural Gas Chemistry*, 2006, 15(3): 235–246.
5. Hu H, Ni Y, Montana V et al. Chemically functionalized carbon nanotubes as substrates for neuronal growth. *Nano Letters*, 2004, 4(3): 507–551.
6. Lijuan K, Qianbin W. Study of carbon nanotubes-hydroxyapatite composite. *International Journal of Stomatology*, 2008, 35(Suppl.): 241–243.
7. de Groot K. Hydroxylapatite as coatings for implants. *InterCeram*, 1987, 4: 38–41.
8. Balani K, Chen Y, Harimkar SP, Dahotre NB, Agarwal A. Tribological behavior of plasma-sprayed carbon nanotube-reinforced hydroxyapatite coating in physiological solution. *Acta Biomaterialia*, 2007, 3: 944–951.
9. Chen Y, Zhang YQ, Gan CH, Zheng CY, Yu G. Carbon nanotube reinforced hydroxyapatite composite coatings produced through laser surface alloying. *Carbon*, 2006, 44(1): 37–45.
10. Chen Y, Gan CH, Zhang TH, Yu G, Bai PC, Kaplon A. Laser surface alloyed? Carbon-nanotube reinforced hydroxyapatite composite coatings. *Applied Physics Letters*, 2005, 86: 251905.
11. Bai Y, Neupane MP, Park IS, Lee MH, Bae TS, Watari F, Motohiro U. Electrophoretic deposition of carbon nanotubes–hydroxyapatite nanocomposites on titanium substrate. *Materials Science and Engineering C*, 2010, C30: 1043–1049.
12. Aryal S, Bhattarai SR, Remant Bahadur KC, Khil MS, Lee D-R, Kim HY. Carbon nanotubes assisted biomimetic synthesis of hydroxyapatite from simulated body fluid. *Materials Science and Engineering A*, 2006, 426: 202–207.
13. Yu W. Preparation of HA/MWNTs composite coating on the surface of NiTi alloy. Master degree thesis, Tianjin University, Tianjin, China, 2008, vol. 33, pp. 37–38.
14. Huijuan H, Fan Z, Yafei Z, Changjian L. Electrophoretic deposition of HA/CNTs composite coating. *Journal of Xiamen University (Natural Science)*, 2006, 45(5): 593–595.

15. Xu JL, Liu F, Wang FP, Yu DZ, Zhao LC. Microstructure and corrosion resistance behavior of ceramic coatings on biomedical NiTi alloy prepared by micro-arc oxidation [J]. *Applied Surface Science*, 2008, 254: 6642–6647.
16. Forsgren J, Svahn F, Jarmar T, Engqvis H. Formation and adhesion of biomimetic hydroxyapatite deposited on titanium substrates [J]. *Acta Biomaterialia*, 2007, 3: 980–984.
17. Byon E, Jeong Y, Takeuchi A, Kamitakahara M, Ohtsuki C. Apatite-forming ability of micro-arc plasma oxidized layer of titanium in simulated body fluids [J]. *Surface and Coatings Technology*, 2007, 201: 5651–5654.
18. Liu SM, Yang XJ, Cui ZD, Zhu SL, Wei Q. One-step synthesis of petal-like apatite/titania composite coating on a titanium by micro-arc oxidation [J]. *Materials Letters*, 2011, 65(6): 1041–1044.
19. Nie X, Leyland A, Matthews A. Deposition of layered bioceramic hydroxyapatite/ TiO_2 coatings on titanium alloys using a hybrid technique of micro-arc oxidation and electrophoresis. *Surface and Coatings Technology*, 2000, 125: 407–414.
20. Li Y, Lee IS, Cui FZ, Choi SH. The biocompatibility of nanostructured calcium phosphate coated on micro-arc oxidized titanium [J]. *Biomaterials*, 2008, 29: 2025–2032.
21. Kar A, Raja KS, Misra M. Electrodeposition of hydroxyapatite onto nanotubular TiO_2 for implant applications[J]. *Surface and Coatings Technology*, 2006, 201: 3723–3731.
22. Ni JH, Shi YL, Yan FY, Chen JZ, Wang L. Preparation of hydroxyapatite-containing titania coating on titanium substrate by micro-arc oxidation[J]. *Materials Research Bulletin*, 2008, 43: 45–53.
23. Tingda J, Chunping Z. *Collagen Protein*. Chemical Industry Press, Beijing, China, 2001.
24. Buehler MJ. Molecular architecture of collagen fibrils: A critical length scale for tough fibrils. *Current Applied Physics*, 2008, 8: 440–442.
25. Tong S, Jingyan W. *Biochemistry*. Higher Education Press, Beijing, China, 2000.
26. Qiyi Z. The fabrication and bioactivity study of Ca-P coatings on titanium surface. Doctoral dissertation, Sichuan University, Sichuan, China, 2003, p. 110.
27. Yanli C. Formation of HA/COL composite coatings on the surface of Ti and NiTi alloy. Master degree thesis, Tianjin University, Tianjin, China, 2005, p. 22.
28. Yanli C, Chunyong L, Shengli Z, Zhenduo C, Yang X. Formation of bonelike apatite-collagen composite coating on the surface of NiTi shape memory alloy. *Scripta Materialia*, 2006, 54(1): 89–92.
29. Fan YW, Duan K, Wang RZ. A composite coating by electrolysis-induced collagen self-assembly and calcium phosphate mineralization. *Biomaterials*, 2005, 26: 1623–1632.
30. Schliephake H, Scharnweber D, Dard M, Rossler S, Sewing A, Huttmann C. Biological performance of biomimetic calcium phosphate coating of titanium implants in the dog mandible. *Journal of Biomedical Materials Research*, 2003, 64: 225–34.
31. Hu K, Yang X, Cai Y, Cui Z, Wei Q. Preparation of bone-like composite coating using a modified simulated body fluid with high Ca and P concentrations. *Journal of Material Science: Materials in Medicine*, 2009, 20: 131–134.
32. Yanli X, Longxiang L, Zhiwang G, Changjian L. Electrochemical codeposition of HA/Collagen coating and preliminary in vitro test. *Electrochemistry*, 2007, 13(3): 238–241.

33. Sun T, Lee W-C, Wang M. A comparative study of apatite coating and apatite/collagen composite coating fabricated on NiTi shape memory alloy through electrochemical deposition. *Materials Letters*, 2011, 65(17–18): 2575–2577.

34. Teng S-H, Lee E-J, Park C-S, Choi W-Y, Shin D-S, Kim H-E. Bioactive nanocomposite coatings of collagen/hydroxyapatite on titanium substrates. *Journal of Materials Science: Materials in Medicine*, 2008, 19: 2453–2461.

35. de Jonge LT, Leeuwenburgh SC, van den Beucken JJ, te Riet J, Daamen WF, Wolke JG, Scharnweber D, Jansen JA. The osteogenic effect of electrosprayed nanoscale collagen/calcium phosphate coatings on titanium. *Biomaterials*, 2010, 31: 2461–2469.

36. Yanli X. Preparation and characterization of the HA/Collagen composite coating and its formation mechanism. Master degree thesis, Xiamen University, Fujian, China, 2005.

37. Ou K-L, Chung R-J, Tsai F-Y, Liang P-Y, Huang S-W, Chang S-Y. Effect of collagen on the mechanical properties of hydroxyapatite coatings. *Journal of the Mechanical Behavior of Biomedical Materials*, 2011, 4(4): 618–624.

38. Kai H. Study on preparing HA/COL composite coating by chemically biomimetic growth and electrochemical deposition. Master degree thesis, Tianjin University, Tianjin, China, 2007, p. 38.

39. Yang XJ, Liang CY, Cai YL, Hu K, Wei Q, Cui ZD. Recombinant human-like collagen modulated the growth of nano-hydroxyapatie on NiTi alloy. *Materials Science and Engineering C*, 2009, 29: 25–28.

40. Zhu X, Eibl O, Scheideler L, Geis-Gerstorfer J. Characterization of nano hydroxyapatite/collagen surfaces and cellular behaviors. *Journal of Biomedical Materials Research Part A*, 2006, 79(1): 114–127.

41. Jin Q, Wang XM, Wang XF, Li XD, Ma JF. The efficacy of collagen-hydroxyapatite composite membrane on bone regeneration. *West China Journal of Stomatology*, 2011, 29(1): 21–26.

42. Wei J, Li YB, Zuo Y, Peng XL, Zhang L. Development of biomimetic needle-like apatite nanocrystals by a simple new method. *Journal of Materials Science and Technology*, 2004, 20(6): 665–667.

43. Li CM, Jin HJ, Botsaris GD et al. Silk apatite composites from electrospun fibers. *Journal of Materials Research*, 2005, 20(12): 3374–3384.

44. Junwu X, Yingchun Z, Yanyan L, Yi Z, Fangfang X. An asymmetric coating composed of gelatin and hydroxyapatite for the delivery of water insoluble drug. *Journal of Materials Science: Materials in Medicine*, 2009, 20: 889–896.

45. Pan M, Kong X, Cai Y, Yao J. Hydroxyapatite coating on the titanium substrate modulated by a recombinant collagen-like protein. *Materials Chemistry and Physics*, 2011, 126: 811–817.

46. Jin Z, Yuping Z, Qianqian G, Yunhui Z, Xiaoyan Y, Kangde Y. Crosslinking of electrospun fibrous gelatin scaffolds for apatite mineralization. *Journal of Applied Polymer Science*, 2011, 119(2): 786–793.

47. Gaihre B, Khil MS, Kang HK, Kim HY. Bioactivity of gelatin coated magnetic iron oxide nanoparticles: In vitro evaluation. *Journal of Materials Science: Materials in Medicine*, 2009, 20(2): 573–581.

48. Rohanizadeh R, Al-Sadeq M, LeGeros RZ. Preparation of different forms of titanium oxide on titanium surface: Effects on apatite deposition. *Journal of Biomedical Materials Research*, 2004, 71A: 343–352.

49. Bian WG, Peng L, Liang FH et al. Bone morphogenetic protein-2 and gel complex on hydroxyapatite-coated porous titanium to repair defects of distal femur in rabbits. *Chinese Journal of Orthopaedic Trauma*, 2007, 9(6): 550–554.
50. Brunette DM, Tengvall P, Textor M, Thomsen P. *Titanium in Medicine [M]*. Springer-Verlag, Berlin, Germany, 2001.
51. Kokubo T, Kim HM, Kawashita M. Novel bioactive materials with different mechanical properties. *Biomaterials*, 2003, 24: 2161–2175.
52. Naka K, Kobayashi A, Chujo Y. Effect of anionic 4.5-generation polyamidoamine dendrimer on the formation of calcium carbonate polymorphs. *Bulletin of the Chemical Society of Japan*, 2002, 75: 2541–2546.
53. Lee CC, MacKay JA, Frechet JMJ, Szoka FC. Designing dendrimers for biological applications. *Nature Biotechnology*, 2005, 23(12): 1517–1526.
54. Ottaviani MF, Bossmann S, Turro NJ, Tomalia DA. Characterization of starburst dendrimers by the EPR technique. 1. Copper complexes in water solution. *Journal of American Chemical Society*, 1994, 116: 661–671.
55. Tomalia DA, Esfand R. Dendrons, dendrimers and dendrigrafts. *Chemistry & Industry*, 1997, 11: 416–420.
56. Sönke S, Tomalia DA. Dendrimers in biomedical applications Reflections on the field. *Advanced Drug Delivery Reviews*, 2005, 57(15): 2106–2129.
57. Jevprasesphant R, Penny J, Jalal R, Attwood D, McKeown NB, D'Emanuele A. The influence of surface modification on the cytotoxicity of PAMAM dendrimers. *International Journal of Pharmaceutics*, 2003, 252(1–2): 263–266.
58. Gallardo-Moreno AM, Pacha-Olivenza MA, Fernandez-Calderon MC, Perez-Giraldo C, Bruque JM, Gonzalez Martin ML. Bactericidal behaviour of Ti6Al4V surfaces after exposure to UV-C light. *Biomaterials*, 2010, 31(19): 5159–5168.
59. Wang L, Erasquin UJ, Zhao M, Ren L, Zhang MY, Cheng GJ, Wang Y, Cai C. Stability, antimicrobial activity, and cytotoxicity of poly(amidoamine) dendrimers on titanium substrates. *ACS Applied Material and Interfaces*, 2011, 3: 2885–2894.
60. Khopade AJ, Khopade S, Jain NK. Development of hemoglobin aquasomes from spherical hydroxyapatite cores precipitated in the presence of half-generation poly(amidoamine) dendrimer. *International Journal of Pharmaceutics*, 2002, 241(1): 145–154.
61. Yan S-J, Zhou Z-H, Zhang F, Yang S-P, Yang L-Z, Yu X-B. Effect of anionic PAMAM with amido groups starburst dendrimers on the crystallization of $Ca_{10}(PO_4)_6(OH)_2$ by hydrothermal method. *Materials Chemistry and Physics*, 2006, 99(1): 164–169.
62. Zhou Z-H, Zhou P-L, Yang S-P, Yu X-B, Yang L-Z. Controllable synthesis of hydroxyapatite nanocrystals via a dendrimer-assisted hydrothermal process. *Materials Research Bulletin*, 2007, 42: 1611–1618.
63. Bosman AW, Janssen HM, Meijer EW. About dendrimers: Structure, physical properties, and applications. *Chemical Reviews*, 1999, 99: 1665–1688.
64. Oliveira JM, Kotobuki N, Marques AP, Pirraco RP, Benesch J, Hirose M, Costa SA, Mano JF, Ohgushi H, Reis RL. Surface engineered carboxymethylchitosan/poly(amidoamine) dendrimer nanoparticles for intracellular targeting. *Advanced Functional Materials*, 2008, 18(12): 1840–1853.
65. Oliveira JM, Sousa RA, Kotobuki N, Tadokoro M, Hirose M, Mano JF, Reis RL, Ohgushi H. The osteogenic differentiation of rat bone marrow stromal cells cultured with dexamethasone-loaded carboxymethylchitosan/poly(amidoamine) dendrimer nanoparticles. *Biomaterials*, 2009, 30(5): 804–813.

66. Esfand R, Tomalia DA. Poly(amidoamine) (PAMAM) dendrimers: From biomimicry to drug delivery and biomedical applications. *Drug Discovery Today*, 2001, 6: 427–436.
67. Gulati A. The articles and references in this volume are specially dedicated to blood substitutes. Preface. *Advanced Drug Delivery Reviews*, 2000, 40: 129–130.
68. Chung YC, Chen CY. Antibacterial characteristics and activity of acid-soluble chitosan. *Bioresource Technology*, 2008, 99: 2806–2814.
69. Seol YJ, Lee JY, Park YJ, Lee YM, Young-Ku, Rhyu IC, Lee SJ, Han SB, Chung CP. Chitosan sponges as tissue engineering scaffolds for bone formation. *Biotechnology Letter*, 2004, 26: 1037–1041.
70. Agrawal P, Strijkers GJ, Nicolay K. Chitosan-based systems for molecular imaging. read more. *Advanced Drug Delivery Reviews*, 2010, 62(1): 42–58.
71. Thanou M, Verhoef JC, Junginger HE. Oral drug absorption enhancement by chitosan and its derivatives. *Advanced Drug Delivery Reviews*, 2001, 52: 117–1126.
72. Jumaa M, Furkert FH, Müller BW. A new lipid emulsion formulation with high antimicrobial efficacy using chitosan. *European Journal of Pharmaceutics and Biopharmaceutics*, 2002, 53(1): 115–123.
73. Wang XH, Li DP, Wang WJ, Feng QL, Cui FZ, Xu YX, Song XH, van der Werf M. Crosslinked collagen/chitosan matrix for artificial livers. *Biomaterials*, 2003, 24: 3213–3220.
74. Sun LM, Berndt CC, Gross KA, Kucuk A. Material fundamentals and clinical performance of plasma-sprayed hydroxyapatite coatings: A review. *Journal of Biomedical Materials Research*, 2001, 58(5): 570–592.
75. Huang Y, Song L, Huang T, Liu X, Xiao Y, Wu Y, Wu F, Gu Z. Characterization and formation mechanism of nano-structured hydroxyapatite coatings deposited by the liquid precursor plasma spraying process. *Biomedical Materials*, 2010, 5(5): 054113.
76. Zhitomirsky I. Cathodic electrophoretic deposition of ceramic and organoceramic materials—Fundamental aspects. *Advances in Colloid Interface Science*, 2002, 97(1–3): 279–317.
77. Hahn B-D, Park D-S, Choi J-J, Ryu J, Yoon W-H, Choi J-H, Kim H-E, Kim S-G. Aerosol deposition of hydroxyapatite–chitosan composite coatings on biodegradable magnesium alloy. *Surface and Coating Technology*, 2011, 205: 3112–3118.
78. Hahn B-D, Park D-S, Choi J-J, Ryu J, Yoon W-H, Kim K-H, Park C, Kim H-E. Dense nanostructured hydroxyapatite coating on titanium by aerosol deposition. *Journal of American Ceramic Society*, 2009, 92: 683–687.
79. Pang X, Zhitomirsky I. Electrodeposition of composite hydroxyapatite–chitosan films. *Materials Chemistry and Physics*, 2005, 94: 245–251.
80. Song L, Gan L, Xiao Y-F, Wu Y, Wu F, Gu Z-W. Antibacterial hydroxyapatite/chitosan complex coatings with superior osteoblastic cell response. *Materials Letter*, 2011, 65: 974–977.
81. Tang X-J, Gui L, Lü X-Y. Hard tissue compatibility of natural hydroxyapatite/chitosan composite. *Biomedical Materials*, 2008, 3(4): 44115–44124.

Index